PLANTATION AND SPICES CROPS

PLANTATION AND SPICES CROPS

Editor

Chadwick Hillam

scitus
academics

Plantation and Spices Crops

Edited by **Chadwick Hillam**

This Book is Printed on Acid-Free Paper. ♾

ISBN: 9781681179384
Library of Congress Control Number: 2017949077

SCITUS Academics LLC,
616, Corporate Way, Suite 2, 4766,
Valley Cottage, NY 10989

www.scitusacademics.com

Printed in United States of America.

Preface

A plantation is a large-scale farm that specializes in cash crops and cultivated on an extensive scale in a large contiguous area, owned and managed by an Individual or a company. The crops include tea, coffee, rubber, cocoa, coconut, arecanut, oil palm, palmyrah, cashew, cinchona etc. These plantation crops are high value commercial crops of greater economic importance and play a vital role in a nation economy. Probably the single most important factor a plantation has on the local environment is the site where the plantation is established. If natural forest is cleared for a planted forest then a reduction in biodiversity and loss of habitat will likely result. In some cases, their establishment may involve draining wetlands to replace mixed hardwoods that formerly predominated with pine species. If a plantation is established on abandoned agricultural land, or highly degraded land, it can result in an increase in both habitat and biodiversity. A planted forest can be profitably established on lands that will not support agriculture or suffer from lack of natural regeneration. However, the prospects of spice crop production are bright because of its high market demand. The most common type of spice crops grown are onion, garlic, ginger, black pepper, hot pepper, celery, shallots, leek, bay leaf, and tanglad. Among these crops however, only onion, garlic, ginger, black pepper, and hot pepper are grown commercially because of their high market demand and usefulness. Spices are primarily used as food seasoning. In addition to their culinary uses, spices are also utilized as flavoring agent in beverages, active ingredient in syrubic medicines, coloring agent of textiles, and an important constituent in cosmetic and perfume products.

This volume 'Plantation and Spices Crops' presents an overview of the importance of plantation crops and spice plants, as well as the current status of the cultivation of these crops. It covers plant production aspects, such as climate and water requirements, planting, fertilizers, irrigation, weeding and hoeing, and harvesting, processing, and pest and disease management. This book is intended for advanced graduate students and practitioners of agriculture and horticulture, and those who are involved in the production and promotion of plantation crops, spice plants, and essential oil and medicinal plants.

Table of Contents

CHAPTER 1

Plantation vs. Natural Forest: Matrix Quality Determines Pollinator Abundance in Crop Fields

Hisatomo Taki[1], Yuichi Yamaura[2], Kimiko Okabe[1] & Kaoru Maeto[3]

[1] *Department of Forest Entomology, Forestry and Forest Products Research Institute, 1 Matsunosato, Tsukuba, Ibaraki 305-8687, Japan,*

[2] *Division of Environmental Resources, Graduate School of Agriculture, Hokkaido University, Nishi 9 Kita 9, Kitaku, Sapporo, Hokkaido 050-8589, Japan,*

[3] *Laboratory of Insect Science, Graduate School of Agricultural Science, Kobe University, Rokkodai, Nada, Kobe, Hyogo 657-8501 Japan.*

ABSTRACT

In terrestrial ecosystems, ecological processes and patterns within focal patches frequently depend on their matrix. Crop fields (focal patches) are often surrounded by a mosaic of other land-use types (matrix), which may act as habitats for organisms and differ in terms of the immigration activities of organisms to the fields. We examined whether matrix quality affects wild pollinator abundance in crop fields, given that the species (*Apis cerana*) generally nest in the cavities of natural trees. We examined fields of a pollination-dependent crop surrounded by plantations and natural forests, which comprised the matrix. Our analysis revealed a clear positive effect of the natural forest on the pollinator abundance, but the plantation forest had little effects. These indicate that agricultural patches are influenced by their matrix quality and the resulting crop pollinator abundance, suggesting the importance of matrix management initiatives such as forest restoration surrounding agricultural fields to improve crop production.

INTRODUCTION

In terrestrial ecosystems, ecological processes and patterns within focal patches frequently depend on their dominant matrix1,2,3. Crop fields (focal patches) are often surrounded by a mosaic of other land use types (matrix), which may act as habitats for organisms and differ in terms of the immigration activities of organisms to the fields[4]. Both theoretical and empirical studies have shown that improving the quality of the surround land use types enhances biodiversity as well as a range of ecosystem functions provided by different species in agro–ecosystems[5,6]. For example, a heterogeneous landscape comprising a matrix of semi-natural or natural land cover alternating with agricultural land generally has a greater biodiversity than a homogeneous landscape[7].

Of the world's terrestrial surfaces, forests comprise the largest proportion of semi-natural or natural land, occupying approximately 30% of the total land area worldwide[8]. In some countries, plantations account for a large proportion of forest area. For example, 12% of the land cover in the UK consists of forest, with 68% of the forest area being plantation, while in Japan, 68% of the land area is covered by forest, with 42% of that being plantation8. Plantation forests, typically consisting of one or a few tree species, are grown as even-aged monocultures, intensively managed, and harvested on relatively short rotations. These characteristics raise concerns that plantation forests may negatively impact forest biodiversity[9]. It is also possible for plantation forests, depending on the specific management practices and region, to play a role in conserving biodiversity[10]. However, for example, the multiple taxa in tropical forests in the Amazon demonstrate that plantation forests are often less diverse than natural forests[11].

Animal pollinators are essential providers of crop pollination services[12]. Several crop pollinator species depend on forests for their food and nest resources, as seen in both temperate and tropical regions (e.g.[13,14]). One such important forest-dependent pollinator species in Asia is *Apis cerana*, which is widespread in temperate and tropical countries[15] and plays important roles in crop pollination services therein[16]. This generalist honeybee species contributes to the pollination of a variety of crops, including apple[17], cauliflower, cabbage[18], radish[19], litchi[20] and cardamom[21]. Wild populations of *A. cerana* are highly dependent on forests for nesting and food resources, given that they mainly nest in tree cavities[22,23] and collect pollen and nectar from various species of trees in temperate and tropical regions[24,25]. Therefore, monocultured, even aged plantation forests that are intensively managed and harvested on relatively short rotations, might limit the available habitat of *A. cerana*.

The main goal of this study was to examine whether the quality of a forested matrix affects the abundance of *A. cerana* in a crop field in central Japan. We examined an agro-ecosystem of buckwheat (*Fagopyrum esculentum*) surrounded by conifer plantations and natural broad-leaved forests as the matrix. The crops in this region are pollinated by wild insect species, including a native honeybee (*A. cerana*), and a managed introduced honeybee (*A. mellifera*)[26]. After examining this system for three years, we predicted that the abundance of *A. cerana* in the buckwheat fields would show a stronger positive correlation with the total area of surrounding natural forests compared to that of surrounding plantation forests[27]. Here, we tested this prediction and analyzed the potentially negative influence of the presence of *A. mellifera* on the abundance of *A. cerana*. *A. mellifera* is a domesticated and dominant managed honeybee in the study region. No wild populations of *A. mellifera* have yet been reported in mainland Japan. It is suspected that *A. mellifera* has been unable to establish wild populations in part due to the presence of the Asian giant hornet (*Vespa mandarinia*), the only known hornet species to have evolved mass predation of other social bees and wasps, whereas *A. cerana* possesses a unique thermal defense against these hornets[28].

RESULTS

The deviance information criteria (DICs) were lowest for the models incorporating the area of natural and plantation forests at 1500-m and 2000-m scales, respectively (Table 1). The model with the area of natural forest set at a 1000–1500-m scale and that with the plantation forest area set at a 1000–2500-m scale were well supported. We then used the best-fit model to estimate the effects of the three covariates (the abundance of *A. mellifera* and the areas of natural and plantation forests) on the abundance of *A. cerana*. The values of $R^\wedge$ for all estimated values were less than 1.01 (Table 2; Supplementary Table S1) indicating that this model was well converged[29]. The results showed that the effect of natural forest was significant and that those of the plantation forest and *A. mellifera* were not significant (Table 2), although the areas of surrounding natural and plantation forests positively affected the abundance of *A. cerana*, while the presence of *A. mellifera* showed a negative effect. Random year effects indicated that *A. cerana* was most abundant in the third year, and the lowest population numbers were observed in the second year. Spatial random effects showed that sites within 1 km were positively correlated (Fig. 1).

Table 1: Deviance information criteria (DIC) of 64 models with different areas of natural and plantation forests.

Table 1 | Deviance information criteria (DIC) of 64 models with different areas of natural and plantation forests.

		Plantation								
		500m	1000m	1500m	2000m	2500m	3000m	3500m	4000m	mean
Natural forest	500m	4.99	2.57	3.77	4.17	4.73	4.5	4.86	5.18	4.35
	1000m	3.3	1.49	1.37	1.41	1.79	1.8	0.98	1.4	1.69
	1500m	1.19	0.77	0.55	0*	0.8	0.67	1.14	1.5	0.83
	2000m	1.75	1.52	2.27	0.58	1.87	2.46	3.58	4	2.25
	2500m	2.36	2.72	3.36	2.52	3.22	3.41	4.15	4.43	3.27
	3000m	2.92	3.34	4.49	5.29	4.74	4.16	4.65	5	4.32
	3500m	3.76	2.07	4.85	5.21	5.36	4.62	5.23	4.87	4.5
	4000m	5.76	1.2	4.07	6.23	6.27	5.94	6.09	6.11	5.21
	mean	3.25	1.96	3.09	3.17	3.6	3.45	3.84	4.06	

Models whose DIC differences to the best model (DIC = 0.0) smaller than 2.0 were highlighted by grey. *Used in parameter estimation.

Table 2: Estimates of the relevant model parameters in the best-fit spatial regression model.

Table 2 | Estimates of the relevant model parameters in the best-fit spatial regression model.

	Range of estimates*			
Variable	2.5%	50%	97.5%	$\hat{R}$
year[1]	−1.00	0.14	1.44	1.00
year[2]	−1.72	−0.54	0.72	1.00
year[3]	−0.70	0.42	1.69	1.00
$\alpha_{intercept}$	−0.13	1.32	2.58	1.00
$\alpha_{mellifera}$	−0.37	−0.18	0.01	1.00
α_{nat_for}	0.60	1.11	1.71	1.00
α_{plant_for}	−0.15	0.18	0.53	1.00
τ	0.93	3.92	12.82	1.00
ϕ	0.18	1.01	4.65	1.00

*Showed in percentile.

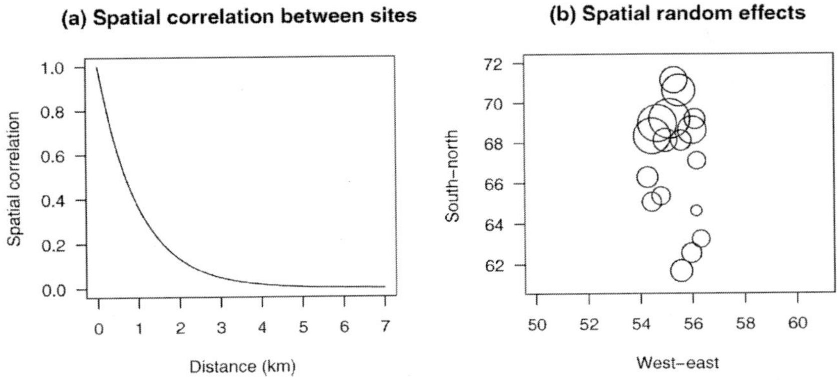

Figure 1: Spatial effects modeled by the spatial regression model. (a) Spatial correlation was depicted using the spatial decay function: exp[-(φd_{ij})], where $\varphi = 1.01$ (estimated median value). (b) Spatial random effects of each site are indicated by relative circle size.

DISCUSSION

Foraging distance of pollinator species is one key factor to influence the spatial scale of certain landscape factors[30,31]. The abundance of *A. cerana* in the crop fields was most strongly correlated with the area of natural forest in the matrix within the 1500-m scale. The wild honeybee species nests in cavities of tree species that are mainly found in forests[22], and uses the flower resources of various forest tree, shrub and herb species for food[24]. Thus, forests provide nest resources (cavities) as well as food resources (flowers)[32]. These resources are relatively fewer in the plantation forests, where the main species of planted conifers are *C. japonica* and *C. obtusa*. Plantation trees are carefully managed and do not provide the types of cavities that are preferred as nesting sites for the wild honeybees. Indeed, previous studies have revealed that tree cavities and understory vegetation are more available in natural forests than in conifer plantations that are grown as even-aged monocultures and are usually harvested on relatively short rotations[33,34].

Although the influence was little, analysis of the relationship between the abundances of *A. cerana* and *A. mellifera* showed that their abundances were negatively related, which could be caused by, for example, competition for limiting resources. Taki et al.[27] reported a negative correlation between the abundances of *A. cerana* and of *A. mellifera* based on two years of observations of buckwheat flowers in the same study region. Additionally, Nagamitsu & Inoue[24] compared pollen sources for *A. cerana japonica* and *A. mellifera* in a primary beech forest in Honshu, Japan. They found that most pollen sources were shared between the two honeybee species and that seasonal variation in the pollen utilization of the two honeybees was similar. These results suggest that there may be some conflict between the two honeybee species, although further research on resource use and behavioral competition for buckwheat flowers between the two bee species is necessary.

In summary, *A. cerana* abundance showed a relatively strong positive correlation with the natural forest matrix compared to the plantation matrix. Thus, replacing natural forest with plantation might have adverse effects on biodiversity conservation and restoration[9,35]. Possible approaches to forest management in a plantation-dominated landscape may include natural regeneration and retention of original native trees[10]. In the case of Japan, where 68% of land is covered by forest and 42% of which comprises plantations of mainly coniferous species[8], public demand for the re-establishment of native broadleaved species has been increasing[36]. Such public involvement must take place in developed and developing countries that continue to practice afforestation. Initiatives to restore monoculture plantations to naturally regenerated forests would initiate positive feedbacks for forest dependent organisms, which may act as ecosystem service providers in agricultural ecosystems.

The patch/matrix dichotomy is a broad oversimplification for many species in human-altered landscapes; however, improving the quality of matrix contexts will likely result in improved species conservation[37]. Therefore, the recognition and consideration of how to manage matrices when making land use decisions is important for the future of the world's terrestrial ecosystems[38,39]. Our results indicate that land management that includes focal agricultural fields and surrounding land use types is important to improving crop yields, and that forest management can contribute to improving agriculture.

METHODS

Study region and crops

The study was conducted in the mountainous region of Hitachiota, Ibaraki Prefecture, central Japan (36°30′–36°40′ N, 140°23′–140°31′ E). Landscapes in the study region consist of agricultural fields (buckwheat fields, paddies, orchards, and tea fields), with the dominant surrounding semi-natural to natural lands comprising cedar plantations and deciduous forests. The original dominant canopy species in the forested areas were deciduous broad-leaved trees such as konara oak (*Quercus serrata*), mizunara oak (*Quercus crispula*) and Japanese beech (*Fagus crenata*), and the main species in the conifer plantations are Japanese cedar (*Cryptomeria japonica*) and Japanese cypress (*Chamaecyparis obtusa*)[40].

Buckwheat is a characteristic crop grown by small local landholders in the region. We conducted a field study over three consecutive years, with 15 field sites examined in 2007 and 2009 and two additional sites in 2008 for a total of 17 sites. The timing of sowing and harvesting is similar in all fields across the region. The sowing of each field is typically completed during the second week of August, and harvesting takes place in mid October. Consequently, the blooming time of each study site was synchronized. No fertilizers, pesticides, or other agrochemicals are used in these fields.

Insect samplings

We sampled *A. cerana* and *A. mellifera* visiting the flowers of common buckwheat for food resources (both pollen and nectar) on sunny days during the blooming period in September (Fig. 2). In each buckwheat field, we used insect nets to capture honeybees visiting flowers within a 1.5 × 1.5-m quadrat approximately 2 m in from the southern edge of the field to standardize sampling among the field sites. In each field, all insects visiting flowers were sampled for 15 min between 9:00 and 12:00. This method was based on previous examinations of flower-visiting insects and pollen availability at different times of the day[41,42,43]. A single person (H. T.) conducted the samplings, which included three trials of 15 min each for each study field between 9:00 and 12:00, and the order of each trial was randomized.

Forested matrix

We measured the areas of tree plantations and natural forests surrounding the study crop sites at radii of 500, 1000, 1500, 2000, 2500, 3000, 3500, and 4000 m from the centers of sampling points in the study fields. The radii were selected based on the reported foraging ranges of *A. cerana*[44]. The tree plantation and natural forest areas were delineated from a 1:50,000 digital vegetation map published by the Ministry of the Environment of Japan in 1999. This vegetation map was updated with more recent large-scale deforestations and land-use changes through the interpretation of 2.5-m spatial resolution panchromatic SPOT 5 satellite imagery (Spot Image, Toulouse, France) in 2004 using ArcGIS 9.2 software (ESRI Inc., Redlands, CA, USA).

Figure 2: Wild honeybees (*Apis cerana*) visiting buckwheat flowers (*Fagopyrum esculentum*).

Data analysis

We examined the effects of three covariates (the abundance of *A. mellifera* and the areas of natural and plantation forests) on the abundance of *A. cerana* as a response variable over three years under a Bayesian framework using WinBUGS Ver. 1.4.3[45], R2WinBUGS Ver. 2.1–13[46], and R Ver. 2.10.1[47]. We assumed the abundance of *A. cerana* to be a Poisson-distributed variable and used a log-link function. Year was used as a random effect (i.e. $year_i$; $i = 1, 2, 3$), and the three covariates as simple linear terms. Given that we used spatial sampling data, it was necessary to consider spatial autocorrelation among sites (e.g.[48]). Thus, we used a random site variable as an additional covariate ($site_j$; $j = 1, ..., 17$) using a Bayesian Gaussian kriging model[49] with GeoBUGS Ver. 1.2[50]. Following Thomas et al.[50], we centered the mean for each site to zero ($\mu = 0$) using a hierarchically centered model. In this spatial model, between-area correlation is modeled using an exponential decay function: $\exp[-(\varphi d_{ij})^{\kappa}]$, where d_{ij} is distance between sites i and j, φ is the rate of decline of correlation with distance, and κ is set to 150. We used uninformative vague priors for estimated parameters[51]: $year_i \sim Norm(0, var_{year})$;

$1/var_{year} \sim$ Gamma(0.001, 0.001); $\boldsymbol{\alpha} \sim$ Norm(0, 1000); $\varphi \sim$ Uniform(0.15, 5); $site_j \sim$ Norm(0, var_{site}); $1/var_{site} \sim$ Gamma(0.001, 0.001), where $\boldsymbol{\alpha}$ is the intercept and estimates the three covariates (i.e. the slope of a simple term), var_{site} is the variation of random site effects (inverse of var_{site} is called τ). For the sites surveyed in only 1 or 2 years, the abundance of *A. cerana* in unsampled years was defined as NA, and the abundance of *A. melifera* was replaced with the average value of 5.7[52]. The range of φ was chosen such that the between-site correlation was within [0, 0.47] at 5 km. The three covariates were standardized before the analysis. We ran three simulations of 100,000 iterations with different initial values, discarded the first 1000 and thinned by 50. This process was conducted with eight different radii of natural and plantation forest areas, producing a total of 8 × 8 models, which were compared using deviance information criteria (DICs)[51].

ACKNOWLEDGEMENTS

We thank the buckwheat growers in Hitachiota for allowing us to conduct our field samplings. We thank Y. Mitsuda for statistical help, S. Sugiura for comments on an earlier version of the manuscript, and M. Akiyama, K. Matsuura, K. Watanabe and K. Aoki for assistance with the selection of study fields. This study was supported by the Global Environment Research Funds (E-0801 and S-9) of the Ministry of the Environment, Japan.

REFERENCES

1. Ricketts, T. H. The matrix matters: Effective isolation in fragmented landscapes. *Am. Nat.* **158**, 87–99 (2001).
2. Vandermeer, J. & Carvajal, R. Metapopulation dynamics and the quality of the matrix. *Am. Nat.* **158**, 211–220 (2001).
3. Prevedello, J. A. & Vieira, M. V. Does the type of matrix matter? A quantitative review of the evidence. *Biodivers. Conserv.* **19**, 1205–1223, 10.1007/s10531-009-9750-z (2010).
4. Tscharntke, T., Klein, A. M., Kruess, A., Steffan-Dewenter, I. & Thies, C. Landscape perspectives on agricultural intensification and biodiversity – ecosystem service management. *Ecology Letters* **8**, 857–874 (2005).
5. Donald, P. F. & Evans, A. D. Habitat connectivity and matrix restoration: the wider implications of agri-environment schemes. *Journal of Applied Ecology* **43**, 209–218, 10.1111/j.1365-2664.2006.01146.x (2006).
6. Perfecto, I. & Vandermeer, J. in *Year in Ecology and Conservation Biology 2008 Vol. 1134 Annals of the New York Academy of Sciences* 173–200 (Blackwell Publishing, 2008).
7. Benton, T. G., Vickery, J. A. & Wilson, J. D. Farmland biodiversity: is habitat heterogeneity the key? *Trends in Ecology & Evolution* **18**, 182–188, 10.1016/s0169-5347(03)00011-9 (2003).

8. FAO. *State of the world's forests 2007*. (Food and Agriculture Organization of the United Nations, 2007).
9. Brockerhoff, E., Jactel, H., Parrotta, J., Quine, C. & Sayer, J.Plantation forests and biodiversity: oxymoron or opportunity?*Biodivers. Conserv.* **17**, 925–951 (2008).
10. Hartley, M. J. Rationale and methods for conserving biodiversity in plantation forests. *For. Ecol. Manage.* **155**, 81–95 (2002).
11. Barlow, J. *et al.* Quantifying the biodiversity value of tropical primary, secondary, and plantation forests. *Proceedings of the National Academy of Sciences of the United States of America* **104**, 18555–18560, 10.1073/pnas.0703333104 (2007).
12. Klein, A. M. *et al.* Importance of pollinators in changing landscapes for world crops. *Proceedings of the Royal Society of London B* **274**, 303–313 (2007).
13. Ricketts, T. H., Daily, G. C., Ehrlich, P. R. & Michener, C. D.Economic value of tropical forest to coffee production. *Proceedings of the National Academy of Sciences of the United States of America* **101**, 12579–12582 (2004).
14. Kremen, C., Williams, N. M. & Thorp, R. W. Crop pollination from native bees at risk from agricultural intensification. *Proceedings of the National Academy of Sciences of the United States of America***99**, 16812–16816, 10.1073/pnas.262413599 (2002).
15. Ruttner, F. *Biogeography and taxonomy of honeybees*. (Springer-Verlag, 1988).
16. Kevan, P. G. *The Asiatic hive bee: apiculture, biology and role in sustainable development in tropical and subtropical Asia*. 315 (Enviroquest Ltd., 1995).
17. Verma, L. R. & Dulta, P. C. Foraging behavior of*Apis cerana indica* and *Apis mellifera* in pollinationg apple flowers.*Journal of Apicultural Research* **25**, 197–201 (1986).
18. Verma, L. R. & Partap, U. Foraging behavior of*Apis cerana* on cauliflower and cabbage and its impact on seed production.*Journal of Apicultural Research* **33**, 231–236 (1994).
19. Partap, U. & Verma, L. R. Pollination of radish by *Apis cerana*.*Journal of Apicultural Research* **33**, 237–241 (1994).
20. Abrol, D. P. Diversity of pollinating insects visiting litchi flowers (*Litchi chinensis* Sonn.) and path analysis of environmental factors influencing foraging behaviour of four honeybee species.*Journal of Apicultural Research* **45**, 180–187 (2006).
21. Kuriakose, G., Sinu, P. A. & Shivanna, K. R. Domestication of cardamom (Elettaria cardamomum) in Western Ghats, India: divergence in productive traits and a shift in major pollinators. *Annals of Botany* **103**, 727–733, 10.1093/aob/mcn262 (2009).
22. Sasaki, M. *Wonders of the Japanese honeybee – Biology of Northernmost Apis cerana -*. (Kaiyusha, 1999).
23. Yoshida, T. *Rearing method and ecology of Japanese honeybee*. (Tamagawa University Press, 2000).
24. Nagamitsu, T. & Inoue, T. Differences in pollen sources of*Apis cerana* and *Apis mellifera* at a primary beech forest in central Japan.*Journal of Apicultural Research* **38**, 71–78 (1999).
25. Corlett, R. T. Pollination in a degraded tropical landscape: a Hong Kong case study.*Journal of Tropical Ecology* **17**, 155–161 (2001).
26. Taki, H., Okabe, K., Makino, S., Yamaura, Y. & Sueyoshi, M.Contribution of small insects to pollination of common buckwheat, a distylous crop. *Annals of Applied Biology* **155**, 121–129 (2009).
27. Taki, H. *et al.* Effects of landscape metrics on *Apis* and non-*Apis* pollinators and seed set in common buckwheat. *Basic and Applied Ecology***11**, 594–602, 10.1016/j.baae.2010.08.004 (2010).
28. Ono, M., Igarashi, T., Ohno, E. & Sasaki, M. Unusual thermal defense by a honeybee against mass attack by hornets. *Nature***377**, 334–336 (1995).
29. Gelman, A. & Hill, J. *Data analysis using regression and multilevel/hierarchical models*. (Cambridge University Press, 2007).
30. Osborne, J. L. *et al.* Bumblebee flight distances in relation to the forage landscape.*Journal of Animal Ecology* **77**, 406–415,

31. Zurbuchen, A. *et al.* Maximum foraging ranges in solitary bees: only few individuals have the capability to cover long foraging distances. *Biol. Conserv.* **143**, 669–676, 10.1016/j.biocon.2009.12.003 (2009).
32. Klein, A. M., Cunningham, S. A., Bos, M. & Steffan-Dewenter, I.Advances in pollination ecology from tropical plantation crops. *Ecology* **89**, 935–943 (2008).
33. Woodley, S. J., Johnson, G., Freedman, B. & Kirk, D. A. Effects of timber harvesting and plantation development on cavity-nesting birds in New Brunswick. *Can. Field-Nat.* **120**, 298–306 (2006).
34. Aubin, I., Messier, C. & Bouchard, A. Can plantations develop understory biological and physical attributes of naturally regenerated forests? *Biol. Conserv.* **141**, 2461–2476, 10.1016/j.biocon.2008.07.007 (2008).
35. Hansen, A. J., Spies, T. A., Swanson, F. J. & Ohmann, J. L.Conserving biodiversity in managed forests. *BioScience* **41**, 382–392 (1991).
36. MAFF. MAFF's biodiversity strategy. 40 (Ministry of Agriculture, Forestry and FisheriesJapan, Tokyo, 2007).
37. Prugh, L. R., Hodges, K. E., Sinclair, A. R. E. & Brashares, J. S. Effect of habitat area and isolation on fragmented animal populations. *Proceedings of the National Academy of Sciences of the United States of America* **105**, 20770–20775, 10.1073/pnas.0806080105 (2008).
38. Franklin, J. F. & Lindenmayer, D. B. Importance of matrix habitats in maintaining biological diversity. *Proceedings of the National Academy of Sciences of the United States of America* **106**, 349–350, 10.1073/pnas.0812016105 (2009).
39. Foley, J. A. *et al.* Global consequences of land use. *Science* **309**, 570–574, 10.1126/science.1111772 (2005).
40. Taki, H. *et al.* Effects of reforestation age on moth assemblages in plantations and naturally regenerated forests. *Insect Conservation and Diversity* **3**, 257–265 (2010).
41. Namai, H. Pollination biology and reproductive ecology for improving genetics and breeding of common buckwheat, *Fagopyrum esculentum* (2). *Fagopyrum* **10**, 35–46 (1990).
42. Bjorkman, T. Role of honey bees (Hymenoptera, Apidae) in the pollination of buckwheat in eastern North America. *J. Econ. Entomol.* **88**, 1739–1745 (1995).
43. Cawoy, V. *et al.* Morph differences and honeybee morph preference in the distylous species *Fagopyrum esculentum* Moench. *Int. J. Plant Sci.* **167**, 853–861 (2006).
44. Dyer, F. C. & Seeley, T. D. Dance dialects and foraging range in three Asian honey bee species. *Behavioral Ecology and Sociobiology* **28**, 227–233 (1991).
45. Lunn, D. J., Thomas, A., Best, N. & Spiegelhalter, D. WinBUGS – a Bayesian modelling framework: concepts, structure, and extensibility. *Statistics and Computing* **10**, 325–337 (2000).
46. Sturtz, S., Ligges, U. & Gelman, A. R2WinBUGS: a package for running WinBUGS from R. *Journal of Statistical Software.* **12**, 1–16 (2005).
47. R: a language and environment for statistical computing. Ver. 2.10.1 (R Foundation for Statistical Computing, Vienna, Austria. http://www.R-project.org/, 2009).
48. Lichstein, J. W., Simons, T. R., Shriner, S. A. & Franzreb, K. E.Spatial autocorrelation and autoregressive models in ecology. *Ecological Monographs* **72**, 445–463 (2002).
49. Diggle, P. J., Tawn, J. A. & Moyeed, R. A. Model-based geostatistics. *Journal of the Royal Statistical Society. Series C (Applied Statistics)* **47**, 299–350 (1998).
50. Thomas, A., Best, N., Lunn, D., Arnold, R. & Spiegelhalter, D. *GeoBUGS user manual. Ver. 1.2.* (2004).
51. McCarthy, M. A. *Bayesian methods for ecology*. (Cambridge University Press, 2007).
52. Kéry, M. *Introduction to WinBUGS for ecologists: a Bayesian approach to regression, ANOVA, mixed models and related analyses.* (Academic Press, 2010).

CHAPTER 2

Oil Palm Plantations in Indonesia: The Implications for Migration, Settlement/Resettlement and Local Economic Development

Suseno Budidarsono, Ari Susanti and Annelies Zoomers

[1] *Utrecht University, Faculty of Geoscience, International Development Studies, The Netherlands*

[2] *World Agroforestry Center (ICRAF), Southeast Asia Regional Programme, JL, CIFOR, Situ Gede, Sindang Barang, Bogor, Indonesia*

[3] *Gadjah Mada University, Faculty of Forestry, Jl. Agro No. 1 Bulaksumur, Yogyakarta, Indonesia*

1. INTRODUCTION

Palm oil is the world's most traded vegetable oil: in August 2012, the share of palm oil (including kernel oil) in world supply was 37.6% [1]. Palm oil is extracted from the fruit of the oil palm tree (*Elaeisguineensis*); the main products are crude palm oil (CPO) and palm kernel oil (PKO). In terms of land use, the oil palm tree is more efficient than any other oil crop [2], and in economic terms palm oil is highly competitive. The value chain of palm oil and its derivatives has a strong degree of vertical integration [3], and its production costs are relatively low compared to other vegetable oils. It is therefore seen as one of the cheapest and most attractive vegetable oils traded on the world market [4, 5].

The palm oil sector provides income and employment for a significant number of individuals in developing countries [6]. A study of the Indonesian palm oil industry carried out as part of a global study under the coordination of the Australian National University, concluded that palm oil developments have had a positive impact on the incomes and living standards of all involved [7]. According to an assessment carried out in Sumatra, oil palm plantations have high labour requirements and show high return to labour [8].

Palm oil, being a multi-purpose vegetable oil, offers good prospects for further expansion. There is a growing demand from the commercial food and oleo-chemical industries that use oil palm in processed foods, cosmetics, soaps, pharmaceuticals, industrial and agro-chemical products, and as a feedstock for bio-diesel. The growing worldwide interest in bio-diesel as an alternative to fossil fuel is expected to lead to the further expansion oil palm plantation [9,10,11]. Tilman and colleagues [12] assert that this might lead to a food–energy–environment 'trilemma'. Even though oil palm trees are not a problem (they are 'green'), the rapid expansion of oil palm plantations across Southeast Asia, and particularly Indonesia and Malaysia, could cause the destruction of rainforests, as well as a lot of social problems, including food security challenges [13]. There are concerns that oil palm expansion will lead to the loss of biodiversity and the conversion of forest area [14 - 16] and the aggravation of social conflicts. According to Colchester and colleagues [17], who analysed the situation on and around six oil palm plantations, local communities face serious problems with the companies and there are many land conflicts. There is a widespread feeling in the communities of being cheated by the companies, and of being pushed into agreements through false promises without having a voice in decision-making. To the extent that people are employed, labour conditions are often not favourable. According to Marti [18], who conducted research on the labour conditions of plantation workers, much will depend on 'whether Indonesia's policy-makers intend to keep a large labouring class in low-paid, low-skill jobs as the rest of the country develops, or whether the country anticipates inviting millions of workers from even less fortunate countries to work on their plantations in future.'

With these controversies in mind, this chapter provides an overview of the pros and cons of oil palm development in Indonesia, paying attention to changing government policies and focusing on the implications of increasing oil palm investments for migration, settlement/resettlement and local economic development.

Since the 1970s, the Indonesian government has stimulated oil palm expansion in various ways, initially in the form of plantations. For a considerable time, the government played a direct role in stimulating investments in oil palm plantations through state agencies (with direct interventions in service provision, institutional support, agricultural extension, access to land and capital, etc.). Plantation development policies were carried out in close relation with other policy objectives, namely population redistribution through resettlement schemes or transmigration to stimulate the development of the outer islands (Sumatra, Kalimantan and Papua) and revitalizing the by then huge transmigration settlements that had often failed to produce more than rice and subsistence crops [19] and enhancing regional development (i.e., the increase in agricultural production, employment generation etc.); and the political objective of promoting national integrity and increasing national security [20].

This chapter is based on empirical data collected in Riau province (one of the booming oil palm producing areas that has undergone rapid expansion in the last 10 years) as well as on desk research. The field data collection focused on the economic activities related to oil palm development. The desk study included an analysis of legal and technical documents related to historical records of oil palm development in Indonesia in general and in Riau province in particular. These documents were collected in libraries and government offices, and from companies and online sources.

After providing some background information on palm oil investments, we give a historical overview of how oil palm expansion took place in Indonesia in close relation to changing policies. This is followed by an assessment of the implications of rapid oil palm expansion for migration, settlement/resettlement and local economic development.

2. INVESTMENT IN PALM OIL PRODUCTION, SETTLEMENT DEVELOPMENT AND RURAL ECONOMY

Investments in palm oil production require large-scale plantations to be economical and the establishment of palm oil processing units (mills) close to the plantations, as the oil palm fruits have to be processed within 48 hours after harvesting. Once investments in palm oil production have been made, the plantations are made to last as long as possible, or at least for one production cycle (25 years). A large-scale oil palm plantation requires a significant amount of labour to establish the plantation, maintain the palm trees and harvest the fruits. A detailed study on palm oil production in Indonesia noted that the labour requirements of a plantation during one production cycle (25 years) vary between 59 and 144 person-days (pds) per ha per year, which is an average of 91 pds per year[21]. Although a mill requires less labour, managers, technical staff and labourers have to be on hand. However, in many (if not all) cases, large-scale plantations are established in scarcely populated areas, and even in 'empty' frontier areas. Large-scale plantations are therefore always accompanied by settlement/resettlement, for which staff and labour facilities must be established [22]. For example, a block of 1000 ha of oil palm trees needs 91,000 pds/year. Assuming that the optimum working time for one person is 250 pds per year, 364 employees and labourers will be needed. As most of them will live with their families, significant provisions have to be made.

The establishment of large-scale oil palm plantations and processing units is an important economic activity that strongly influences the rate of land development in a region. It speeds up the development of infrastructure (e.g. the construction of roads to open up less accessible areas, and the provision of health and educational facilities) and stimulates the growth of the local economy through income spending by the plantation workers and the expenses of the company. Economic diversification will take place in the surrounding area, for example from subsistence agriculture to market-oriented cash production. The most fundamental effect of cash-cropping is that it increases the separation between the consumption and the production activities of the household. This in turn tends to change the pattern of family and communal dependencies and to strengthen the role of the nucleus family as a productive enterprise [23]. Another effect is the emergence of non-agricultural activities (trade, home industry and services). Urban employment opportunities subsequently increase.

Empirical research carried out in Riau province assessing the development impact of large-scale oil palm plantations on the local economy [24] revealed that the investment in palm oil production strongly induced local economic growth. The study found that marginal propensity to consume locally (MPCL) of the community living from oil palm plantation was 0.8415. This means that about 84% of additional income was spent locally. The percentage of money spent locally that becomes local income (PSY) was 71%; nearly three quarters of the necessary inputs needed could be provided locally. Based on these figures, the calculated income multiplier effect was 2.48 (1/(1 – (MPCL x PSY)). This means that an autonomous IDR 1 million change in income (or investment expenditures) in the area results in a change of IDR 2.48 million. The multiplier effect refers to the increase in final income arising from any new injection of spending.

3. OIL PALM PLANTATION DEVELOPMENT IN INDONESIA: A BRIEF REVIEW

3.1. The Early Stage of Palm Oil Development

The oil palm (*Elaeis guineensis*) originates from the tropical rainforest of West Africa, and was originally cultivated mostly by independent small farmers with landholdings of up to 7.5 ha. Palm oil has been commercially traded to Europe since 1811 [25, 26]. In Indonesia, oil palms have been cultivated commercially since 1911, when they were first developed in the east coast area of Sumatra under Dutch administration [2]. While the tree was cultivated successfully in this area in large plantations, the native population did not replace their coconut palms with this new palm species. They planted it only for decorative purposes.

In the east coast region of Sumatra, palm oil production (CPO) grew dramatically, from 181 tons in 1919 to 190,627 tons of CPO and 39,630 tons of kernel oil in 1937 [27]. This was in line with the oil palm plantation development in the area. The size of the first estate, which was established in 1910–14, was 2,620 ha. The plantation area increased to 6,920 ha in 1919, and steadily increased until 1936, when the total area planted amounted to 75,000 ha, of which 63,234 ha were in a productive stage. In this regard, Deasy [28] wrote that:

... the oil palm industry in Sumatra during the last two decades has bordered on the phenomenal. Production of palm oil in that island has increased from a few thousand metric tons in the early twenties of this century to almost two hundred thousand metric tons in 1937. In 1920, Sumatra accounted for less than one-half of one per cent of the world's annual export of palm oil; by 1937 over 40 per cent of the total export of that commodity was originating in the island. From an insignificant crop with an output far below that of most other Sumatran estate crops in 1920, the oil palm by 1937 had risen to a position of joint supremacy with rubber in the agricultural economy of the island.

The east coast of Sumatra became the site of one of the most intensive and successful pursuits of foreign agricultural enterprise or plantation [29]. Bio-physically, the area is quite suitable for oil palm growing, with high rainfall (minimum 1600 mm/year) and a tropical climate within 10° of the equator. Land and labour, the most important inputs, were available. The land was made available by getting long-term land leases from the local indigenous rulers through the colonial government [29, 30]. In the 1870s, the Dutch government started to withdraw from direct involvement in economic production, and increasingly concentrated on creating incentives that would stimulate the private initiative. As Breman [30] wrote:

The 1870 Agrarian Act, which officially put an end to the forced cultivation system in Java and started transition to an era of unbridled liberalism, indicated the orientation of the new policy: the archipelago's natural resources were henceforth to be made accessible to capitalist interests in the mother country.

The establishment of the palm oil industry in Sumatra started at the right time. The demand for palm oil was growing enormously worldwide, due to the increasing uses of palm oil products. Palm oil had already been traded to Europe and the United States since the early nineteenth century. Industrial development in Great Britain stimulated the demand for oil, a demand that was partly satisfied through trade with the West African coast. However, the increased mechanization of industry created a much larger need for lubricants, and palm oil proved useful in the manufacture of grease. Of equal importance, though, was the value of palm oil in the making of soap, candles and medicinal ointments, the increased use of which was also a result of the Industrial Revolution [26]. Scientific research supported the production of oil palm by modern, well-equipped estates and mills [25]. As a result, oil palm plantations in the Sumatra region became five times more productive than in Africa. The plantation development culture acquired from the cultivation and processing of latex rubber was a good foundation on which to introduce the large-scale cultivation of palm oil. The emergence of corporate capitalism since the 1850s, especially during the 'Tobacco Deli era' in the east coast of Sumatra, provided an ideal environment for large-scale plantation investments [29]. The capital market, labour market and remuneration systems had already developed. There was a banking system created by the tobacco planters association. Plantations that found themselves short of labour could easily acquire more through labour suppliers in Malaysia and Singapore.

Export-oriented plantation culture in Sumatra East Coast provided local employment that also attracted immigrants, which brought about more economic activity in the region. However, there are some criticisms regarding the development of large-scale plantations during the colonial era. The plantations did not lead to balanced economic growth and regional development [31]. The large-scale monocultures of capital-intensive enterprises created little wealth for unskilled workers (migrants) or local farmers, since wage levels were low. Many local farmers (especially shifting cultivators) suffered from land grabbing by the estate companies and the local economy suffered from the siphoning off of the locally produced surpluses (profits, dividends, staff salaries) to the head offices in urban centres and abroad. Moreover, the highly specialized plantations often suffered from instabilities on the world market, while having high fixed costs (for machinery, buildings, staff and maintenance). Together with the low purchasing power of the local population, this caused a low demand for urban goods and services and, as a result, a weak development of local market centres. There were consequently few opportunities for the local diversification of production and employment. Historical studies by Stoler [29] and Breman [30] to some degree provide a similar nuance from the labour perspectives.

However, those observations do not negate the fact that during the early stage of oil palm plantation, the east coast of Sumatra underwent a high population growth. As shown in Table 1, the annual population growth rate in the area in 1900–30 was 10.4%. Chinese and Javanese workers were the largest groups among the plantation workers (54% and 34%, respectively in 1902; [30]). It seems that the region was quite attractive for people to move to. The share of local people was assumed to be 56%, with an immigrant population of 44% in 1930.

According to Stoler [29] and Breman [30] investments in the plantation agriculture on the east coast of Sumatra (and the emergence of a plantation society) induced local development. The large number of people coming to and residing in the region created a market for consumer goods. To some degree, agglomeration economies evolved in the region. It is also noted that plantation support services, such as banking systems, capital and labour markets, were introduced, established and worked well. Within about five decades, the east coast of Sumatra region had been transformed from a frontier area into a well-developed and integrated plantation area. However, no study has looked at how the investments in such large-scale and modern plantation activities led to settlement development in the surrounding area, how the rural service centres developed and grew into urban centres, or, more in general, what kind of transformation took place in the region because of the new technology and the increase in population caused by the inflow of different ethnic groups.

Table 1. Population of the east coast of Sumatra, 1900–30: population, proportion by ethnic group and annual growth rate

[i] - Sources: Adapted from Het Oostkust van Sumatra Instituut, 1938.

[ii] - Note: referring to Breman [30]

[iii] - 1. Includes Dutch, Belgian, German, Swiss, French, British and Austrian

[iv] - 2. Consists of local community and immigrants: Javanese, Banjarese, Baweanese, Bataks, Malay, Gayo, Mandailing

[v] - 3. Consists of Thai, Indians

Year	European 1)	Inlanders 2)	Chinese	Other 3)	Total
Population					
1900	2,079	306,035	103,768	9,208	421,090
1905	2,667	450,941	99,236	15,573	568,417
1915	5,200	681,000	132,000	14,320	832,520
1920	7,882	1,042,930	134,750	11,992	1,197,554
1930	11,079	1,470,395	192,822	18,904	1,693,200
	Proportion (%)				
1900	0.5%	72.7%	24.6%	2.2%	
1905	17.1%	79.3%	17.5%	2.7%	
1920	0.7%	81.8%	15.9%	1.7%	
1930	0.7%	87.1%	11.3%	1.0%	
Annual growth rate 1900–30	14.9%	13.1%	3.0%	3.6%	10.4%
Annual growth rate 1900–20	14.7%	12.7%	1.6%	1.6%	9.7%
Annual growth rate 1920–30	4.5%	4.6%	4.8%	6.4%	4.6%

3.2. Palm Oil Production after Independence and Recent Development

Not much has been written about plantation development - or more specifically oil palm plantations – in the period from World War II until the independence of Indonesia, or during the transitional phase after independence. In general terms, Benjamin Higgins, a financial advisor for the Indonesian government in the early 1950s under the UN Technical Assistance Programme, wrote that the plantation industry was facing serious problems, which was shown by the fact that the area under plantation crops was still only two thirds of the pre-war level [32]. Figures showed that palm oil exports amounted to only 109,000 tons in 1960 [33], compared to 240,000 tons of CPO and kernel oil in 1937 [27]. However, since the earliest national reconstruction plan the Indonesian government has included plantation agriculture in its development policy. In 1955, the

Ministry of Transmigration submitted a 5-year plan to the National Planning Bureau to resettle 400,000 families from overpopulated Java to unsettled areas in the outer islands (mainly associated with traditional export commodities from plantation agriculture, such as rubber, coffee, tea, etc.). According to Higgins, the objectives of the transmigration programme were not achieved, mainly due to a lack of assured financing, inadequate manpower and a lack of technical education [32].

Looking at time series data of oil palm plantation in Indonesia (Figure 1), it appears that Indonesian oil palm plantation increased significantly only after the 1970s. This is related to the New Order government policy for the agricultural sectors, which included plantation development.

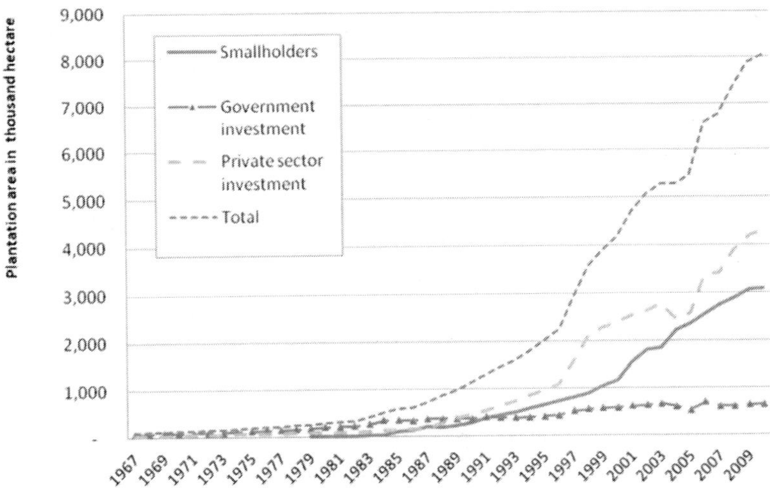

Figure 1. The development of oil palm plantation area in Indonesia, by actors. Source: Series of Statistik Perkebunan Indonesia (estate statistics)

Much has been published about the post-1970s development of the palm oil industry in Indonesia, focusing on topics such as germ palm development [34], competitiveness [7], socio-economic improvement [19], agrarian issues and social transformation [5, 17], and even human rights [18].

As mentioned, the Indonesian government began to put sustained effort into promoting tree plantation crops in the late 1970s. It established a scheme called Nucleus Estate Scheme (NES) (*Perkebunan Inti Rakyat*; PIR), whereby state-owned plantation companies (the 'nucleus') helped farmers (namely plasma farmers) to grow oil palm. The plantation companies provided seedlings, technical assistance and financing to small holders. Their crops would be purchased by companies' mills [35]. The policy was not implemented for solely plantation development; it was linked and integrated with other policy objectives: population redistribution through resettlement schemes or transmigration (i.e. by moving people from densely populated regions to scarcely populated areas), socio-economic progress (regional development, increased agricultural production, employment generation) and political consolidation (promotion of national integrity and security) [36]. Thus, the initial programme consisted basically of direct state investments through state-owned companies (PTPN) and was integrated with government-sponsored transmigration programmes to provide a labour force for the new plantations [22]. This integration was embryonic for smallholder engagement in state-led agribusiness; the emergence of smallholder oil palm planters constituted a spread effect of plantation development led by the government, and, most importantly, it was the time when settlement development was started in the surrounding of large-scale oil palm plantation.

Analysing the evolution of oil palm plantation development since the 1970s, McCarthy [5] differentiates into three phases of development: (1) the New Order state developmental period (the late 1970s to 1994), which was characterized by direct state intervention; (2) the transitional period towards private initiative through the KKPA model (1994–98); and (3) the 'laissez-faire' period (the term used by McCarthy to name the *Reformasi* era) since 1998.

First phase: During the New Order period, the government pursued a developmental agenda that combined the aim of ensuring political and macro-economic stability by financing infrastructure and providing subsidies derived from oil revenues [5]. There were direct interventions by the state that enabled state-owned companies to have more access to land and capital; institutional support was also provided [37]. The introduction of the Nucleus Estate Scheme (the PIR) in combination with the transmigration programme (the PIR-Trans) took place in this period. In general, in terms of area, the ratio between the nucleus estate and the smallholder plasma was 20:80; the nucleus estate held 20% of the total area and the remaining 80% was owned by smallholders – plasma farmers with technical assistance from the nucleus estate. Under the PIR-Trans scheme, trans-migrants played significant roles as labour for the plantation (crop care and harvesting) and constitute an important component during the establishment years. The government provided financial support to smallholders to establish their plantations and to finance their living and housing expenses, while the nucleus estate (the company or investors) was responsible for extension services and for collecting and processing the fruit bunches. The government also facilitated access to land (mostly state forest land and village lands), developed some infrastructure and granted credit at concessionary rates for plantation development. The PIR-Trans schemes, which existed between 1986 and 1994, benefited smallholders, as 'plasma' farmers. The state provided access to 'free land' and concessionary credit in exchange for submission to a particular agribusiness model, namely the inclusion of smallholders in peripheral areas. The farmers would obtain fully private rights over their holdings upon settlement of the oil palm development loan. By incorporating trans-migrants, new settlements were established in the area surrounding the plantation, besides those of the local people.

Second phase: In the transitional period towards private–community initiatives (the KKPA model), the government changed its policy by seeking to encourage private sector initiatives, facilitate foreign direct investment and accelerate estate crop development. The new orientation was in response to the World Bank's criticism of the on-going state support of smallholder oil palm schemes. It advised the government to officially abandon its direct subsidizing role and leave oil palm development to the market. The government ignored this advice for some time, before changing its policy as a result of increasing pressure on the state budget, as well as donor advocacy for a more direct social–private partnership model. Accordingly, the next phase in the development of oil palm schemes marked a new milestone in an on-going trend towards state withdrawal.

The new scheme was known as the KKPA (*Koperasi Kredit Primer untuk Anggota*; Primary Cooperative Credit for Members) and covered the period 1995–98. It involved a more direct private–community 'partnership' model, with the plantation firm being responsible for nearly all of the project, working directly with the participating farmers to resolve land problems, and providing training and extension. During this period the door for foreign direct investment in large-scale plantations was opened. Local communities, including settlements of trans-migrants, which had often been unsuccessful to move beyond the production of rice and subsistence crops, could be transformed into oil palm plantations, depending on the agreement between both parties. More independent smallholder oil palm farmers emerged in that time, and led to more spontaneous migration into the oil palm area.

Third phase: McCarthy [5] asserts that in the 'laissez-faire' period (1998 onwards), the shift to decentralization, public–private partnerships between market actors and the government, and social–private partnerships between market actors and communities have affected Indonesian policy. The year 1998 was, in fact, a key moment in the transition from a developmental approach to a more neoliberal, market-driven model [5]. A series of policy changes provided for the development of 'community plantations' (*perkebunan rakyat*) under various partnership (*kemitraan*) models (see the two ministerial decrees: the Decree of Forestry and Estate Ministry No. 107/Kpts-II/1999 and the Decree of Agricultural Ministry No 26/2007). Existing estates entered into partnerships with large, capital-intensive companies willing to invest in labour-intensive oil palm projects [19]. During this period, farmers, who were initially in the PIR-Trans schemes, managed to gain access to oil palm technology and improve their incomes. They were eventually able to access investment capital, because, once they had paid off their credit, they obtained land certificates, which could be used as collateral for borrowing money from local banks to expand production. At a time of rising oil palm prices, many of these new landowners used these accumulated assets to rapidly expand their holdings. During the later years of the oil palm boom (i.e. prior to 2008), these actors were joined by successful KKPA farmers, who were (to a limited degree) using incomes from productive oil palm holdings to invest in upgrading unproductive land into oil palm plantations. The result was spontaneous frontier development on the margins of already existing palm oil plantations. This means that the spread effect of oil palm plantation was more prevalent in this period.

Summarizing, the state agribusiness-driven policy has transformed rural areas, making Indonesia the world's largest oil palm producer. The area devoted to palm oil production had doubled to 5.5 million ha by the year 2000 [38], and by 2009 Indonesia was the world's leading oil (CPO and PKO) producer with 24.5 million tons as compared to Malaysia with 22.1 million tons [39]. The introduction of the oil palm was initially associated with direct state investments via state-owned companies. After implementing various types of schemes, the state introduced the estate-transmigration programme; both models of palm oil development led to frontier development. Settlements were officially established in newly opened areas, in the expectation that economic activities would slowly develop. During the second and the third period, both spontaneous migration and the number of independent oil palm smallholders increased in the area surrounding the existing oil palm plantations.

4. OIL PALM PLANTATION DEVELOPMENT IN RIAU PROVINCE

Riau Province is located in the centre and on the eastern coast of Sumatra along the Strait of Malacca, the busy international shipping route connecting the Indian Ocean with the South Chinese Sea and the Pacific Ocean. Riau is also quite near Singapore, one of the biggest trading centres in Southeast Asia (see Figure 2).

Riau is rich in natural resources, particularly petroleum, natural gas, coal, forest, and rubber, oil palm and fibre plantations. There are huge deposits of oil and natural gas beneath the ground, making the province the country's largest producer of oil: 80% in the early 1970s [41] and 50% in 2006 [42, 43]. In 2010, there were 2.1 million ha of oil palm plantations [44] and also two giant pulp and paper companies. The economic potential attracts people from the surrounding areas and even from other islands (Java, Kalimantan) who hope to find a better life. Net migration to Riau has rapidly increased over time (Table 2).

Riau is divided into 10 regencies and 2 autonomous cities. Until 2004 the province included the Riau Islands, a large group of small islands located east of Sumatra and south of Singapore, but in July 2004 these islands were split off as a separate province.

According to the 2010 census, the population of Riau province was 5.54 million. Population growth was far above national average: the annual population growth rate was 4.35% in 1970-90, while the national average was only 1.49%. In 2000–10, the rate it was 3.9%, while the national figure was only 1.9%. This population growth reflects the economic potential of the province. Table 3 presents the detailed population growth by regency and city.

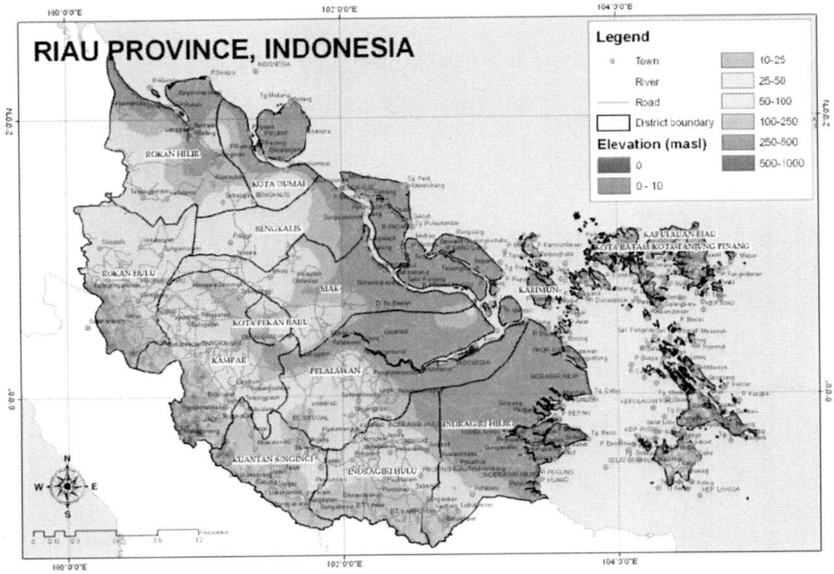

Figure 2. Topographical map of Riau. Source: ICRAF database

Table 2. Net migration to Riau and Sumatra, 1971–2005

[i] - Source: BPS 2011

	1971	1980	1985	1990	1995	2000	2005
Riau	175,498 (11%)	269,732 (11%)	221,654 (11%)	561,364 (22%)	714,828 (29%)	1,370,491 (72%)	1,127,824 (64%)
Sumatra	1,547,401	2,384,232	2,095,077	2,552,764	2,457,843	1,903,777	1,762,728

As shown in Table 3, the regency of Pelalawan – where timber plantations for the pulp and paper industry, and also oil palm plantations are located – had the highest population growth rate (6.8%). Siak, Kampar, Rokan Hulu and Rokan Hilir, which have significant areas of oil palm plantations, also experienced a high population growth (above 4%). Although there is no big difference in population between the districts, several districts (such as the city of Pekanbaru, Kampar Regency and Inderagiri Hilir Regency) have quite high concentrations compared to other districts. However, migrants are found almost everywhere. Many come from Java, and use the social networks that have been established through trans-migrants who settled before them [45].

Table 3. Area and population of Riau by regency and city, 2000 and 2010

[i] - *) Regency established in 2009; previously part of Bengkalis Regency. No data are available.

[ii] - Source: BPS, 2011

Regency / city	Area (km^2)	Population		
		Census 2000	Census 2010	Annual growth
Pekanbaru (city)	633	585,440	903,902	4.3%
Dumai (city)	2,039	173,188	254,337	3.8%
Kuantan Singingi Regency	5,235	216,732	291,044	2.9%
Indragiri Hulu Regency	7,611	247,306	362,961	3.8%
Indragiri Hilir Regency	13,633	555,701	662,305	1.8%

Pelalawan Regency	12,482	152,949	303,021	6.8%
Siak Regency	8,216	238,786	377,232	4.6%
Kampar Regency	10,814	447,157	686,030	4.3%
Bengkalis Regency	11,932	520,241	674,755	2.6%
Rokan Hulu Regency	7,225	265,686	475,011	5.8%
Rokan Hilir Regency	8,852	352,299	552,433	4.5%
Meranti Islands Regency*)	no data	no data	no data	–
RIAU PROVINCE	88,673	3,755,485	5,543,031	3.9%

Since the early 1980s, Riau Province has been the primary target for oil palm plantation development as part of Indonesia's agricultural development policy. The first large-scale oil palm plantation was established in Rokan Hulu Regency by PTPN (a state-owned plantation enterprise). The current Rokan Hulu Regency was part of Kampar Regency in the 1980s; like many other Indonesian provinces, Riau experienced the formation of new regencies in the era of post-Soeharto decentralization. This process, known as *pemekaran*, led to an increase in the number of regencies. Until 1999, Riau consisted of 5 regencies and 2 cities; it now has 10 regencies and 2 cities.

The establishment of oil palm plantations was initially realised through direct state investments. Then the private sector slowly took a more important role, especially after 1998. As mentioned, the establishment of oil palm plantations took place in the frontier area and in tandem with a resettlement programme to provide labour. Because the plantations provide higher wages and better profit, they attract people who seek to improve their livelihood and try their luck by engaging in oil palm growing. It is estimated that migrants comprise around 24% of the total population (about 67% of this group arrived in the context of the transmigration programmes [46]). This migration process stimulated oil palm to expand (occupying the surrounding area), and the decision of spontaneous independent farmers to cultivate oil palm has made the oil palm sector in the province boom (Table 4). Looking at the age distribution of the plantations, the table shows that the largest proportion of newest (still immature) oil palm plantations are under the smallholder system, followed by the private estates. It means that the recent development of oil palm plantation in Riau was done by private sector and smallholder farmers.

A study on livelihoods in Riau revealed that many of the migrants are well-off and are able to buy the land of local people and plant oil palms.

However, it also involves the encroachment of state forest land. For example, small-scale farmers have entered the Tesso Nilo National Park to establish oil palm plantations [46, 47].

Table 4. Oil palm plantations by actors in Riau, 2010

[i] - Sources: Dinas Perkebunan Provinsi Riau (2011) Statistik Perkebunan Provinsi Riau 2010. Pekanbaru

	Area (ha)			
	Immature	Productive stage	Not-productive	Total area
State-owned estates	1,000	78,546	-	79,546
Private estates	147,162	758,402	385	905,949
Smallholdings	318,969	780,959	17,625	1,117,553
	467,131	1,617,907	18,010	2,103,048

In only two decades, Riau overtook North Sumatra as the leading province in palm oil production (Figure 3). The rapid expansion of oil palm plantations has been in line with the economic attractiveness and profitability of this perennial crop. Independent smallholders are becoming increasingly dominant. They benefit from a direct link with the plantation companies, which provide technical assistance, good planting material, fertilizers and delivery contracts to the mill.

The rapid expansion of oil palm plantations is clearly illustrated in Riau's economic performance. Table 5 presents the distribution of oil palm plantations and the economic performance of the various regencies.

As shown in Table 5, each regency in Riau has oil palm plantations, with the planted area ranging from 12% in the capital city of Pekanbaru to 58.5% in Rokan Hulu Regency. In 2000–10, Riau underwent relatively high population growth. Indragiri Hilir Regency (which is dominated by peat land) had the lowest population growth rate. Rokan Hulu Regency (which has mostly mineral soil and the largest oil palm plantation in the province) had the highest population growth (see also Table 3).

Table 5. Oil palm plantation and economic performance in Riau, by regency

[i] - Sources: adapted from the respective [Regencies' statistics from 2011]

[ii] - Notes:

[iii] - 1. Regency established in 2009; previously part of Bengkalis Regency. No data are available.

[iv] - 2. Gross regional domestic product in constant prices (hence 2000 price).

[v] - 3. Annual average 2004–10.

[vi] - 4. Annual average 2008–10.

Regency / city	Area (km^2)	Population			Oil palm plantation (2010)		Economic performance		
		Population in 2010	Population density (per km^2)	Population growth rate 2000–10	Planted area (ha)	% to total area	GRDP per capita 2010 [2]	GRDP Growth [3]	Economic growth rate [4]
Kuantan Singingi	5,235	291,044	56	2.9%	121,709	23.4%	10,649.44	8.28	7.40
Indragiri Hulu	7,611	362,961	48	3.8%	118,538	15.4%	11,088.16	7.30	6.75
Indragiri Hilir	13,633	662,305	49	1.8%	213,541	15.5%	10,157.36	7.55	7.47
Pelalawan	12,482	303,021	24	6.8%	184,110	14.8%	10,321.78	7.21	7.10
Siak	8,216	377,232	46	4.6%	232,857	28.3%	10,123.38	7.45	7.37
Kampar	10,814	686,030	63	4.3%	353,792	32.4%	6,772,80	7,40	7,29
Rokan Hulu	7,225	475,011	66	5.8%	422,613	58.5%	5,395,28	7,06	6,02
Bengkalis	11,932	674,755	57	2.6%	177,130	21.0%	6,862,21	7,68	7,40
Rokan Hilir	8,852	552,433	62	4.5%	237,743	26.5%	7,439,10	7,69	7,57
Meranti	no dat	no data	no	no data	-	-	8,049	7,29	6,29

Islands 1)	a		data				,62		
Pekan baru (city)	633	903,902	1.428	4.3%	8,080	12.8 %	10,078,26	9,72	8,90
Dumai (city)	2,039	254,337	125	3.8%	32,935	16.2 %	8,221,24	8,60	8,56
RIAU	88,673	5,543,031	63	3.9%	2,103,048	23.6 %	8,782,70	7,99	8,90

Two economic indicators are used to assess the economic performance of all regencies in Riau in relation to oil palm plantation in the region, namely gross regional domestic product (GRDP) and economic growth. GRDP is used to measure the size of a region's economy. It is basically the aggregate of gross value added of all resident producer units in the region (using GRDP per capita as an approximation of the value of goods produced per person in a region); economic growth is measured on the basis of increase in the capacity of an economy to produce goods and services from one period of time to another.

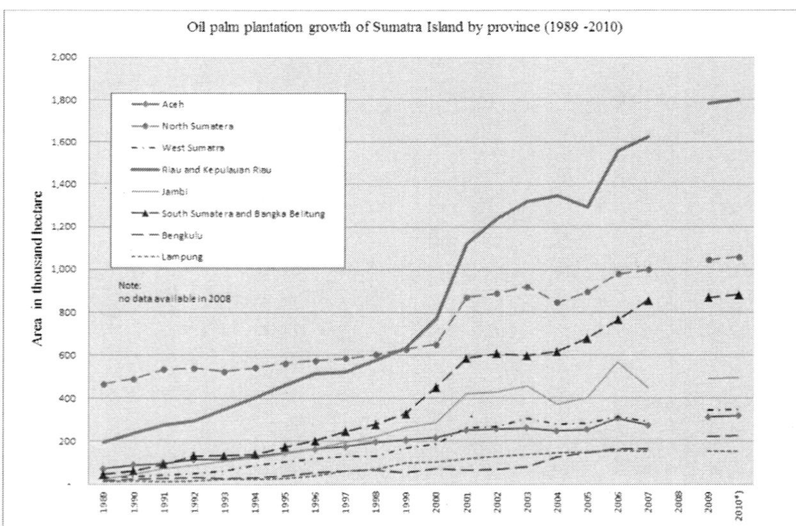

Figure 3. Oil palm plantation development in Sumatra Island by province (1989-2010)

In general, Riau performed well economically. GRDP per capita ranged from IDR 5.4 million in Rokan Hulu to IDR 11.0 million in Indragiri Hulu. The growth rate of GRDP in 2004–10 was almost 8%. In addition to oil palm, other sectors (e.g., mining in Rokan Hilir; and industry in Pelalawan and Bengkalis) also contribute to GRDP. However, looking at oil palm plantations, a recent economic assessment of small-scale oil palm cultivation revealed that returns to land ranged from IDR 92 million to IDR 143 million per ha in a 25-year cycle, and returns to labour from IDR 122,000 to IDR 178,000 per person per day. Returns to land of smallholding oil palm plantations ranged from IDR 125 million to IDR 266 million; returns to labour ranged from IDR 67,000 to IDR 297,000 [48]. It is therefore not too surprising that net migration was high, especially in sparsely populated area (where the agricultural wage rate were relatively high). Especially here, land use conversion was accelerated by the attraction of more people.

For the development of settlements, we looked at the Village Potential Statistics (PODES / *PotensiDesa*), a village-level census that is carried out three times per decade. PODES is administered by BPS (Indonesia Statistical Body) and serves to collect socio-economic information from all 69,000 Indonesian rural villages and urban neighbourhoods. The survey is based on responses of the village heads and includes a wide range of indicators, from population characteristics to infrastructure, economic activities and social life. According to PODES, from 1996 (roughly 10 years after the introduction of large-scale oil palm plantations through direct government intervention), some 100 settlement units were being prepared to become definite village administrative units in Riau. These settlements were previously part of the resettlement scheme linked to the agricultural development programme, and some of them were part of the oil palm development programme. In Kampar Regency, where large-scale oil palm plantation was first introduced, it was noted that there were 53 new settlements being prepared to become definite village administrative units. There were also 32 settlements (UPT; *Unit Permukiman Transmigrasi*) that still fall under the authority of the Transmigration Office. These settlements will also become definite village administrative unit in the near future.

Thus, the number of settlements in Riau also increased significantly. 'Settlements' in this context refers to *desa* (settlements in rural areas) and *kelurahan* (settlements in urban neighbourhoods), the *desa/ kelurahan* being the lowest level of government administration in Indonesia. Using the PODES data from 1996, 2005 and 2010, the number of *desa* in ex-Kampar Regency (Rokan Hulu, Kampar and Pelalawan regencies) increased from 309 in 1996 to 492 in 2011. In the same period, the number of *kelurahan* increased from 8 to 24. The figures indicate that significant changes have taken place over a 15-year period. Even though other factors may also play a role, the multiplication and rapid growth of the *desa* and *kelurahan* cannot be explained without taking into account the rapid expansion of oil palm plantations and related activities.

Making an assessment of the socio-economic impact of oil palm adoption, we see that people living in the immediate surroundings of oil palm estates often have considerable benefits [49]. Village-level assessment showed that villages that adopt oil palm as a major source of income tend to perform well on indicators of physical, financial and human capital. At the household level, an assessment showed that more than 18% of the households had increased their income (in real terms) by 200–300 per cent after 5 years of practising oil palm cultivation. About 35% had increased their income between 400 and 1300 per cent after 5–10 years of engagement, and about 45% of the households that had practised oil palm cultivation for more than 10 years had increased their income by 2200 to over 25,000 per cent. Such positive income effects make oil palm expansion extremely difficult to stop.

In policy debates on and cost–benefit analyses of the effects of oil palm expansion, much attention is often paid to the direct effects, that is, the effects on potential incomes and/or environmental effects. However, little or no attention is paid to the indirect population effects, how to deal with the massive population inflow, and whether oil palm production will in the long run be able to sustain urban populations.

5. CONCLUDING REMARKS

It is not difficult for policy makers to show that oil palms are an economically rentable crop with a huge potential for further economic growth. In addition to national demands, the growing worldwide interest in biofuels as an alternative to fossil fuels will increase demand for its feedstock and lead to the expansion of oil palm plantations in climatically suitable regions.

On the basis of a cost–benefit analysis of various crops, oil palm will probably continue to be seen as a highly profitable crop with interesting possibilities for being promoted as a source of 'green' development. The Indonesian population has increasingly been attracted by this crop, as it provides them with opportunities to benefit and multiply their incomes, which will in itself provide capital for improving consumption and having a good life.

At the same time, however, Riau shows us that there are environmental costs (deforestation, invasion into peat land areas etc.) and that oil palm expansion is accompanied by rapid immigration and urbanization. Even though policy attempts are made to control land conversion or to stop deforestation, much of what is happening today cannot easily be regulated. It is not the direct effects (i.e. the expansion of plantations and production) but the indirect effects and multipliers that bring into question the long-term sustainability of the development model. The establishment of new settlements, rapid urbanization and continuing immigration will require additional employment opportunities. And along with the growing population and the conversion of rice lands into oil palm fields, food security issues will increasingly become a problem. Rather than making quick money from oil palm production, the Indonesian government should make efforts to better control the indirect effects and especially to investigate the problem how to make oil palm-based economies more sustainable and equitable in the longer term. Without interventions – and with today's laissez-faire approach – further oil palm expansion will soon lead to the depletion of natural resources and an increase in social tensions as a result of unemployment and food insecurity.

ACKNOWLEDGEMENT

This article is based on the Agriculture beyond Food (AbF) programme, funded by Netherlands Organisation for Scientific Research (NWO) in collaboration with the Royal Dutch Academy of Sciences (KNAW).

REFERENCES

1. USDA (United States Department of Agriculture). 2012. Oil Seeds: World Market and Trade.Washington DC. Foreign Agricultural Services. Circular Series FOP 08–12, August 2012.Table 03.
2. Corley, R. H. V., and Tinker, P.B.H. 2003, The Oil Palm, 4th ed. Blackwell Science. Ltd, p. 1.
3. World Bank and IFC. 2011. The World Bank Group Framework and IFC Strategy for Engagement in the Palm Oil Sector. DRAFT FOR CONSULTATIONS.
4. Tan, K.T., Lee K.T., Mohamed, A.R., Bhatia, S., 2009. Palm oil: Addressing issues and towards sustainable development. Renewable and Sustainable Energy Reviews 13 (2009) 420–427. Elsevier.
5. McCarthy, John F. 2010. Processes of inclusion and adverse incorporation: oil palm and agrarian change in Sumatra, Indonesia. Journal of Peasant Studies, 37: 4, 821–850. DOI: 10.1080/03066150.2010.512460 (accessed 5 January 2011).
6. German, L., Schoneveld, G., Skutsch, M., Andriani, R.; Obidzinski, K., Pacheco, P., Komarudin, H., Andrianto, A., Lima, M.; Dayang Norwana, A.A.B. 2010. The local social and environmental impacts of biofuel feedstock expansion, A synthesis of case studies from Asia, Africa and Latin America. Infobrief No. 34. Center for International Forestry Research (CIFOR), Bogor, Indonesia.
7. Barlow, C., Zen, Z. and Gondowarsito, R., 2003. The Indonesian Oil Palm Industry. Oil Palm Industry Economic Journal. Vol. 3. No 1. Malaysia Palm Oil Board.

8. Tomich, T. P., van Noordwijk, M., Budidarsono, S., Gilison, A., Kusumanto, T., Mudiyarso, D., Stolle, F., and Fagi, A.M., 2001. Agricultural Intensification, Deforestation, and the Environment: Assessing Tradeoffs in Sumatra, Indonesia. In D.R. Lee and C. B. Barret (eds). Trade-offs and Synergies? Agricultural Intensification, Economic development, and Environment. Wallingford, UK: CAB International.
9. Edwards, R., Mulligan, D. and Marelli, L. 2010. Indirect land use change from increased biofuels demand: Comparison of models and results for marginal biofuels production from different feedstock. Final Report. Luxemburg. European Commission Joint Research Centre – Institute for Energy.
10. Koh, L.P. and Ghazoul J. 2008. Biofuels, biodiversity, and people: Understanding the conflict and finding opportunities. Biological Conservation. 141. pp. 2450–2460.
11. Germer, J. and Sauerborn, J., 2008. Estimation of the impact of oil palm plantation establishment on greenhouse gas balance. Environment, Development and Sustainability. Vol. 10, No. 6, pp. 697–716, DOI:10.1007/s10668-006-9080-1.
12. Tilman, David, Robert Socolow, Jonathan A. Foley, Jason Hill, Eric Larson, Lee Lynd, Stephen Pacala, John Reilly, Tim Searchinger, Chris Somerville, Robert Williams. 2009. Beneficial Biofuels – the Food, Energy, and Environment Trilemma. SCIENCE. vol. 325. 2009. P. 270. AAAS.
13. Pye, Oliver. 2009. Palm Oil as a Transnational Crisis in South-East Asia, Austrian Journal of South-East Asian Studies, 2 (2), Vienna, Austria. Society for South-East Asian Studies.
14. Koh, L.P., 2007. Potential Habitat and Biodiversity Losses from Intensified Biodiesel Feedstock production. Conservation Biology. Vol. 21, No. 5. pp. 1373–1375.
15. Fitzherbert, E.B., Matthew J. Struebig, A. Morel, F. Danielsen, Carsten A. Brvhl, Paul F. Donald and Ben Phalan. 2008. How will oil palm expansion affect biodiversity? Trends in Ecology and Evolution. Vol. 23, No. 10. 538–545.
16. Buckland, H. 2005. The Oil for Ape Scandal: How Palm Oil Is Threatening Orang-Utan Survival, Friends of the Earth, the Ape Alliance, the Borneo Orangutan Survival Foundation, The Orangutan Foundation (UK) and the Sumatran Orangutan Society.
17. Colchester, M., Jiwan, N., Andiko, Sirait, M., Firdaus, A. Y,, Surambo, A. and Pane, H. 2006. Promised Land: Palm Oil and Land Acquisition in Indonesia – Implications for Local Communities and Indigenous Peoples. Monograph. Forest People Programme (UK) and Sawit Watch (Indonesia).
18. Marti, Serge. 2008. Losing Ground. The human rights impacts of oil palm plantation expansion in Indonesia. A Report. Friends of the Earth, LifeMosaic and Sawit Watch.
19. Zen, Z., Barlow, C. and Gondowarsito, R., 2006. Oil Palm in Indonesian Socio-Economic Improvement – A Review of Options. Oil Palm Industry Economic Journal. Vol. 6. No 1. Malaysia Palm Oil Board.
20. Government of Indonesia, 1981. Transmigration Programme: an Overview. Paper presented by the government of Indonesia at the IGGI meeting, Amsterdam, May 1981.
21. Budidarsono, S., Khasanah, N., Ekadinata, A., Rahayu, S., Dewi, S., Suharto, R. and van Noordwijk, M. (eds) 2011. Reducing carbon emissions associated with oil palm plantations in Indonesia: accounting for greenhouse gas emissions over the full life cycle on peat and mineral soils and building capacity for and industry response to emerging environmental regulation in European markets. Research Report. Jakarta. Indonesian Palm Oil Commission and World Agroforestry Centre.
22. Jelsma I. et al. (2009). Smallholder oil palm production system in Indonesia: lessons from the NESP Ophir project. Wageningen: Wageningen University.
23. Michael Bunce. 1982. Rural Settlement in an Urban World. London. Croom Helm Ltd.
24. Syahza, Almasdi (2004). Development Impact of Palm Oil Plantation on Rural Economic Multiplier Effect in Riau Province). Jurnal Ekonomi, Th.X/03/ November/2005, Jakarta. Fakultas Ekonomi Universitas Tarumanagara, Jakarta.

25. Poku, Kwasi. 2002. Small-scale Palm Oil Processing in Africa. FAO Agricultural Services Bulletin No 148. Rome. Food and Agricultural Organization of United Nations.
26. Sanders, J. 1982. Palm Oil production on the Gold Coast in Aftermath of Slave Trade: Case Study of the Dante. The International Journal of African Historical Studies, Vol. 15 No. 1. pp. 49–63.
27. Het Oostkust van Sumatra Instituut, 1938. Data Deli 1883–1938. Mededeling No. 26.
28. Deasy, George F. 1948. Localization of Sumatra's Oil Palm Industry. Economic Geography, Vol. 18, No. 2 (Apr., 1942), pp. 153–158. Massachusetts. Clark University.
29. Stoler, Ann L., 1985. Capitalism and Confrontation in Sumatra's Plantation Belt, 1870–1979. London. Yale University Press.
30. Breman, Jan, 1989. Taming the Coolie Beast – Plantation Society and the Colonial Order in Southeast Asia. New Delhi. Oxford University Press.
31. Gilbert, A. and Gugler, J. 1992. Poverty and Development: Urbanization and the Third World. 2nd edition. London, Oxford University Press, pp. 50–52.
32. Higgins, B. 1956. Indonesia's Development Plan and Problems. Pacific Affairs, Vol. 29. No. 2 (June 1956) pp. 107–125. University of British Columbia.
33. Graham, E. and I. Floering, 1984. The modern plantation in the Third world. Great Britain. Croom Helm Ltd.
34. Pamin, K., 1998. A hundred and fifty years of Oil Palm development in Indonesia: from Bogor Botanical Garden to the Industry. In Proceedings of 1998 International Oil Palm Conference: Commodity of the past, today, and the future (eds A. Jatmika et al.), pp. 3–23, Indonesian Oil Palm Research Institute, Medan, Indonesia.
35. Bangun, Derom, 2006, Indonesia Palm Oil Industry, Conference Paper. The National Institute of Oilseed Products Annual Convention, 21–25 March 2006, Sheraton Wild Horse Pass, Phoenix, Arizona, USA. Available at http://www.oilseed.org/pdf/am_2006_materials/Bangun_Text.pdf (accessed 12 July 2012)
36. Hoshour, C. (1997). Resettlement and the politization of ethnicity in Indonesia. Bijdragen tot de Taal, Land, en Volkenkunde, Riau in Transition 153(4).
37. Humphrey, J. and Schmitz, H., 2000. Governance and upgrading: linking industrial cluster and global value chain research. Institute of Development Studies, Sussex, IDS Working Paper 120.
38. Casson, A., 2000. The hesitant boom: Indonesia's oil palm sub-sector in an era of economic crisis and political change. Bogor. CIFOR. P. 8.
39. FAO. FAOSTAT – Agriculture. Interactive online statistics http://www.fao.org/crop/en/
40. BAPPENAS (National Planning Body), 1975. Rencana Pembangunan Lima Tahun II (2nd Five Year Development Plan). Jakarta. BAPPENAS. Book D, p. 69.
41. Statistics Indonesia. Petroleum and Natural Gas Production, 1996–2010. Jakarta. Badan Pusat Statistik Republik Indonesia.
42. BPS dan BAPPEDA Provinsi Riau. 2008. Riau Dalam Angka 2007. Pekanbaru. BPS Provinsi Riau. Table 7.2.1.
43. Dinas Perkebunan Provinsi Riau. 2011. Statistik Perkebunan Provinsi Riau 2010. Pekanbaru. Dinas Perkebunan.
44. Susanti, A. and Burgers, P. 2012. Oil palm expansion in Riau Province, Indonesia: serving people, planet, profit? Background paper for European Report on Development. Confronting scarcity: managing water, energy and land for inclusive and sustainable growth. Belgium: European Union.
45. BPS dan BAPPEDA Provinsi Riau. 2011. Riau Dalam Angka 2010. Pekanbaru. Pekanbaru. BPS Provinsi Riau.
46. Derkzen, M.L. 2011. Conserve or convert? Forest as the fuel of oil palm – an assessment of rural livelihood and their strategies to cope in an oil palm environment, Riau, Indonesia. Field Work Report. International Development Studies, Utrecht University.

47. Heijman, S. 2010. Is the forest reduced to just an economic resources in an era of rapid oil palm expansion? Field Work Report. International Development Study, Utrecht University.
48. Budidarsono, S., Rahmanulloh, A., Sofiyuddin, M. 2012. Economics Assessment of Palm Oil Production. Technical Brief No. 26: Oil Palm Series. Bogor, Indonesia: World Agroforestry Centre (ICRAF) Southeast Asia Regional Program. 6 pp.
49. Budidarsono, S., Dewi, S., Sofiyuddin, M., Rahmanulloh, A. 2012. Socioeconomic Impact Assessment of Palm Oil Production. Technical Brief No. 27: Oil Palm Series. Bogor, Indonesia. World Agroforestry Centre (ICRAF) Southeast Asia Regional Program.

CHAPTER 3

Establishment of in vitro Culture of Rubber (Hevea brasiliensis) from Field-derived Explants: Effective Role of Silver Nanoparticles in Reducing Contamination and Browning

Moradpour M[1*], Aziz MA[1,2] and Abdullah SNA[1,2]

[1]*Laboratory of Plantation Crops, Institute of Tropical Agriculture, Universiti Putra Malaysia, 43300 Serdang, Selangor Darulehasan, Malaysia*

[2]*Department of Agriculture Technology, Faculty of Agriculture, Universiti Putra Malaysia, 43300 Serdang, Selangor Darulehasan, Malaysia*

ABSTRACT

The development of in vitro methods to produce high quality clonal planting materials of c for replanting and new planting is highly essential. The establishment of in vitro cultures of challenging woody plant species like H. brasiliensis has frequently been hampered by persistent microbial contamination, phenolic production, phase change and low response to media. Therefore, the overall goal of this study was to establish an efficient in vitro propagation method for H. brasiliensis through solving those persistent problems. In determining the most suitable method of reducing explant contamination, different concentrations of sodium hypochlorite, mercuric chloride and **nano** silver (NS) at different immersion times were assessed on explants derived from three physiological leaf stages (bronze, light green and stable green leaf stages) of preculture and non-preculture treated plants. The highest percentage of survival (94.44%) was from explants derived from the light green leaf stage of preculture treated **plants** immersed in 10 ppm NS for 20 min. Different types of adsorbents which were silver nitrate, activated charcoal and NS were assessed in controlling browning of in vitro cultures of H. brasiliensis. NS at 4 ppm significantly produced the highest percentage of explant survival (87.03%) with a low percentage of browning (10%). In determining the most suitable medium for in vitro culture of H. brasiliensis explants, Murashige and Skoog (MS) medium, Woody Plant medium (WPM) and Enjalric and Carron (MB) medium in combination with 3%, 6% and 9% sucrose concentrations were assessed. After 16 weeks of culture, the highest percentage of shoot formation (80%) was on MB medium supplemented with 6% sucrose.

Keywords: Micropropagation; ***Hevea brasiliensis;*** Sterilization; Phenolic compounds; Hevea leaf development

INTRODUCTION

Hevea brasiliensis is commonly the rubber tree, is a tree belonging to the family Euphorbiaceae. It is the most economically important member of the genus Hevea. It is of major economic importance because the milky latex extracted from the tree is the primary source of natural rubber. *H. brasiliensis* is still traditionally propagated by grafting clonal axillary buds onto unselected seedlings as stock plant to maintain intraclonal heterogeneity for both vigour and productivity [1]. New methods of propagation are highly recommended in order to eliminate the possible incompatibility between rootstock and scion, as well as to reduce the cost and time required for plant production. Micropropagation is a promising technique for large scale multiplication of selected clones. It has the advantages of small space requirements and reduced costs when compared to conventional vegetative propagation techniques. Although gross experimentation was conducted in the past for standardizing *in vitro* technologies in rubber, there had been many setbacks in commercializing these procedures [2,3]. A number of aspects inherent in the explant **tissue** are responsible for the delay in commercialization. These include: (i) the release of phenols; (ii) contamination by bacteria and fungi; (iii) their recalcitrant status; (iv) reduced axillary branches; (v) lack of sufficient juvenility; and above all (vi) increased sensitivity of *in vitro* raised plantlets towards environmental attributes. There are some methods and chemicals available to control *in vitro* contamination. The efficiency of some of these methods is low, and some of the chemicals are too toxic. The laboratory-based sterilization of explant cannot fulfil its potential to control *in vitro* contamination until it includes reference to the natural environment in which plants evolved. In fact, interaction with the environment is a key factor affecting the establishment of explants in plant tissue culture. Previous studies have shown that **physiological** aging negatively affects the micropropagation capacity of *H. brasiliensis* from microcuttings [4,5]. This phase change phenomenon has been observed in many studies and is associated with a noticeable decline in the potential of micropropagation or somatic embryogenesis of most arborescent species [6-9]. The second difficulty in *in vitro* establishment of Hevea species is the rapid browning and/ or necrosis of the explants. These problems are partly caused by oxidation of polyphenols which are abundant in Hevea species [10]. Also, inhibition of shoot organogenesis and necrosis of the explants are associated with considerable leaking of exudates into the culture medium [11]. There are no preliminary reports on reduction of browning in this plant material. To optimise an efficient tissue culture method for Hevea rubber, explant preparation is one of the important steps that need to be given attention in order to overcome the problem of microbial contamination and **browning**. Besides understanding the mother plant and its interaction with the environment it is also important that new types of sterilants are introduced to combat the problems faced during *in vitro* culture establishment. The antibacterial effects of Ag salts have been noticed since antiquity [12]. Nano silver is a new and non-toxic material which shows high capability in eliminating microorganisms, e.g., fungus, bacteria and viruses through the release of tiny particles of silver that could destroy the microorganisms [13]. The **detrimental** effects of nano silver have been shown on more than six hundred microorganisms [14]. There are no preliminary reports on reduction of contamination and browning using silver nanoparticles in *in vitro* culture of *Hevea*. This study attempts to establish an efficient *in vitro* culture of *H. brasiliensis* from field-derived explants through determining the most suitable method of reducing contamination and browning, as well as determining a suitable medium for promoting shoot formation (**Figure 1**).

MATERIALS AND METHODS

Plant materials

One to two year-old grafted plants of *H. brasiliensis* clone RRIM 2025, grown in polybags were obtained from Rubber Industry Smallholder Development

Authority (RISDA) nursery in Jelebu, Negri Sembilan, Malaysia. The plants were maintained in the Department of Agriculture Technology, Faculty of Agriculture, Universiti Putra Malaysia. The plants were regularly pruned to enhance lateral shoot development for use in this study. In order to assess the effects of environmental conditions on the degree of microbial infection of the Hevea tissues, the grafted plants were divided into two groups, with preculture **treatment** and without preculture treatment (control). The preculture treatment was applied by spraying the plants with a mixture consisting of 0.1% white oil, 0.2% benomyl and 90 mg L-1 Welgro fertilizer. Then, explants consisting of shoot tips and axillary buds were taken from branches at different leaf developmental (physiological) stages (Figure 2) indicated by the leaf colour: (1) bronze leaf stage, (2) light green leaf stage and (3) stable green leaf stage of the preculture and non-preculture treated plants.

Effects of sterilant type and concentration, preculture treatment and leaf developmental stage on overcoming contamination of *H. brasiliensis* explants

Sodium hypochlorite (NaOCl) at 10%, 20%, 30% and 40%, mercuric chloride ($HgCl_2$) at 0.1% and 0.2% and NS at 2.5, 5, 10, 50 and 200 ppm were freshly prepared prior to use and dispensed in jam jars. *Hevea* explants were harvested from the bronze, light green and stable green leaf stage of both preculture treated plants and the control. The explants were harvested at 10-12 days after pruning for the bronze coloured leaf stage, 23-25 days after pruning for the light green leaf stage and 40-45 day after pruning for the stable green leaf stage. All expanded leaves were removed and the explants were excised into 1 cm pieces and washed thoroughly until all latex and rubber particles were removed. The explants were washed in 70% v/v ethanol for 30 seconds. Then, the explants were **immersed** in the different concentrations of NaOCl for 10, 20 and 30 min, in $HgCl_2$ for 5 and 10 min and in NS for 5, 20 and 60 min, added with a few drops of Tween-20. They were shaken throughout in the sterilant and washed 5-6 times with sterile distilled water. Explants without exposure to the sterilants were considered as the control. Finally, the cut ends of the explants were trimmed and the explants were cultured on 40 mL of MS medium containing 3% sucrose, 0.3% Gelrite without hormone supplementation. Each treatment combination was repeated three times and each replication per treatment consisted of 6 explants. At the end of 4 weeks of culture, the percentage of explant survival was recorded.

Control of browning of *in vitro* cultures of *H. brasiliensis*

To reduce the occurrence of browning of cultures, different adsorbents which were silver nitrate ($AgNO_3$), activated charcoal (AC) and NS at various concentrations were evaluated. The explants were derived from the light green leaf stage of preculture treated plants and sterilized with 10 ppm NS for 20 min (best **sterilization** procedure determined above). MS basal medium was supplemented with 2, 4 and 6 ppm of $AgNO_3$; 1, 3 and 5 ppm of AC and 4, 8, 12 and 16 ppm of NS. The explants were cultured on 10 mL aliquots of the respective media contained in 25 x 100 mm vials. Explants placed on hormone-free MS medium were considered as the control. Each treatment was replicated three times and each replication per treatment consisted of 6 explants. The percentage of browning and percentage of explant survival were recorded at the end of 5 weeks of culture.

Effects of basal medium and sucrose concentration on shoot induction

Low response to media is another hindrance in the tissue culture of *H. brasiliensis* explants. In this experiment, full strength of WPM, MS and MB media were prepared in combination with 3%, 6% and 9% sucrose. The explants were derived from the light green leaf stage of preculture treated plants and sterilized with 10 ppm NS for 20 min (best sterilization procedure determined above). To overcome browning, the media were incorporated with 0.5% AC (best adsorbent treatment determined above). The explants were cultured on 40 mL of solid medium contained in 100 mL Pyrex Erlenmeyer flasks. Each treatment was replicated three times and each replication per treatment consisted of 18 explants. The explants were subcultured on fresh medium of the same constituents at 4-week intervals. The percentage of shoot formation was recorded at the end of 16 weeks of culture.

Culture conditions

Cultures were incubated on racks in an air-conditioned culture room with the temperature maintained at 25 ± 2°C and a relative humidity of 50 to 70%. Illumination was provided by 80 watt fluorescent lamps (Philips TL 80 W/55T) at an intensity of 45 µmol m-2 s-1 for 16 hours photoperiod.

Figure 1: Hevea brasiliensis.

Figure 1: Hevea brasiliensis.

Statistical analysis

Statistical analyses were performed using the statistical analysis system (SAS) program (Version 9.2). Data were subjected to one-way analysis of variance (ANOVA) for mean comparison and significant differences were calculated according to Duncan's multiple range test (DMRT). Differences at P=0.05 were considered statistically significant.

RESULTS

Effects of NaOCl in combination with preculture treatment and leaf developmental stage

Table 1 shows the effects of different treatments of NaOCl on survival of explants derived from the three leaf developmental stages of preculture and non-preculture treated plants. It is obvious that explants derived from the three leaf developmental stages of preculture treated plants generally

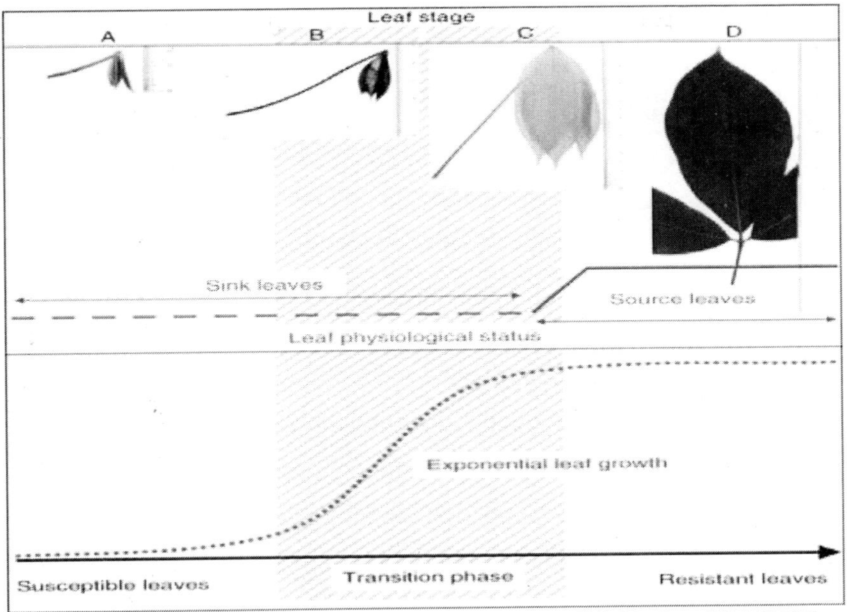

Figure 2: Relationship between exponential leaf growth, transition phase from completely susceptible leaf stage to completely resistant leaf stage, and the short physiological step from sink to source leaves. This qualitative comparison attempts to visualize the different developmental processes that turn a susceptible rubber tree leaf into a completely age-resistant organ (Lieberei, 2007).

Figure 2: Relationship between exponential leaf growth, transition phase from completely susceptible leaf stage to completely resistant leaf stage, and the short physiological step from sink to source leaves. This qualitative comparison attempts to visualize the different developmental processes that turn a susceptible rubber tree leaf into a completely age-resistant organ (Lieberei, 2007).

showed moderate to high percentage of explant survival. Increase in the concentration of NaOCl and duration of immersion decreased the percentage of survival of explants taken from all leaf developmental stages. Explants from the bronze leaf stage of preculture treated plants showed up to 79.62% survival after sterilization in 20% NaOCl for 20 min, while explants from the bronze leaf stage of non-preculture treated plants exhibited a maximum of 57.40% survival after sterilization in 20% NaOCl for 20 min. Meanwhile, explants from the stable green leaf stage of preculture treated plants exhibited a maximum of 62.96% explant survival after sterilization in 20% NaOCl for 20 min, while explants from the stable green leaf stage of non-preculture treated plants attained the highest survival of 51.85% also after sterilization in 20% NaOCl for 20 min as well. Overall, NaOCl was effective to some extent the increase in immersion time resulted in loss of explants due to browning followed by leaching of chlorophyll from the explants. In order to use NaOCl as a sterilizing agent for *H. brasiliensis* explants, it is therefore recommended to take explants from the bronze leaf stage of preculture treated plants and sterilized with 20% of NaOCl for 20 min.

Effects of $HgCl_2$ in combination with preculture treatment and leaf developmental stage

Table 2 shows the effects of different treatments of $HgCl_2$ on survival of explants derived from the three leaf developmental stages of preculture and non-preculture treated plants. The survival of explants was noticeably high (42.59%) for explants derived from the stable green leaf stage of preculture treated plants whereas the highest survival was only 27.77% for explants derived from the stable green leaf stage of non-preculture treated plants, both after sterilization in 0.1% of $HgCl_2$ for 10 min. Meanwhile the highest survival recorded for explant derived from the bronze leaf stage of preculture treated plants was 38.88% after sterilization in 0.1% of $HgCl_2$ for 10 min, while the highest from the bronze leaf stage of non-peculture treated plants was 37.03% survival after immersion in 0.1% of $HgCl_2$ for 5 min. In the case of explants derived from the light green leaf stage of preculture treated plants the highest survival was 31.47%, while the highest survival of explants taken from the light green leaf stage of non-preculture treated plants reached only 18.51%, both after exposure to 0.2% of $HgCl_2$ for 5 min.

Effects of Nano silver in combination with preculture treatment and leaf developmental stage

`Table 3 shows the effects of different treatments of NS on survival of explants from the three leaf developmental stages of preculture and non-preculture treated plants. It is very clear that explants derived from the light green and bronze leaf stages of preculture and nonpreculture treated plants generally showed high survival. Increase in the concentration of NS and duration of immersion decreased the percentage of survival of explants from all three leaf developmental stages of preculture and non-preculture treated plants. The survival of explants from the light green leaf stage reached 94.44% for preculture treated plants and 88.88% for non-preculture treated plants after sterilization in 10 ppm of NS for 20 min. Explants taken from the bronze leaf stage of preculture treated plants attained the highest survival of 85.18% following immersion in 10 ppm of NS for 5 min, while explants from the bronze leaf stage of non-preculture treated plants reached 77.77% of survival after sterilization in 10 ppm NS for 20 mim. Explants from the stable green leaf stage of preculture treated plants could also attain 77.77% survival after sterilization in 50 ppm of NS for 20 min. This study revealed that NS was more potent than NaOCl and $HgCl_2$ for effective disinfection of field-derived shoot tip and axillary bud explants of *H. brasiliensis*.

Control of browning of cultures of *H. brasiliensis*

Figure 3 shows that the effect of $AgNO_3$, AC and NS in controlling the browning of cultures of *H. brasiliensis* after 5 weeks of culture. NS at 4 ppm significantly produced the highest percentage of explant survival (87.03%) with a low percentage of browning (10%). In contrast, explants cultured on solid MS medium supplemented with 2 ppm $AgNO_3$ showed the highest percentage of **browning** (70%) compared to the rest of the treatments. The present study also showed that NS at 12 ppm and 16 ppm could totally prevent the incidence of browning of Hevea explants but with significant reduction in the survival of explants compared to NS at 4 ppm. **Figure 4A** shows shoot of *H. brasiliensis* free of contamination and browning.

Effects of basal medium and sucrose concentration on shoot induction

The results showed poor shoot formation (18%) on WPM medium with 6% sucrose. Meanwhile, there was no shoot formation on WPM medium with 3% and 9% sucrose and MS medium at all concentrations of sucrose tested (**Figure 5**). On the contrary, there was 80% shoot formation on MB medium supplemented with 6% sucrose, which differed significantly compared to the other treatments. This was followed by 34% shoot induction on MB medium with 9% sucrose and 28% on MB medium with 3% sucrose. It can clearly be seen that MB medium was most effective on induction of shoots from the fieldderived *H. brasiliensis* explants compared to WPM and MS medium, with the percentage of shoot formation being most enhanced in the presence of 6% sucrose **Figure 4B** shows healthy and well- formed regenerants on MB medium supplemented with 6% sucrose.

DISCUSSION

A successful tissue culture initiation step often begins with an effective explant sterilization technique and in the micropropagation of *H. brasiliensis* the establishment of an axenic culture is one of the most **challenging** steps. In this study, NaOCl, $HgCl_2$ and NS were assessed as sterilants on explants derived from different leaf developmental stages of preculture and non-preculture treated plants of *H. brasiliensis*. The use of increasing strength of NaOCl and $HgCl_2$ or extending the time of immersion in the sterilants resulted in the loss of explants due to browning followed by leaching of chlorophyll from the explants. The results indicated the necessity of using the correct concentration and immersion time in NaOCl and $HgCl_2$ as well as a suitable leaf developmental stage and with preculture treatment in order to overcome the contamination of H. brsiliensis explants *in vitro*. Debergh and Maene [15] proposed a scheme focusing on the preculture phase to obtain healthier explants and uniform responses. Growth, morphogenesis and rates of propagation *in vitro* can be improved by applying appropriate environmental and chemical pretreatments to the stock plants. To our knowledge, there are no reports so far investigating on the effects of NaOCl and $HgCl_2$ in combination with leaf developmental stage and preculture treatment on overcoming **contamination** of field-derived explants of *H. brasiliensis*. In the assessment of NS as a sterilant, the results revealed that it was more effective than NaOCl and $HgCl_2$ in reducing contamination of *H. brasiliensis* explants. Increasing the strength of NS or extending the time of immersion in the sterilant did not cause browning and leaching of chlorophyll from the explants. The highest percentage of survival (94.44%) was attained when explants derived from the light green leaf stage of preculture treated plants were sterilized in 10 ppm of NS for 20 min. Chen and Schluesener [16], Mahna et al. [17] and Razali et al. [18] also suggested the use of NS for disinfection of *in vitro* plant species they worked on. Abdi et al. [14] and Abdi [19] reported on the application of 100 ppm NS for 60 min as the most effective decontamination procedure for *Valeriana officinalis* explants. However in this study, a low concentration of 10 ppm NS at an immersion time of 20 min was highly effective in reducing microbial contamination and increasing the survival of *H. brasiliensis* explants. The mechanism of action of NS in terminating the microorganisms is not known clearly. However, it has been shown that NS releases silver ions (Ag+) slowly which can destroy the cell structure of microorganisms [20]. Dibrov et al. [21] stated that the effects of Ag++ on microorganisms may be due to their chemosmotic activity. These researchers showed that Ag+ affects the phospholipids and destroy the cell membrane of microorganisms. Ag+ also may substitute the sulphur in the –SH groups of cell membrane of microorganisms and destroy them. The destructive effects of Ag+ are related to the production of active silver containing organic compounds. These compounds can attract the microorganisms and destroy their structure. The use of nano-sized silver particles as antimicrobial agent has become more common as technological advances make their production more economical. One of the potential applications of silver is in the management of plant diseases. Since silver displays multiple modes of inhibitory action on microorganisms, it may be used for controlling various plant pathogens in a relatively safer way compared to synthetic fungicides [22]. Results obtained in this study including the overall low percentage of microbial contamination of explants derived from the different phenological stages after sterilization in NS reconfirmed the antifungal activity of NS as reported by Jo et al. [22].

Another most frequently encountered problem during micropropagation of *H. brasiliensis* is the occurrence of browning. Several authors have reported on methods of controlling **phenolic** compounds exudation from the species [23-25]. This study revealed the effectiveness of nano silver particles as an antioxidant and polyphenol adsorbent compared to silver nitrate and activated charcoal on explants of *H. brasiliensis.* Earlier investigations revealed that the dark exudate was very difficult to eradicate by pretreatment washes [26]. Ziv and Halvey [27] found that phenoloic compounds inhibited shoot formation in *Strelitzia reginae*and *Geranium* species and also caused browning and necrosis of *Kangarof paws.* Reports have shown that in other woody species when activated charcoal was added to the medium, with or without dark periods at the beginning of the experiment, it eliminated browning and was beneficial to shoot formation [28,29]. However, in our study on rubber though the explant survival reached 65% in the presence of 3 ppm activated charcoal the percentage of browning was still high at 40%. Meanwhile, evaluation on effects of $AgNO_3$ showed even higher percentage of browning (70%) with an explant survival of less than 30% upon application of 2 ppm $AgNO_3$. Earlier, Housti [10] reported that $AgNO_3$ showed only a slight positive effect on organogenesis of *H. brasiliensis* and a very slight antibrowning effect on callus of this species. It was also evident from our study that shoots of*H. brasiliensis* produced in the presence of activated charcoal and silver nitrate were not vigorous and showed symptoms of chlorosis. Interestingly, our study shows that nano silver is highly effective on the explant survival, behaviour and response to organogenesis. Transmission electron microscopy (TEM) images (**Figure 6**) of the silver nanoparticles used in this study indicated the extremely minute size of the particles. A study by Zhang et al. [30] reported that Ag+ release rates were found to depend on the primary particle size and concentration of NS, when other environmental factors (e.g. dissolved oxygen and protons) were kept constant.. To our knowledge, there are no reports so far investigating the effects of nano silver as an antioxidant and polyphenol adsorbent in overcoming browning of field-derived *H. brasiliensis* explants.

Table 1: Effects of NaOCl, preculture treatment and leaf developmental stage on percentage of survival of *H. brasiliensis* explants after 4 weeks of culture. Means followed by the same letter in the same column are not significantly different based on DMRT (p=0.05).

Time (min)	NaOCl concentration (%)	Mean percentage of explant survival					
		Bronze Leaf Stage		Light Green Leaf Stage		Stable Green Leaf Stage	
		Control	Preculture	Control	Preculture	Control	Preculture
Control	Control	0.0 ± 0.0f	1.85 ± 3.2^{h}	0.0 ± 0.0^{d}	5.55 ± 5.55^{g}	0.0 ± 0.0^{e}	3.7 ± 3.2^{e}
10	10	33.32 ± 5.55^{bcd}	44.44 ± 5.56^{cde}	1.85 ± 3.2^{d}	11.10 ± 5.55^{fg}	9.25 ± 6.41^{de}	5.55 ± 5.55^{e}
	20	40.73 ± 3.21^{b}	49.99 ± 5.55^{bcd}	11.10 ± 5.55^{c}	16.66 ± 5.55^{ef}	9.25 ± 6.41^{de}	18.51 ± 3.2^{cd}
	30	31.47 ± 8.48^{bcd}	53.7 ± 3.2^{bc}	16.66 ± 5.55^{c}	24.07 ± 3.2^{de}	12.96 ± 13.97^{cd}	22.22 ± 11.11^{c}
	40	16.66 ± 5.55^{e}	33.32 ± 5.55^{f}	12.96 ± 3.2^{c}	33.32 ± 5.55^{bc}	22.21 ± 5.55^{bc}	24.07 ± 3.2^{c}
20	10	37.03 ± 3.2^{bcd}	55.55 ± 5.55^{b}	29.62 ± 3.2^{b}	38.88 ± 5.55^{b}	29.62 ± 3.21^{b}	38.88 ± 5.55^{b}
	20	57.40 ± 3.21^{a}	79.62 ± 3.21^{a}	51.85 ± 3.2^{a}	55.55 ± 5.55^{a}	51.85 ± 3.2^{a}	62.96 ± 3.2^{a}
	30	38.88 ± 5.55^{bc}	46.29 ± 3.21^{bcde}	27.77 ± 5.55^{b}	49.99 ± 5.55^{a}	27.77 ± 5.55^{b}	44.44 ± 5.56^{b}
	40	27.77 ± 5.55^{d}	33.32 ± 5.55^{f}	11.10 ± 5.55^{c}	16.66 ± 5.55	16.66 ± 5.55^{cd}	22.21 ± 5.55^{c}
30	10	29.62 ± 8.48^{cd}	38.88 ± 5.55^{ef}	24.07 ± 3.2^{b}	37.03 ± 3.2^{b}	12.96 ± 3.2^{cd}	38.88 ± 0.0^{b}
	20	33.32 ± 5.55^{bcd}	42.59 ± 8.48^{def}	12.96 ± 3.2^{c}	25.92 ± 3.2^{cd}	20.36 ± 3.21^{bcd}	27.77 ± 5.55^{c}
	30	27.77 ± 5.55^{d}	33.32 ± 5.55^{f}	0.0 ± 0.0^{d}	12.96 ± 3.2^{fg}	11.10 ± 5.55^{de}	5.55 ± 5.55^{e}
	40	11.10 ± 5.55^{e}	22.21 ± 5.55^{g}	1.85 ± 3.2^{d}	5.55 ± 5.5^{g}	12.96 ± 3.2^{cd}	11.10 ± 5.55^{de}

Table 2: Effects of $HgCl_2$, preculture treatment and leaf developmental stage on percentage of survival of *H. brasiliensis*explants after 4 weeks of culture. Means followed by the same letter in the same column are not significantly different based on DMRT (p=0.05).

Times (min)	$HgCl_2$ Concentration (%)	Mean percentage of explant survival					
		Bronze Leaf Stage		Light Green Leaf Stage		Stable Green Leaf Stage	
		Control	Preculture	Control	Preculture	Control	Preculture
Control	Control	0.0 ± 0.0^{d}	5.55 ± 5.55^{b}	0.0 ± 0.0^{c}	1.85 ± 3.2^{c}	0.0 ± 0.0^{c}	0.0 ± 0.0^{c}
5	0.1	37.03 ± 3.2^{a}	24.07 ± 23.13^{ab}	14.81 ± 3.2^{ab}	27.77 ± 0.0^{a}	11.10 ± 5.55^{b}	25.92 ± 3.20^{b}
	0.2	24.07 ± 3.2^{b}	18.52 ± 6.41^{ab}	18.51 ± 3.21^{a}	31.47 ± 8.48^{a}	22.21 ± 5.55^{a}	20.36 ± 3.21^{b}
10	0.1	12.96 ± 3.2^{c}	38.88 ± 24.21^{a}	11.10 ± 5.55^{b}	27.77 ± 5.55^{a}	27.77 ± 5.55^{a}	42.59 ± 8.48^{a}
	0.2	0.0 ± 0.0^{d}	1.85 ± 3.2^{b}	1.85 ± 3.2^{c}	14.81 ± 6.41^{b}	5.55 ± 5.55^{bc}	5.55 ± 5.55^{c}

Table 3: Effects of NS, preculture treatment and leaf developmental stageon percentage of survival of H. brasiliensisexplants after 4 weeks of culture. Means followed by the same letter in the same column are not significantly different based on DMRT (p=0.05).

Times (min)	NS concentration (ppm)	Mean percentage of explant survival (%)					
		Bronze Leaf Stage		Light Green Leaf Stage		Stable Green Leaf Stage	
		Control	Preculture	Control	Preculture	Control	Preculture
Control	Control	0.0 ± 0.0^{f}	0.0 ± 0.0^{i}	0.0 ± 0.0^{h}	0.0 ± 0.0^{h}	0.0 ± 0.0^{h}	0.0 ± 0.0^{h}
5	2.5	49.99 ± 11.11^{cd}	55.55 ± 5.55^{f}	33.32 ± 5.55^{ef}	46.29 ± 3.21^{e}	16.66 ± 5.55^{g}	22.21 ± 5.55^{g}
	5	61.11 ± 11.11^{bc}	66.66 ± 5.55^{de}	40.73 ± 3.21^{de}	66.66 ± 5.55^{cd}	22.21 ± 5.55^{fg}	35.18 ± 3.2^{ef}
	10	66.66 ± 11.11^{ab}	85.18 ± 8.48^{a}	44.44 ± 5.5^{d}	90.73 ± 8.48^{a}	33.32 ± 5.55^{cd}	46.29 ± 3.21^{cd}
	50	75.92 ± 13.97^{ab}	81.47 ± 6.41^{ab}	62.96 ± 3.2^{bc}	77.77 ± 5.55^{b}	38.88 ± 5.55^{c}	49.99 ± 5.55^{c}
	200	61.10 ± 5.55^{bc}	72.21 ± 5.55^{bcd}	38.88 ± 5.55^{de}	49.99 ± 5.55^{e}	38.88 ± 5.55^{c}	51.85 ± 3.2^{c}
20	2.5	62.96 ± 3.20^{abc}	64.81 ± 3.2^{def}	55.55 ± 5.55^{c}	61.10 ± 5.55^{d}	29.62 ± 3.21^{def}	33.32 ± 5.55^{ef}
	5	68.51 ± 8.48^{ab}	72.21 ± 5.55^{bcd}	61.10 ± 5.55^{bc}	66.66 ± 5.55^{cd}	31.47 ± 3.21^{cde}	38.88 ± 5.55^{de}
	10	77.77 ± 5.55^{a}	77.77 ± 5.55^{abc}	88.88 ± 5.55^{a}	94.44 ± 5.56^{a}	51.85 ± 3.2^{b}	70.36 ± 3.21^{a}
	50	64.81 ± 8.48^{abc}	61.10 ± 5.55^{ef}	44.44 ± 5.56^{e}	66.66 ± 5.55^{cd}	53.7 ± 3.2^{b}	77.77 ± 5.55^{a}
	200	18.51 ± 3.21^{e}	18.51 ± 3.21^{h}	27.77 ± 5.55^{fg}	44.44 ± 5.56^{ef}	66.66 ± 5.55^{a}	61.10 ± 5.55^{b}
60	2.5	64.81 ± 3.20^{abc}	66.66 ± 5.55^{de}	44.44 ± 5.56^{d}	66.66 ± 5.55^{cd}	24.07 ± 3.2^{efg}	27.77 ± 5.55^{fg}
	5	68.51 ± 6.41^{ab}	68.51 ± 6.41^{cde}	57.40 ± 3.21^{c}	72.21 ± 5.55^{bc}	24.06 ± 6.41^{efg}	38.88 ± 5.55^{de}
	10	61.11 ± 11.11^{bc}	66.66 ± 5.55^{de}	66.66 ± 5.55^{b}	75.92 ± 3.2^{bc}	49.99 ± 5.55^{b}	44.44 ± 5.56^{cd}
	50	38.88 ± 11.11^{d}	44.44 ± 5.56^{g}	22.21 ± 5.55^{g}	37.03 ± 3.2^{f}	29.62 ± 3.21^{def}	44.44 ± 5.56^{cd}
	200	0.0 ± 0.0^{f}	0.0 ± 0.0^{i}	0.0 ± 0.0^{h}	16.66 ± 5.55^{g}	5.55 ± 5.55^{h}	3.7 ± 6.41^{h}

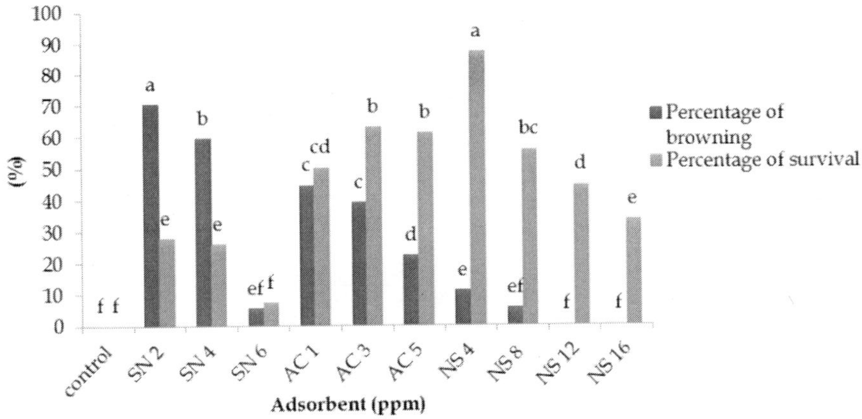

Figure 3: Effects of different adsorbents on control of browning of in vitro culture of H. brasiliensisafter 5 weeks of culture. Values represent means for 18 explants per treatment per replication. The same letters represent no significant difference between treatments according to one-way analysis of variance (P=0.05).

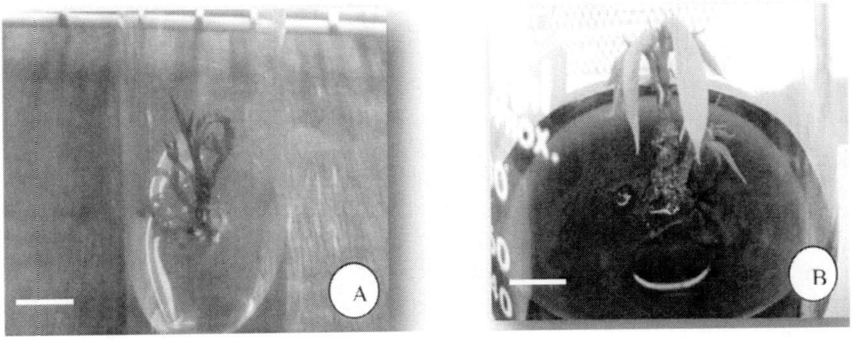

Figure 4:*H. brasiliensisshoots* after NS treatment (bar=0.75 cm), (A) shoot without contamination and browning (bar=0.28 cm), (B) healthy and well-formed regenerants (bar=0.75 cm).

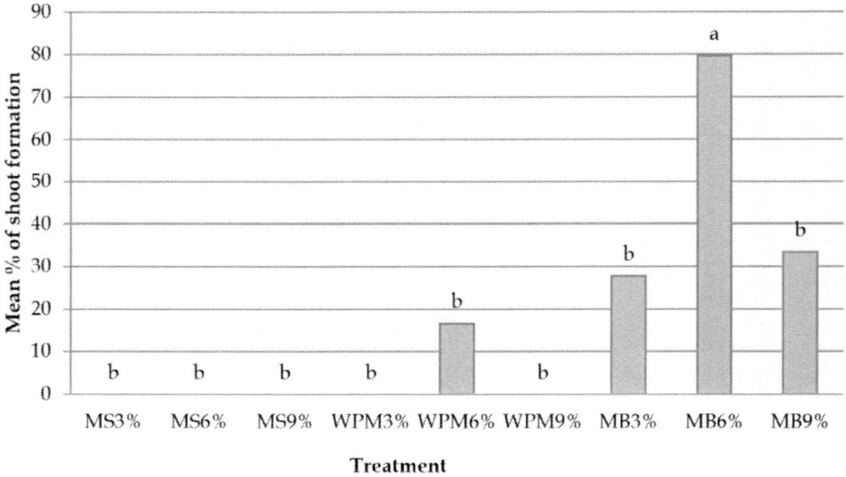

Figure 5: Effects of basal medium and sucrose concentration on shoot induction of *H. brasiliensis*explants after 16 weeks of culture. Medium formulations: MS, WPM and MB; Sucrose concentrations: 3%, 6% and 9%. Values represent means for 18 explants per treatment per replication. The same letters represent no significant difference between treatments according to one-way analysis of variance (P=0.05).

In this study, MB medium was comparatively found to be more suitable than WPM and MS media on *in vitro* shoot induction of *H. brasiliensis*. MB medium had earlier been specifically suggested for *in vitro* culture of *Hevea* by Enjalric and Carron [31]. There are numerous reports on the better performance of woody plant species grown on low salt media including WPM, however it was not ideally the case with *H. brasiliensis.* This study showed that the mean percentage of shoot formation from cultures grown on MB medium supplemented with 3%, 6% and 9% sucrose was better than that obtained on WPM with 6% sucrose. WPM contains much less macro elements in comparison with MB medium. Earlier, Seneviratne et al. [32] showed better growth and survival of*H. brasiliensis* axillary bud on WPM supplemented with 10% sucrose in comparison with MS medium [33-35].

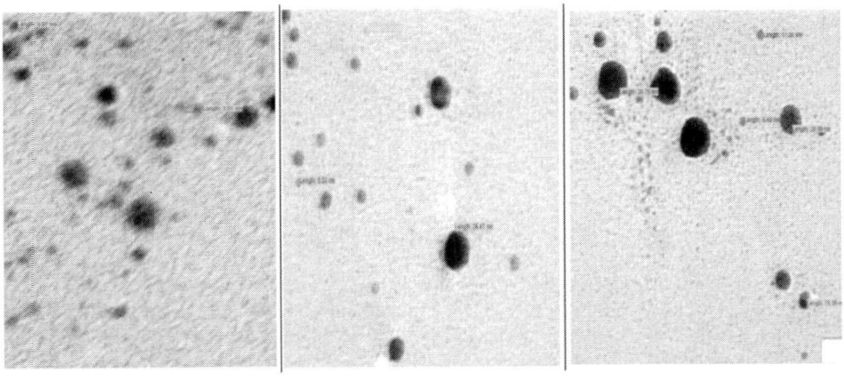

Figure 6: Transmission electron microscopy (TEM) image of silver nanoparticles. Scale bars: 50, 100 and 200 nm..

CONCLUSION

This research study indicates that the successful establishment of invitro culture of *H. brasiliensis* requires the understanding of the species leaf developmental stages, the interaction of the explant source with the environment and the effective application of the non-toxic silver nano particles in reducing microbial contamination and browning of Hevea explants. An efficient initial culture establishment developed in this study thus opens the way for future application of the microcutting technique for propagation as well as genetic improvement of *H. brasiliensis.*

ACKNOWLEDGEMENTS

The authors are thankful to the Department of Agriculture Technology, Faculty of Agriculture, for the laboratory facilities provided.

REFERENCES

1. HuaYW, Huang TD, Huang H (2010) Micropropagation of self-rooting juvenile clones by secondary somatic embryogenesis inHeveabrasiliensis. Plant Breeding 129:202-207.
2. Carron MP, DAuzac J (1992) Biochemical and histological features of somatic embryogenesis in Heveabrasiliensis. Indian J Natural Rubber Res 5:7-17.

3. Thulaseedharan A, Jayasree PK, Venkatachalam P (2000) Biotechnological approaches for crop improvement in rubber. In: Biotechnology in Horticultural and Plantation Crops. Malhotra Publishing House New Delhi 323-351.
4. Lardet L, Enjalric F, Carron MP (1990) Culture d'apexd'Heveabrasiliensis (Mull. Arg.). Influence du stademorphogenetiqueet de l'age des explants. ComptesRendus de l'Académie des Sciences Paris 310:195-202.
5. Perrin Y, Lardet L, Enjalric F, CarronMP (1994) Rajeunissement de clones matures d'Heveabrasiliensis (Müll.Arg.) parmicrogreffage*in vitro*. Can J Plant Sci 74:623-630.
6. Bonga, JM (1982) Vegetative propagation in relation to juvenility, maturity and rejuvenation. In Tissue Culture in Forestery. Springer 387-412.
7. Hackett WP (1985) Juvenility, maturation and rejuvenation in woody plants. Hortic Rev 7:109-155.
8. Monteuuis O (1989) Maturation concept and possible rejuvenation of arborescent species. Limits and promises of shoot apical meristems to ensure successful cloning. In Breeding Tropical Trees: Population Structure and Genetic Improvement Strategies in Clonal and Seedling ForestryWinrock International 106-118.
9. vonAderkas P, Bonga JM (2000) Influencing micropropagation and somatic embryogenesis in mature trees by manipulation of phase change, stress and culture environment. Tree Physiol 20:921-928.
10. Housti F, Coupe M, d'Auzac J (1992) Browning mechanisms and factors of influence in in vitro Heveacalli cultures. Indian J Natural Rubber Res 5:1-2.
11. Mederos-Molina S, Trujillo MI (1999) Elimination of browning exudate and in vitro development of shoots in Pistaciavera L. cv. Mateur and PistaciaatlanticaDesf. Culture. ActaSocietatisBotanicorumPoloniae, 68:21-24.
12. Silver S,Phung LT (1996) Bacterial heavy metal resistance: new surprises. Annu Rev Microbiol 50: 753-789.
13. Sondi I,Salopek-Sondi B (2004) Silver nanoparticles as antimicrobial agent: a case study on E. coli as a model for Gram-negative bacteria.J Colloid Interface Sci 275: 177-182.
14. Abdi G, Salehi H, Khosh-Khui M (2008) Nano silver: a novel nanomaterial for removal of bacterial contaminants in valerian (Valerianaofficinalis L.) tissue culture. ActaPhysiol Plant 30:709-714.
15. Debergh PC, Maene LJ (1981) A scheme for commercial propagation of ornamental plants by tissue culture. ScientiaHortic 14:335-345.
16. Chen X,Schluesener HJ (2008) Nanosilver: a nanoproduct in medical application. ToxicolLett 176: 1-12.
17. Mahna N, Vahed SZ, Khani S (2013) Plant in vitro culture goes nano: nanosilver-mediated decontamination of ex vitro explants. J Nanomed and Nanotechnol.
18. Razali M, Nur Aida MP, Habsah M, Hairiyah M, Rohaya MA, et al. (2012) Influence of nanosilver packaging on quality retention of peeled garlic. In VII International Postharvest Symposium 1012:1317-1322.
19. Abdi G (2012) Evaluation on the potential of Nano silver for removal of bacterial contaminants in valerian (Valerianaofficinalis L.) tissue culture. J Biol Environ Sci 6:199-205.
20. Lubick N (2008) Nanosilver toxicity: ions, nanoparticles--or both? Environ SciTechnol 42: 8617.
21. Dibrov P,Dzioba J, Gosink KK, Häse CC (2002) Chemiosmotic mechanism of antimicrobial activity of Ag(+) in Vibrio cholerae. Antimicrob Agents Chemother 46: 2668-2670.
22. Jo YK, Kim BH, Jung G (2009) Antifungal activity of silver ions and nanoparticles on phytopathogenic fungi. Plant Dis 93:1037-1043.
23. Enjalric F, Carron MP, Lardet L (1987) Contamination of primary cultures in tropical areas: the case of Heveabrasiliensis. Bacterial and Bacteria-like Contaminants of Plant Tissue Cultures 225:57-66.
24. Seneviratne P, Wijesekara GAS (1996) The problem of phenolic exudates in in vitro cultures of mature Heveabrasiliensis. J Plantation Crops 24:54-62.

25. Sirisom Y, Te-chato S (2012) The effect of peptone and silver nitrate on in vitro shoot formation in HeveabrasiliensisMuell Arg. J AgricTechnol 8:1509-1516.
26. Barghchi M, Alderson PG (1985) *In vitro* propagation of Pistaciavera L. and the commercial cultivars Ohadi and Kalleghochi. J HorticSci 60: 423-430.
27. Ziv M, Halevy AH (1983) Control of oxidative browning and in vitro propagation of Strelitziareginae. HorticSci 18: 434-436.
28. George EF (1996) Plant propagation by tissue culture. Part 2. Exegetics Ltd. Hardcover USA.
29. Gamborg O, Phillips G (1995) Plant cell, tissue and organ culture: fundamental methods. Springer-Verlag, Heidelberg, New York, USA.
30. Zhang W, Yao Y, Sullivan N, Chen Y (2011) Modeling the primary size effects of citrate-coated silver nanoparticles on their ion release kinetics. Environ SciTechnol 45: 4422-4428.
31. Enjalric F, Carron MP (1982) Microbouturage in vitro de jeunes plants d'Heveabrasiliensis (Kunth., Mull. Arg). ComptesRendus Des Seances de l'Academie Des Science 295: 261-264.
32. Seneviratne P, Flegmann AW, Wijesekera GAS (1996) The effects of the basic medium and the carbohydrate content on shoot cultures of Heveabrasiliensis. J Rubber Res Inst Sri Lanka 78: 60-68.
33. Alderson PG, Dullforce WM (1987) Micropropagation in horticulture, practice and commercial problems. The University of Nottingham Trent Print Unit.
34. Lieberei R1 (2007) South American leaf blight of the rubber tree (Hevea spp.): new steps in plant domestication using physiological features and molecular markers. Ann Bot 100: 1125-1142.
35. Murashige T, Skoog F (1962) A revised medium for rapid growth and bio assays with tobacco tissue cultures. Physiol Plant 15:473-479.

CHAPTER 4

Mitigating Biodiversity Concerns in Eucalyptus Plantations Located in South China

Roger A. Williams

School of Environment and Natural Resources, Ohio State University, Columbus, Ohio, USA

ABSTRACT

China's growing economy and changes in policies that encourage afforestation, particularly in the industrial sector, have led vast areas in south China to be planted with eucalyptus. These large areas of eucalyptus plantings have elicited environmental concerns for two primary reasons. First there is a concern related to the water demand of eucalyptus, in which it is feared these large areas of eucalyptus will deplete aquifers and create shortages in water supplies. The second concern is in regard to the reduction in biodiversity across large landscapes, leading to further ecological demises. This paper proposes two ideas to possibly mitigate some of the biodiversity concerns. The first is the interplanting of alder-leaf birch (Betula alnoides), a native but dwindling species in south China, to enhance biodiversity and encourage it's reestablishment across the landscape. The second is to encourage retention harvests of alder-leaf birch planted within eucalyptus plantations to enhance not only biological diversity but also structural diversity across the landscape. Alder-leaf birch has demonstrated great potential in producing high quality timber and wood for use in furniture manufacturing.

Keywords: Eucalyptus, Alder-Leaf Birch, Biodiversity, Retention Harvests, Mixed Plantations

1. INTRODUCTION

China has experienced a rapid increase in the amount of forest plantations created for the purpose of providing a resource base for the expansion of the wood products sector. China's booming economy has created a high demand for construction materials and paper. During the first decade of the 21st century China developed policies that further enhanced the wood products sector, thus encouraging the large expanse of eucalyptus plantations.

The Chinese forest sector has experienced a transformation over the past decade with a marked increase in the demand by China for wood products. This demand has been the result of the growth in exports as well as the increase in domestic demand. Over the same period environmental concerns have resulted in significant constraints being placed on timber production from China's natural forests, while changes in land tenure and use rights have paved the way for increased private ownership and investment in plantation forestry.

The Chinese government initiated forestry sector reform beginning in 2000 where a series of policies were introduced that encouraged the development of the forest industry sector. The Chinese government released a Forestry Industry Promotion Plan for the years 2010-2012 with the following four main objectives: 1) to assist the leading 100 key state level forestry enterprises; 2) to support policies that encourage leading enterprises to raise debt finance; 3) to continue and increase the subsidizing the purchase of seedlings required for plantations; and 4) to reduce charges for the forest maintenance fund. Additionally, the Development Plan aim was to maintain the growth rate of forestry's output value at 12% per year and raise the output value of the forestry industry from RMB1.44 trillion in 2008 to RMB2.26 trillion by 2012.

China has undertaken a set of policy initiatives aimed at expanding its forest cover, most notably the Industrial Base Development Program. This program promotes the establishment of fast-growing and high-yield timber (FGHY) plantations in selected regions by subsidizing plantation establishment with loan interest subsidies, discounted loans from state banks and extended repayment periods [1]. The program's goal is to plant 13.3 million hectares of fast growing plantations between 2001 and 2015, creating an estate that can supply 130 million $m^3 \cdot yr^{-1}$ of timber for China's domestic market. Private companies are expected to play an important role in the operation of the program by leasing or sub-leasing state-owned and collectively owned land for plantation development.

These policy changes have caused a rapid increase in the amount of forest cover, particularly with plantations comprised of fast-growing species. According to the 7th National Forestry Inventory (2004 to 2008), the forest area of China is 195.5 million hectares, 20.4% of the country's land area. The forest area has increased by 20.5 million hectares since the 6th National Forestry Inventory conducted between 1999 and 2003 [2]. China has the fifth largest forest area of any country in the world with natural forest covering 125 million hectares and plantations totaling 64.5 million hectares [3]. However, China's per capita forest area is only one quarter of the world average, and the percentage of land under forest cover is low by international standards.

The species most commonly used in southern China for FGHY plantations is eucalyptus. The eucalyptus species, endemic to Australia, was a tree that was first introduced into China in 1890 for ornamental purposes, and later developed into large plantations in the 1950's [4]. Since then, it has been used extensively as protection forests and production forests. Much research has been carried out on eucalyptus since the 1980's, which has helped perpetuate its use. As a result, eucalyptus has become a widespread and valuable resource in southern China for commercial timber and provides a variety of oil extracts for medicinal and aromatherapy purposes. The fast growth of eucalyptus also makes this resource readily available for bioenergy use, which is a top priority in China's government agenda as the Renewable Energy Law was implemented beginning in January 2006 [5].

Guangxi, a southern province, has recently realized a sharp increase in the number of eucalyptus plantations due in part to the increased investments by the Chinese government in industrial development and by the forest products companies such as Guangxi Sunway Forest Products, Stora Enso, Sino-Forest Corporation, Guangxi Fenglin Wood Industry Group, Guangxi Huafeng Forestry Inc and Asian Pulp and Paper. Guangxi now is one of the top three provinces accounting for the total area in eucalyptus plantations in China [6].

China now has one of the largest areas of eucalyptus plantations in the world, occupying more than 2.0 million ha of eucalyptus plantations mainly in Guangdong, Guangxi and Hainan provinces [6]. The area of eucalyptus plantations has since increased to 3.13 million hectares in 2010. Fast-growing species and hybrids (e.g., E. urophylla, E. tereticornis, E. grandis, E. urophylla × E. grandis) are most widely planted in South China. These eucalyptus plantations have primarily employed a continuous cropping system, involving clear-cutting, prescribed burning, and mechanical site preparation on the site, and 60% - 70% of them have been managed in short rotation for industrial purposes. The normal rotation period of short-rotation eucalyptus plantation is usually 6 or 7 years.

Plantations of eucalyptus and other exotic tree species have been extensively established in southern China during the past 50 years [7]. The total area of eucalyptus plantations in China was around 460,000 ha in 1993 [8], increasing by 100,000 ha or more per year to an estimated 1.3 million ha in 2000 [9] in response to government afforestation programs and cooperative joint ventures between provincial forestry authorities and foreign investors. More recent estimates place the total planted area of eucalyptus in China to more than 2.5 million ha in 2010 [10], making it second only to Brazil, the country with the largest area of eucalyptus plantations with an estimated 3.1 million ha. In addition to their key role in timber and fiber production, eucalypt plantations in China have an important ecological role in the re-vegetation of degraded lands [11] [12].

2. ENVIRONMENTAL CONCERNS

Investments in the timber and forests by large institutional investors, asset owners and pension funds have increased significantly over the past several years. These investments have the potential to reap handsome returns as investments are made in the biological growth of forests which occurs independently of movements in the financial markets. Forest industries have a similar view of these investments but also look at this investment from the perspective that it is the renewable raw material that provides the basis for the existence of their industry and provides value-added potential to the various products made and sold.

With a heightened awareness of environmental impacts (both negative and positive) of forest management practices, companies must pay more attention to the environmental outcomes of their management activities. Digital and social media have enabled more transparency of company activities than in the past. This includes greater public access to satellite imagery and Google Earth. Greater societal demands and expectations on environmentally responsible practices are becoming more of the norm. Accordingly, there are greater expectations of companies to practice more responsible land stewardship in their management practices. This is becoming more of a requirement of consumers in purchase of various products produced. Subsequently, investors are likewise looking at not only profit margins but the environmental ethics being practiced by said companies.

The introduction of eucalyptus into south China has been controversial. These large areas of eucalyptus plantings have elicited environmental concerns for two primary reasons. First there is a concern related to the water demand of eucalyptus, in which it is feared these large areas of eucalyptus will deplete aquifers and create shortages in water supplies. The second concern is in regard to the reduction in biodiversity across large landscapes, leading to further ecological demises. The fact that eucalyptus is an introduced species to southern China has exacerbated this concern.

Disputes have arisen between larger corporate eucalyptus plantations and the use of land by farmers. The source of these disputes are varied, but tend to focus on issues of land takeover and use by the state, and thereby minimizing farmer's sources of income. There are also concerns over the conversions of large areas of native forests and farmland to eucalyptus plantations, creating less diversity across the landscape and other negative impacts on the environment. Many ecological concerns have been raised regarding the extensive areas planted to eucalyptus, including whether there is adverse effects on floristic composition and species diversity [13]-[16], soil fertility [17], and local hydrology and water use [18] [19]. Large contiguous areas are more susceptible to disease and insects, such as the gall wasp. Many concerns have been raised regarding water usage by eucalyptus trees, a very water-demanding species, upon soil water resources. Many questions in this regard have not been answered, but some literature indicates that this high water usage can be partly mitigated through wider planting spacing [20].

Eucalyptus plantations in several countries have been the subject of criticism because of their high water use and other negative environmental impacts [20]. Examination of the evidence for these claims has usually concluded that well-managed plantations are beneficial rather than detrimental to the environment [21]-[23]. However, some studies indicate that whenever water resources are limited, the area, location and management of plantations must be carefully considered to avoid conflict with other water users [24] [25].

One of the pro-arguments for the vast areas of planted eucalyptus in south China is that it restores a level of biodiversity on lands that have been in agriculture for extensive periods of time. However, one has to be careful about claims that reforestation of any degraded land would offset any loss of species following deforestation [26]. While it is known that the conversion of primary forests to intensive agriculture leads to dramatic losses in biodiversity, there is far less certainty about the conservation to secondary and planted forests [26]-[28]. Biodiversity in eucalyptus plantations has been widely studied since it is one of the major concerns regarding the management of this forest type [13] [16] [29]-[31].

However, depending on forest location and specific management method these studies have found that management of eucalyptus plantations can encourage the regeneration of natural forest biodiversity in the understory and restore degraded land, or it can have an opposite effect by facilitating the recovery of light-demanding grasses and other pioneer weeds, thus suppressing the recruitment of native plants [32]. It is believed that allelopathogenic effects of the eucalyptus litter, which contains a variety of oils and resins from foliage, has an adverse effect on vegetation thus reducing the number and amount of understory species [10] [22].

There are a variety of studies that investigate the allelopathic effects of eucalyptus on understory vegetation [33] and results are contrasting. Some studies find, for example, that eucalyptus inhibits understory herbaceous and woody species growth by producing allelochemicals [34] [35]; while on the other hand, some studies find that the overstory eucalypt trees do not seem to inhibit the colonization of understory woody or herbaceous species [29] [33] [36]. A variety of environmental conditions, particularly related to soil depth, texture, and moisture and climate [33] may play a role in this variation of findings.

3. MITIGATING THE CONCERNS

One possible way to mitigate the biodiversity and other environmental issues/concerns regarding vast areas planted to eucalyptus and still retain high production levels for the growing industries and economy is to develop mixed plantings of eucalyptus with other native fast-growing species. There is much evidence to support the benefits of mixed forest species plantings [37] and the role of mixed forests in maintaining biodiversity and ecosystem function [33] [38] [39]. Many benefits that have been reported with the use of mixed-species plantations include the potential to increase biomass production and carbon sequestration [40]-[45], improve soil fertility and enhance nutrient cycling [41] [46], and protection from pest and disease outbreaks, thus improving risk management [46]-[48]. Even though there often is resistance to utilizing mixed-species plantations for timber production, they show potential to be used as a silvicultural system for growing high value timber and at the same time provide a more diverse range of products [42] [47] [49].

One species that has been receiving much attention lately and is a candidate to consider in mixed plantings with eucalyptus is alder-leaf birch (Betula alnoides Buch.-Ham. ex D. Don) which is indigenous to the northern part of India Peninsula, Myanmar, Indochina Peninsula and south China [50]. This species is fast growing and well formed, and is suitably used to grow large size timber; its wood is commonly used in high-quality furniture making and room decorations [51]. The growth of alder-leaf birch has been shown to be better on shady slopes as compared to sunny slopes [52]. This suggests that alder-leaf birch requires some protection via shady conditions, thus making it potentially suitable to plant in mixtures. Alder-leaf birch also plays important roles in water conservation, long-term maintenance of land fertility and biodiversity of forest ecosystems [50]. It is becoming one of the most important native, broadleaf timber species for plantation establishment in southwestern China [53]-[55]. It is a pioneer species, supporting forest regeneration on marginal lands, clear-cuts, and forest gaps. Because the tree is fast-growing with a straight stem and little tapering, alder-leaf birch is suitable for large-diameter timber cultivation [54]. Alder-leaf birch outperformed other tree species (Mytilaria laosensis, Castanopsis hystrix, Cunninghamia lanceolata, Pinus massoniana) native to south China in terms of height, diameter and total volume growth in plantation situations. Accordingly, alder-leaf birch had the greatest economic gain and potential compared to the other species studied and showed promise as being a valued timber species [56]. The fast growth rate of alder leaf birch allows it to have comparable rotations to eucalyptus, and has wood properties that enable it to become economically competitive with eucalyptus [55].

In a comparative study of alder-leaf birch plantations with native forests, Reference [54] found that 8- to 13-yr-old alder-leaf birch plantations had higher species diversity in the understory compared to the natural native forests. The species richness increased significantly with the increase in plantation age and the understory vegetation changed from pioneer species to more shade-tolerant plants and sciophytes. The fact that alder-leaf birch plantations demonstrated higher diversity than the other native forests indicates that these plantations can facilitate the development of plant species under their canopy in terms of species composition, diversity, and restoring native species diversity. The species could thus not only be a valuable timber species but also be an effective tree species for restoring tree diversity in tropical southwestern China [57]. Alder-leaf birch also shows promise in sequestering carbon in both monocultures and mixed plantings, sequestering 148.42 and 140.33 tC/ha, respectively, at an annual rate of 4.01 and 4.59 t/ha/yr [58] which compares to 103.73 tC/ha reported for mature eucalyptus stands in the Pearl River Delta of south China [59].

Guangxi province is one of the most important areas in China for this species, where the natural populations are fragmented by limestone soil, which is unsuitable for the long-term survival of alder-leaf birch [60] [61]. Within the last two decades, the wide use of this species for furniture and decoration has led to heavy deforestation, consequently resulting in a more fragmented and diminished distribution of this forest type [60]. Therefore, efforts to re-establish this species on the landscape within this province as well as throughout its native range add greater importance in the use of this species in restoration efforts.

4. WILL IT WORK?

The overarching question to this approach is whether mixed plantations of eucalyptus and alder-leaf birch will increase biodiversity, mitigate any soil degradation that may occur from short rotation practices, reduce soil water losses, and be sustainable and economically feasible. While some short term benefits may be sacrificed initially as this system is put into place, the potential long term benefits appear to be promising. Selective thinning of alder-leaf birch has been performed in Yunan province at age 15 years with the resulting value of the lumber produced ranging from RMB 700 - 1500 m^{-3} [62]. Alder-leaf birch was final harvested at age 20 years in an agroforestry system with pineapple producing timber at a value ranging from RMB 100,000 - 225,000 ha^{-1}, depending on tree size, planting spacing and time of harvest [62].

With eucalyptus continually placed on short rotations, concerns regarding the long term soil sustainability arise. In spite of the fact that many of these forests are fertilized at the time of planting, continuous short rotations expose the soil to rain and greater erosion thereby degrading the soil. This could jeopardize the long term sustainability of these systems. Delayed harvests of alder-leaf birch mixed with eucalyptus would continue to provide protection to soil when eucalyptus is harvested. When it comes time to harvest birch, eucalyptus will be in place to continue to provide soil protection. To the extent biodiversity is enhanced is unknown, but the protection of soil is vital to the development of plant understories in the immediate and long term. Protection of soil plus the presence of alder-leaf birch should facilitate the establishment and growth of a more diverse understory.

Changing social values and expectations have led some of the industries in North America to consider alternatives to clearcut harvesting. The recognition of the short term and long term consequences of this practice has resulted in focusing on what can be done to mitigate some of the negative effects of intensive-fiber oriented management. Subsequently many studies have been conducted to determine both the ecological and economic benefits of retaining forest structure through a method referred to as variable retention harvests (i.e., leaving parts of the forest unharvested) for the purpose of retaining structural diversity, increasing biodiversity and the maintenance of ecosystem functions through retaining legacy systems [63]. Forest industries in North America such as Weyerhaeuser Company, TimbeWest and Interfor have adopted this approach on some of their forest land with success over the past decade. The short answer to offsetting costs associated with such a management approach would be to receive a premium for the products. Society as a whole expects greater care be taken of industrial forests and markets presently exist and more can be created where consumers will pay more for sustainable forest practices. New economic models may need to be created by industries for longer term outlook that develops revenue streams from non-timber sources such as eco-tourism, recreation, carbon, biodiversity and botanicals. While ultimately it will be the marketplace that will determine what non-timber values are worth, the forest industries can play a major role in shaping that discussion through their management practices.

American business magnate and philanthropist John D. Rockefeller said "If you want to succeed you should strike out on new paths, rather than travel the worn paths of accepted success". While the current practices of intensively managing large areas of eucalyptus have been successful in supplying the needed wood resource to the Chinese economy, it is necessary to begin looking toward the next success. China has recently been the world leader in afforestation [3], but concerns have arisen regarding the long-term sustainability of the eucalyptus resource and the intensive practices due to the potential impacts upon the soil [64]-[66]. The forest industries should examine other/additional potentials in the growing "green" economy and become more diversified in their product streams. This could include creating carbon offsets through variable retention harvests. A less tangible but possibly more valuable product is the favorable perception of industries by the societal community.

ACKNOWLEDGEMENTS

The author would like to thank Dr. Yuhua Tao, Guangxi Eco-engineering Vocational and Technical College, for her assistance in retrieving some of the Chinese articles for this paper.

REFERENCES

1. Barr, C. and Cossalter, C. (2004) China's Development of a Plantation-Based Wood Pulp Industry: Government Policies, Financial Incentives, and Investment Trends. Int Forest Rev, 6, 267-281. http://dx.doi.org/10.1505/ifor.6.3.267.59977
2. FAO (2009) The Importance of China's Forest Products Markets to the unece region. Geneva Timber and Forest Discussion Paper 57. United Nations Economic Commission for Europe. Timber Section, Geneva, Switzerland, 36 p.

3. FAO (2010) Global Forest Resources Assessment 2010, Country Report, China. FAO Forestry Department FRA2010/ 042, Rome, 101 p.
4. Jiayu, B. and Siming, G. (1996) Eucalyptus Plantations in China. In: Kashio, M. and White, K., Eds., Reports Submitted to the Regional Expert Consultation on Eucalyptus—Volume II, FAO Regional Office for Asia and the Pacific (RAP), RAP Publication, 1996/44.
5. RELaw Assist. (2007) Renewable Energy Law in China—Issues Paper. Baker & McKenzie, the Renewable Energy Generators of Australia (REGA), the Chinese Renewable Energy Industry Association (CREIA) and the Centre for Renewable Energy Development (CRED), 72 p.
6. Wen, Y.G. (2008) Eucalyptus Ecological, Social Issues and Scientific Development. China Forestry Publishing House, Beijing.
7. Qi, S.X. (2002) Eucalyptus in China. 2nd Edition, Chinese Forestry Press, Beijing, 517 p.
8. Bai, J.Y. and Gan, S.M. (1996) Eucalyptus Plantations in China. In: Kashio, M. and White, K., Eds., Reports Submitted to the Regional Expert Consultation on Eucalyptus, October 1993, RAPA Publication 1996/44, FAO Regional Office for Asia and the Pacific, Bangkok, 23-32.
9. FAO (2001) Global Forest Resources Assessment 2000: Main Report. FAO Forestry Paper 140, United Nations Food and Agriculture Organization, Rome, 511 p.
10. Zhang, W.J. (2012) Did Eucalyptus Contribute to Environment Degradation? Implications from a Dispute on Causes of Severe Drought in Yunnan and Guizhou, China. Environmental Skeptics and Critics, 1, 34-38.
11. Corlett, R.T. (1999) Environmental Forestry in Hong Kong: 1871-1977. For Ecol Manag., 16, 93-105.
12. Zhou, G.Y., Morris, J.D., Yan, J.H., Yu, Z.Y. and Peng, S.L. (2002) Hydrological Impacts of Reafforestation with Eucalypts and Indigenous Species: A Case Study in Southern China. For. Ecol. Manage., 167, 209-222. Giambastiani, B.M.S. (2007) Evoluzione Idrologica ed Idrogeologica Della Pineta di san Vitale (Ravenna). Ph.D. Thesis, Bologna University, Bologna.
13. Chen, Q.B. (2001) A Review of Researches on Biodiversity in Eucalyptus Plantations. Chin J Trop Crops, 4, 82-90.
14. Wen, Y.G., Li, X.X., He, T.P. and Zhong, J.B. (2000) Study on the Functions of Biodiversity Conservation of the Coast Protection Forest in Guangxi Province. Prot For Sci Technol, 1, 1-4.
15. Wen, Y.G., Liu, S.R. and Chen, F. (2005) The Ecological Problems and Sustainable Management of Industrial Eucalyptus Plantation. J Guangxi Acad Sci, 21, 13-19.
16. Yu, X.B., Zhong, L.S., Yang, W.D. and Long, T. (1999) Structure of Understory Vegetation in Eucalyptusus Plantations. Chin J Trop Crops, 20, 66-72.
17. Stape, J.L., Dan, B. and Michael, G.R. (2004) Eucalyptus Production and the Supply, Use and Efficiency of Use of Water, Light and Nitrogen across a Geographic Gradient in Brazil. For Ecol Manag, 193, 17-31. http://dx.doi.org/10. 1016/j.foreco .2004.01.020
18. Calder, I.R., Hall, R.L. and Adlard, P.G., Eds. (1992) Growth and Water Use of Forest Plantations. Wiley, New York.
19. Patrick, N.J., Lane, J.M., Zhang, N.N., Zhou, G.Y. and Xu, D.P. (2004) Water Balance of Tropical Eucalyptus Plantations in South-Eastern China. Agric For Meteorol, 124, 253-267. http://dx.doi.org/10.1016/j.agrformet.2004.01.015
20. Morris, J., Zhang, N., Yang, Z., Collopy, J. and Xu, D. (2004) Water Use by Fast-Growing Eucalyptus Urophylla Plantations in Southern China. Tree Physiol., 24, 1035-1044. http://dx.doi.org/10.1093/treephys/24.9.1035
21. Casson, A. (1997) The Controversy Surrounding Eucalypts in Social Forestry Programs of Asia. Economics Division Working Paper 97/1, National Centre for Development Studies, Australian National University, Canberra, 46 p.

22. Davidson, J. (1996) Ecological Aspects of Eucalyptus Plantations. In: White, K., Ball, J. and Kashio, M., Eds., Proceedings: Regional Expert Consultation on Eucalyptus. Rapa Publication 1995/6, FAO Regional Office for Asia and the Pacific, Bangkok, 202 p.
23. Poore, M.E.D. and Fries, C. (1985) The Ecological Effects of Eucalyptus. FAO Forestry Paper No. 59, Food and Agriculture Organization, Rome, 87 p.
24. Calder, I.R., Hall, R.L. and Adlard, P.G., Eds. (1992) Growth and Water Use of Forest Plantations. Wiley, New York.
25. Dye, P.J. (1996) Climate, Forest and Streamflow Relationships in South African Afforested Catchments. Comm. For. Rev., 75, 31-38.
26. Gardner, T.A., Barlow, J., Parry, L.T.W. and Peres, C.A. (2007) Predicting the Uncertain Future of Tropical Forest Species in a Data Vacuum. Biotropica, 39, 25-30. http://dx.doi.org/10.1111/j.1744-7429.2006.00228.x
27. Dunn, R.R. (2004) Recovery of Faunal Communities during Tropical Forest Regeneration. Conserv Biol., 18, 302-309. http://dx.doi.org/10.1111/j.1523-1739.2004.00151.x
28. Wright, S.J. and Muller-Landau, H.C. (2006) The Future of Tropical Forest Species. Biotropica, 38, 207-301.
29. Michelsen, A., Lisanework, N., Friis, I. and Holst, N. (1996) Comparisons of Understory Vegetation and Soil Fertility in Plantations and Adjacent Natural Forests in the Ethiopian Highlands. J Appl Ecol, 33, 627-642. http://dx.doi .org/10.2307/2404991
30. Parrotta, J.A. (1995) Influence of Overstory Composition on Understory Co-lonization by Native Species in Plantations on a Degraded Tropical Site. J Veg Sci, 6, 627-636. http://dx.doi.org/10.2307/3236433
31. Yu, X.B. and Li, W.G. (1997) Ecological Problems and Its Review of Research in Eucalyptus Plantations. Chin J Trop Crops, 18, 60-68.
32. Wen, Y., Duo, Y., Chen, F., Liu, S. and Liang, H. (2010) The Changes of Understory Plant Diversity in Continuous Cropping System of Eucalyptus Plantations. South China. J For Res, 15, 252-258. http://dx.doi.org/10.1007/s10310-010-0179-8
33. Yirdaw, E. and Luukkanen, O. (2003) Indigenous Woody Species Diversity in Eucalyptus globulus Labill. ssp. Globulus Plantations in the Ethiopian Highlands. Biodiversity and Conservation, 12, 567-582.http://dx.doi.org/ 10.1023 /A:1022483700992
34. Basanta, M., Vizcaino, E.D., Casal, M. and Morey, M. (1989) Di-versity Measurements in Shrubland Communities of Galicia (NW Spain). Vegetatio, 82, 105-112. http://dx.doi.org/10.1007/BF00045024
35. Madeira, M.A.V., Andreux, F. and Portal, J.M. (1989) Changes in Soil Organic Matter Characteristics Due to Reforestation with Eucalyptus globulus, in Portugal. Sci Total Environ., 81/82, 481-488. http://dx.doi.org/10.1016/0048-9697(89)90157-5
36. da Silva, J.M.C., Scarano, F.R. and Souza, F.C. (1995) Rege-neration of an Atlantic Forest Formation in the Understory of a Eucalyptus grandis Plantation in South-Eastern Brazil. J Trop Ecol., 11, 47-152.
37. Nichols, J.D., Bristow, M. and Vanclay, J.K. (2006) Mixed-Species Plantations: Prospects and Challenges. For Ecol Manag., 233, 383-390.
38. Barlow, J., Gardner, T.A., Araujo, I.S., Avila-Pires, T.C., Bonaldo, A.B., Costa, J.E., Esposito, M.C., Ferreira, L.V., Hawes, J., Hernandez, M.I.M., Hoogmoed, M.S., Leite, R.N., Lo-Man-Hung, N.F., Malcolm, J.R., Martins, M.B., Mestre, L.A.M., Miranda-Santos, R., Nunes-Gutjahr, A.L., Overal, W.L., Parry, L., Peters, S.L., Ribeiro-Junior, M.A., da Silva, M.N.F., Silva Motta, C. and Peres, C.A. (2007) Quantifying the Biodiversity Value of Tropical Primary, Secondary, and Plantation Forests. Proceedings of the National Academy of Sciences, 104, 18555-18560. http://dx.doi .org/10.1073/pnas.0703333104
39. Ren, H., Li, Z.A., Shen, W.J., Yu, Z.Y., Peng, S.L., Liao, C.H., Ding, M.M. and Wu, J.G. (2007) Changes in Biodiversity and Ecosystem Function during the Restoration of a Tropical Forest in South China. Science in China Series C: Life Sciences, 50, 277-284. http://dx.doi.org/10.1007/s11427-007-0028-y

40. Bauhus, J., van Winden, A.P. and Nicotra, A.B. (2004) Above-Ground Interactions and Productivity in Mixed-Species Plantations of Acacia mearnsii and Eucalyptus globulus. Can. J. For. Res., 34, 686-694. http://dx.doi.org/10.1139/x03-243
41. Binkley, D. (2003) Seven Decades of Stand Development in Mixed and Pure Stands of Conifers and Nitrogen-Fixing Red Alder. Can. J. For. Res., 33, 2274-2279. http://dx.doi.org/10.1139/x03-158

42. DeBell, D.S., Whitesell, C.D. and Schubert. T.H. (1985) Mixed Plantations of Eucalyptus and Leguminous Trees Enhance Biomass Production. USDA For. Serv. Res., Paper PSW-175.
43. Kaye, J.P., Resh, S.C., Kaye, M.W. and Chimmer, R.A. (2000) Nutrient and Carbon Dynamics in a Replacement Series of Eucalyptus and Albizia Trees. Ecology, 81, 3267-3273. Keenan, R., Lamb, D. and Sexton, G. (1995) Experience with Mixed Species Rainforest Plantations in North Queensland. Commun. For. Rev., 74, 315-321. http://dx.doi.org/10.1890/0012-9658(2000)081[3267:NACDIA]2.0.CO;2
44. Parrotta, J.A. (1999) Productivity, Nutrient Cycling, and Succession in Single- and Mixed-Species Plantations of Casuarina equisetifolia, Eucalyptus robusta, and Leucaena leucocephala in Puerto Rico. For Ecol Manage., 124, 45-77.
45. Resh, S.C., Binkley, D. and Parrotta, J.A. (2002) Greater Soil Carbon Sequestration under Nitrogen-Fixing Trees Compared with Eucalyptus Species. Ecosystems, 5, 217-231. http://dx.doi.org/10.1007/s10021-001-0067-3
46. Montagnini, F. (2000) Accumulation in Above-Ground Biomass and Soil Storage of Mineral Nutrients in Pure and Mixed Plantations in a Humid Tropical Lowland. For Ecol Manage., 134, 257-270.
47. Ewel, J.J. (1986) Designing Agricultural Ecosystems for the Humid Tropics. Ann. Rev. Ecol. Syst., 17, 245-271. http://dx.doi.org/10.1146/annurev.es.17.110186.001333
48. FAO (1992) Mixed and Pure Forest Plantations in the Tropics and Subtropics. FAO Forestry Paper 103 (Based on the Work of T.J. Wormald). Food and Agriculture Organization of the United Nations, FAO of the UN, Rome.
49. Montagnini, F., Gonza'les, E. and Porras, C. (1995) Mixed and Pure Forest Plantations in the Humid Neotropics: A Comparison of Early Growth, Pest Damage and Establishment Costs. Commun. For. Rev., 74, 306-314.
50. Guo, J.J., Zeng, J., Zhou, S.L. and Zhao, Z.G. (2008) Isolation and Characterization of 19 Microsatellite Markers in a Tropical and Warm Subtropical Birch, Betula alnoides Buch.-Ham. ex. D. Don. Mol Ecol Resour., 8, 895-897.http://dx.doi.org /10.1111/j.1755-0998.2008.02101.x
51. Zeng, J., Wang, Z., Zhou, S., Bai, J. and Zheng, H. (2003) Allozyme Variation and Population Genetic Structure of Betula alnoides from Guangxi, China. Biochemical Genetics, 41, 66-75. http://dx.doi.org/10.1023/A:1022027832065
52. Su, F. (2011) Effect of Betula alnoides on the Afforestation of Different Exposures. Journal of Anhui Agricultural Sciences, 18, 10858-10859, 10984. (In Chinese)
53. Wang, W.B. and Zhang, J.F. (2004) The Manual of Betula alnoides Plantation Establishment Technology. Yunnan Technology Press, Kunming, 6 p.
54. Wang, W.B., Zhang, J.F., Yang, D.J. and Geng, Y.F. (2011) Comparative Study of Plant Diversity between Betula alnoides Plantations and Adjacent Natural Forests. Taiwan J For Sci, 26, 323-339.

55. Zeng, J., Zheng, H.S. and Wang, B.G. (1998) Betula alnoides—The Fast-Growing and Valuable Timber Tree Species in Tropical and Southern Sub-Tropical Area. Yunnan For Sci Technol, 98, 18-20. (In Chinese)
56. Lu, L.H., Cai, D.X., He, R.M. and Guo, W.F. (2006) Evaluation of Tree Species of Plantation in Southwest Guangxi. Forest Research, 2. (In Chinese)
57. Zhang, J., Yang, D. and Geng, Y. (2012) Plant Diversity among Betula alnoides Plantations and Adjacent Natural Forests. 22-24. In: Heok-Choh, S., Hamid, S.A. and Mei, L., Eds., Asia and the Pacific Workshop—Multinational and Transboundary Conservation of Valuable and Endangered Forest Tree Species, Extended Abstracts

from the Workshop held in Guangzhou, China, 5-7 December 2012. IUFRO World Series Vol. 30, 129 p.

58. Yang, D., Wang, W., Geng, Y. and Qiu, Q. (2008) Comparative Analysis of Ecosystem Carbon Stocks for Pure and Mixed Stands of Betula alnoides. Journal of Fujian College of Forestry, 2, 151-155. (In Chinese)

59. Hui, Z., Guan, D.S. and Song, M.W. (2012) Biomass and Carbon Storage of Eucalyptus and Acacia Plantations in the Pearl River Delta, South China. Forest Ecol Manag, 277, 90-97. http://dx.doi.org/10.1016/j.foreco.2012.04.016

60. Zeng, J., Zou, Y., Bai, J. and Zheng, H. (2003) RAPD Analysis of Genetic Variation in Natural Populations of Betula alnoides from Guangxi, China. Euphytica, 134, 33-41.

61. Zeng, J., Zheng, H.S. and Weng, Q.J. (1999) Geographic Distributions and Ecological Conditions of Betula alnoides Buch.-Ham. ex D.Don in China. Forest Res, 12, 479-484. (In Chinese)

62. Liang, L., Shen, L., Yang, W., Yang, X. and Zhang, Y. (2009) Building on Traditional Shifting Cultivation for rotational Agroforestry: Experiences from Yunan, China. For Ecol Manage., 257, 1989-1994. http://dx.doi.org/10.1016/j.foreco.2008.11.032

63. Serrouya, R., and D'Eon, R. (2004) Variable Retention Forest Harvesting: Research Synthesis and Implementation Guidelines. Knowledge Exchange and Technology Extension Program, Sustainable Forest Management Network, 53 p.

64. Nambiar, E.K.S. (1996) Sustained Productivity of Forests Is a Continuing Challenge to Soil Science. Soil Sci. Soc. Am. J., 60, 1629-1642. http://dx.doi.org/10.2136/sssaj1996.03615995006000060006x

65. Nambiar, E.K.S. (1999) Pursuit of Sustainable Plantation Forestry. S.Afr. For. J., 189, 45-62.

66. Powers, R.F., Scott, D.A., Sanchez, F.G., Voldseth, R.A., Page-Dumroese, D., Elioff, J.D. and Stone, D.M. (2005) The North American Long-Term Soil Productivity Experiment: Findings from the First Decade of Research. For Ecol Manag., 220, 31-50.http://dx.doi.org/10.1016/j.foreco.2005.08.003

CHAPTER 5

Agriculture Intensifies Soil Moisture Decline in Northern China

Yaling Liu[1,2,3], Zhihua Pan[1], Qianlai Zhuang[2,4], Diego G. Miralles[5,6], Adriaan J. Teuling[7], Tonglin Zhang[8], Pingli An[1], Zhiqiang Dong[1], Jingting Zhang[1], Di He[1], Liwei Wang[1], Xuebiao Pan[1], Wei Bai[9] & Dev Niyogi[2,4]

[1]*College of Resources & Environmental Sciences, China Agricultural University, Beijing, 100193, China.*

[2]*Department of Earth, Atmospheric, and Planetary Sciences, Purdue University, West Lafayette, Indiana, 47907, USA.*

[3]*Pacific Northwest National Laboratory, Joint Global Change Research Institute, College Park, Maryland, 20740, USA.*

[4]*Department of Agronomy, Purdue University, West Lafayette, Indiana, 47907, USA.* [5]*Department of Earth Sciences, VU University Amsterdam, Amsterdam 1081 HV, The Netherlands.*

[6]*Laboratory of Hydrology and Water Management, Ghent University, B-9000 Ghent, Belgium.*

[7]*Hydrology and Quantitative Water Management Group, Wageningen University, 6708PB Wageningen, The Netherlands.*

[8]*Department of Statistics, Purdue University, West Lafayette, Indiana, 47907, USA.*

[9]*China Center for Urban Development, Beijing, 100045, China.*

ABSTRACT

Northern China is one of the most densely populated regions in the world. Agricultural activities have intensified since the 1980s to provide food security to the country. However, this intensification has likely contributed to an increasing scarcity in water resources, which may in turn be endangering food security. Based on *in-situ* measurements of soil moisture collected in agricultural plots during 1983–2012, we find that topsoil (0–50 cm) volumetric water content during the growing season has declined significantly ($p < 0.01$), with a trend of −0.011 to −0.015 $m^3\ m^{-3}$ per decade. Observed discharge declines for the three large river basins are consistent with the effects of agricultural intensification, although other factors (e.g. dam constructions) likely have contributed to these trends. Practices like fertilizer application have favoured biomass growth and increased transpiration rates, thus reducing available soil water. In addition, the rapid proliferation of water-expensive crops (e.g., maize) and the expansion of the area dedicated to food production have also contributed to soil drying. Adoption of alternative agricultural practices that can meet the immediate food demand without compromising future water resources seem critical for the sustainability of the food production system.

INTRODUCTION

Droughts are one of the most costly natural hazards in China due to their unusual severity and persistence[1]; they resulted in an average annual loss of $37.9 billion during 1978–2003[2]. While recent studies indicate that globally droughts are not intensifying on average[3], Northern China (NC) is currently facing increasing water scarcity[1,4,5]. In fact, China is one of thirteen countries considered to be facing extreme water shortages, as the water resources per capita is only one quarter of the world average[6]. Additionally, the distribution of water resources in China does not correspond well with the geospatial distribution of water demands. With only 18% of China's water resources, NC accounts for as much as 65% of the arable land[6] and 40% of the population[7]. This imbalance leads to a high exploitation of water in NC, which induces severe ecological and environmental problems derived from the degradation of land and freshwater[8]. In the past three decades, droughts affected an average area of 23.9 million ha per year, and 35.6% of agricultural area in NC[9]. Moreover, recent studies indicate that dry extremes are prone to become more frequent and intense in this region[10,11].

While the soil dry-out in NC has been well documented[1,4,5], the drivers of this dry-out and the role of agricultural intensification remain understudied. Changes in soil water content are affected by a large number of factors, such as soil properties (e.g., soil texture, structure, organic matter, depth, density and salinity), climate (e.g., precipitation, solar radiation, temperature, etc.), topography and land cover. These factors regulate infiltration, permeability, water holding capacity and moisture loss rates[12]. Current agricultural practices, such as crop type selection, agronomic measures, fertilizer application, or irrigation management, are expected to change the water use and the dynamics of soil moisture[13] due to their impacts on the physical and biogeochemical interactions within ecosystems[14,15]. Despite recent analyses of meteorological droughts focusing on trends in rainfall[3,4,5,6,10], and studies of model-simulated or satellite-retrieved soil moisture changes[1,16], there is a lack of ground-truth based observational assessments of soil moisture trends. Moreover, the regional impacts of the intensification of agricultural practices remain poorly understood. Consequently, investigating the mechanisms behind the NC dry-out remains crucial for the improvement of regional agricultural and water resource management, and it can provide useful insights that are transferable to other regions experiencing similar water-stress situations[14,17].

This study aims at understanding soil moisture trends in NC using robust observational evidence, and unravelling the impacts of regional agricultural practices on these trends. We first use satellite remote sensing, long-term discharge measurements and meteorological observations to assess the regional drying during the past three decades. Then we use unique long-term *in-situ* soil moisture records to evaluate observed soil drying in agricultural plots. The effects of agricultural intensification on soil moisture decline are assessed through multiple linear regression analysis using data for fertilizer usage and information on crop types and crop area. The Xinjiang autonomous region and the Qinghai province are excluded from the analysis, as population density is low[7], agriculture is not extensive[18], and there is a lack of historical data.

RESULTS

Satellite observations of surface soil moisture and total terrestrial water storage both point to NC as a hot-spot region of declining water availability at the global scale[16,19] (Fig. 1a,b). Moreover, the growing season river discharges (*Q*) at the outlets of the Yellow, Haihe and Liaohe river basins all show significant negative trends during 1983–2012 (Fig. 1c). This is consistent with the aforementioned satellite-observed regional drying, although dam construction, water withdrawal and groundwater abstraction have also altered the regional water distribution[20]. The Yellow river basin exhibits the largest decline, which is consistent with the reports of the frequent dry-out of the lower reach since 1972[21].

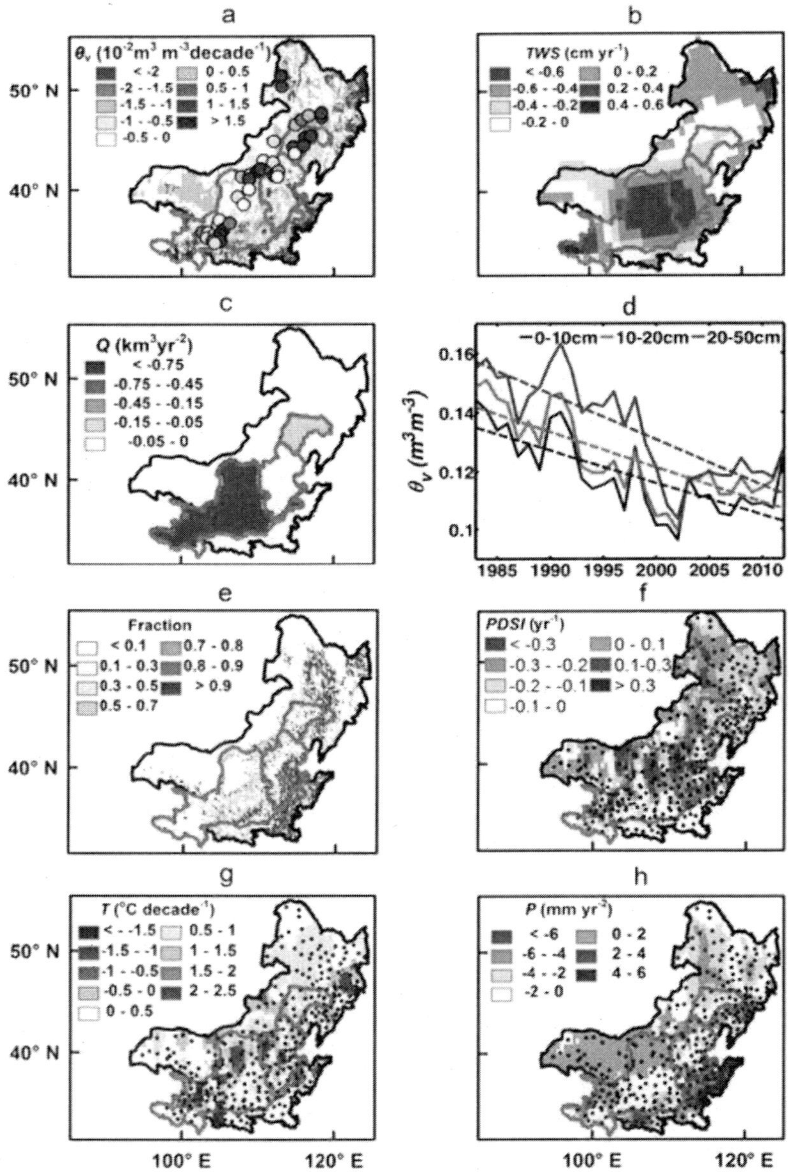

Figure 1: Growing season trends in Northern China (NC) : **a**) Volumetric soil moisture (θ_v) trend during 1983–2010 across NC and at the 40 agricultural meteorological stations (circles on the map) during 1983–2012. The former is derived from CCI-WACMOS soil moisture product16 and the latter is from *in-situ* measurements of the top 0–10 cm soil. The red polygons delineate the boundaries of the Yellow, Haihe and

Liaohe river basins from west to east, respectively. Note that the boundary of Liaohe river basin is only extended to Tieling gauge station. **b**) Terrestrial water storage (*TWS*) trend during 2003–2013. **c**) Discharge (*Q*) trends in the three basins during 1980–2012. **d**) Average θ_v trends in 0–10 cm, 10–20 cm and 20–50 cm soil across the 40 agricultural meteorological stations during 1983–2012. **e**) Fraction coverage of cropland in 2010. **f–h**) Palmer Drought Severity Index (*PDSI*), air temperature (*T*) and Precipitation (*P*) trend during 1984–2007. The black dots in **f–h**) represent the locations of 307 weather stations. All maps are generated via ArcMap10.2, and their projected coordinate systems are Asia North Albers Equal Area Conical.

In situ soil moisture measurements across NC reveal a significant soil moisture decline at dryland crop sites. At the 40 agricultural meteorological stations in NC (Fig. 1a), soil moisture was monitored 3–5 times per month using the gravimetric technique during the growing season (generally April-October). These stations cover rain-fed croplands where a variety of different crop types were grown. Across these 40 stations, we find that the average volumetric soil moisture (θ_v) during growing season decreased significantly in the last three decades (Fig. 1d). We find a significant ($p < 0.01$) θ_v trend of −0.011, −0.012 and −0.015 $m^3 m^{-3}$ per decade in the top 0–10 cm, 10–20 cm, and 20–50 cm of soil (i.e., the fitted θ_v in 20–50 cm of soil decreased from 0.157 $m^3 m^{-3}$in 1983 to 0.112 $m^3 m^{-3}$ in 2012 or by about 29%, see Fig. 1d), respectively. Similar negative trends (−6.2% to −10.4% per decade) are also found in the soil moisture index (SMI, Fig. S1), a measure of soil moisture content as ratio of the total potential storage available to plants22. This analysis thus provides evidence for a notable soil moisture decline at agricultural fields across NC.

The trends in satellite-observed surface soil moisture are generally consistent with the ground-measured top 10 cm soil moisture at the 40 stations (Fig. 1a, correlation coefficient $r = 0.75$). Disagreements in the magnitude of these trends might occur due to the scale-mismatch between the point measurements and the satellite footprint, and also because crops do not necessarily cover the entire area of each pixel18(Fig. 1e). Additionally, the satellite sensors usually detect the conditions in the top 0.5–2 cm of soil, rather than the 0–10 cm of soil sampled by the *in-situ* measurements.

The detected soil moisture decline over the last three decades could have multiple drivers, including climate change. In fact, an overall decline of the Palmer Drought Severity Index23 (PDSI, Fig. 1f) – a widely used metric to monitor meteorological droughts – reflects the simultaneous spatiotemporal increase in air temperature (Fig. 1g, Fig. S2a) and decrease in rainfall (Fig. 1h, Fig S2b) in the region over this period. The most dramatic decline in PDSI occurs in the northeast part of Inner Mongolia, where a pronounced warming and drying trend is also found. Trends in other drought indices such as Standardized Precipitation Index (SPI)[24] and Standardized Precipitation Evapotranspiration Index (SPEI)[25] further support these results (Fig. S3). The negative trends for SPEI are more prominent than that for SPI, suggesting that the increased atmospheric demand for water has played a role as well[25]. These results suggest that climate change has contributed to soil drying in NC[26].

We disentangle the effects of climate change from those of agricultural practices using a multiple linear regression analysis. Note that soil properties, microclimate and topography intrinsically affect soil moisture content[12] and may vary from county to county due to heterogeneity in the environmental conditions; their effects on soil water content may overshadow the effects induced by agricultural practices[12,27]. We define the variable "county effect" (C_m) to account for differences in soil moisture level caused by differences in the environmental conditions across different counties/stations. Other than C_m, results show that among all the meteorological and agricultural explanatory variables considered (see Methods), precipitation (P), fertilizer use rate (F) and crop type (c) are most significant in influencing the soil moisture at the three soil depths considered. Here we take the 20–50 cm soil layer, which comprises a large portion of root zone[28,29] and is directly affected by transpiration, to illustrate the correlated effects of declining soil moisture. The analysis yields slopes of 4.1×10^{-4} ($m^3\ m^{-3}\ mm^{-1}$ yr), -3.0×10^{-4} ($m^3\ m^{-3} kg^{-1}$ ha yr), −0.013 and −0.026, for P, F, crop group 2 (i.e., $c2$, wheat) and group 3 (i.e., $c3$, maize and rapeseed), respectively – see model (1) in Methods. So, for instance, while a 100 mm increase in growing season precipitation could lead to an increase of 0.041 $m^3\ m^{-3}$ in θ_v, a 100 kg ha^{-1} increase in fertilizer use could lead to an 0.030 $m^3\ m^{-3}$ reduction in θ_v over 1983–2006, which is approximately 20% of the mean observed θ_v(Fig. 1d).

The model explains 81% of variation in θ_v, indicating that the changes in soil moisture can be reasonably well captured by the explanatory variables considered. A large fraction (60%) of the total variation explained by the model (81%) is due to C_m; however, the average variation derived from C_m adjusted by the degree of freedom ($df = 39$) is small. In terms of the relative contribution, P, F and c account for 32%, 38% and 30%, respectively, of the the remaining model variation (40% of the 81% explained by the model). The 32% relative contribution of P indicates that climate change induced soil moisture decline is substantial. We do not find T has significant impact on θ_v. This suggests that θ_v is more sensitive to P, and that the impact of water supply overshadows that of atmospheric evaporative demand increase induced by warming.

The positive covariance between P and θ_v is expected, as θ_v responds directly to changes in P. However, the considerable contribution of F and c to θ_v decline is of interest and may be less straightforward. In order to further explore this negative correlation, we examine data gathered at the Wuchuan Field Experiment Station, where the relations between soil moisture and fertilizer loads were monitored from 2008–2010. In this experiment, the effect of fertilizer was monitored in isolation as fertilizer use was the only changing agricultural variable, and other agricultural practices such as tilling method, plant variety and plant density were the same across different treatments and along time. Results show a clear rise in field water consumption with increased fertilizer use, by 4%–54% depending on the specific fertilizer treatment (Table S1, S2, Fig. 2a). In addition, the choice of crop types also leads to significant differences in terms of θ_v. Taking the crop group 1 (i.e., $c1$, soybean and potato) as baseline, regression results indicate that the choice of crop group $c2$, and especially $c3$, has negative effects due to their higher water consumption. This is consistent with the findings from previous field experiments conducted in NC[30,31,32,33,34,35] that indicate that crop water consumption during the growing season follows the order: $c1$ (326–452 mm) < $c2$ (450–500 mm) < $c3$ (398–568 mm). These results suggest that the increased fertilizer use (Fig. 2b) and the associated agricultural practices such as crop area expansion (Fig. 3a) in NC[9,18]– with a disproportionate proliferation of high water-consuming crops (e.g., maize, Fig. 3b) – have contributed to the widespread soil moisture decline in the region.

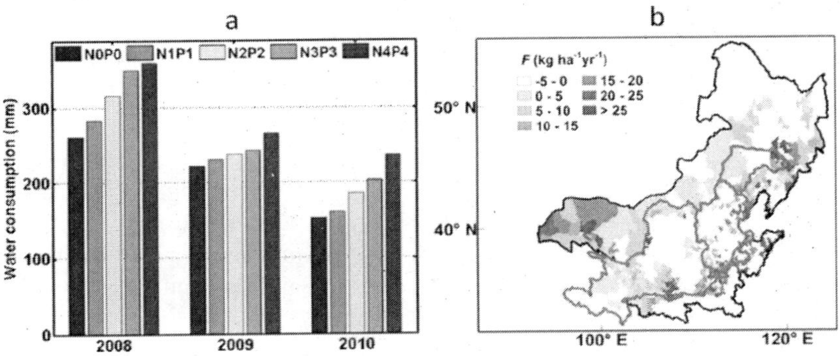

Figure 2: Fertilizer use experiment and fertilizer use trend in Northern China a) water consumption in N0P0-N4P4 fertilizer use treatments between 2008–2010, which is derived from fertilizer use experiment in Wuchuan Field Experiment Station, where N0P0–N4P4 stand for increasing fertilizer use treatments that correspond to Table S1; and **b**) fertilizer use (F) trend during 1983–2006 across the region. Note that area without fertilizer use data is left vacant. The map in **b**) is generated via ArcMap10.2, and the projected coordinate system is Asia North Albers Equal Area Conical.

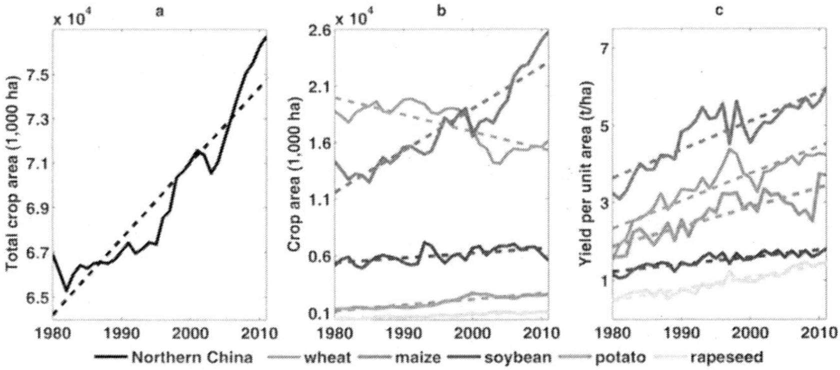

Figure 3: Changes in crop area and yield per unit area in Northern China during 1980–2011 : **a**) total crop area in the region; **b**) planting area for each major crop; and **c**) yield per unit area for each major crop.

The factors that impact discharge decline in NC are also assessed for the three river basins based on multiple linear regression models. Results are similar for each of the three basins. In the following, we focus on the Yellow river basin, due to its larger coverage and dense population. Our regression model for the Yellow river (see model (2) in Methods section) explains 82% of the long-term variation in discharge. P has a significant positive effect on Q (slope of 0.12 km^3 mm^{-1}), whereas F and A_w (area covered by wheat plantations) have significant negative effects (−0.29 km^3 kg^{-1} ha and −0.06 km^3 10^{-3} ha^{-1}, respectively). The crop area of maize (A_{mz}) is not considered to have significant effects on Q (i.e., $p = 0.97$) despite the high correlation between A_{mz} and Q ($r = -0.76, p < 10^{-5}$). This occurs due to the "Simpson Paradox"36, in which two strongly dependent variables like F and A_{mz} ($r = 0.92$) cannot co-exist in the model. Replacing A_w with A_{mz} in the model, we find A_{mz} also has significant negative effect on Q ($p < 0.001$). Overall, the effects of P, F, acreage of maize and wheat on Q are consistent with those on θ_v. This finding suggests that extensive agriculture intensification in the Yellow, Haihe and Liaohe river basins – where 26.6%, 48.2% and 34.1% of the area is covered by cropland in 201018 – could partly explain the Q declines.

A pair-wise experiment conducted at the Wuchuan Agricultural Meteorology Observation Station further supports the hypothesis that agricultural intensification accelerates soil moisture decline. For nearly three decades (1983–2009), soil moisture has been monitored at two contiguous sites: a pristine pasture and an agricultural site. Measurements reveal a significant ($p < 0.05$) negative trend of soil moisture in the first 50 cm of soil for the cropland (for all crop types), contrasting with the slightly positive – but statistically significant ($p < 0.05$) – trend observed in the pristine pasture (Fig. 4). In potato, naked oats and spring wheat fields, θ_v decreased by 0.011, 0.011 and 0.006 $m^3 m^{-3}$ per decade, respectively. For instance, the fitted θ_v decreased from 0.103 $m^3\ m^{-3}$ in year 1983 to 0.074 $m^3\ m^{-3}$ in 2009 at the potato field. However, in the pristine pasture – where trends reflect larger scale climatic change only – there was an increase of 0.003 $m^3\ m^{-3}$ per decade. Since no significant differences in soil texture, topography and microclimate exist between the cropland and pasture sites, differences in θ_v may be attributed to the agricultural activities.

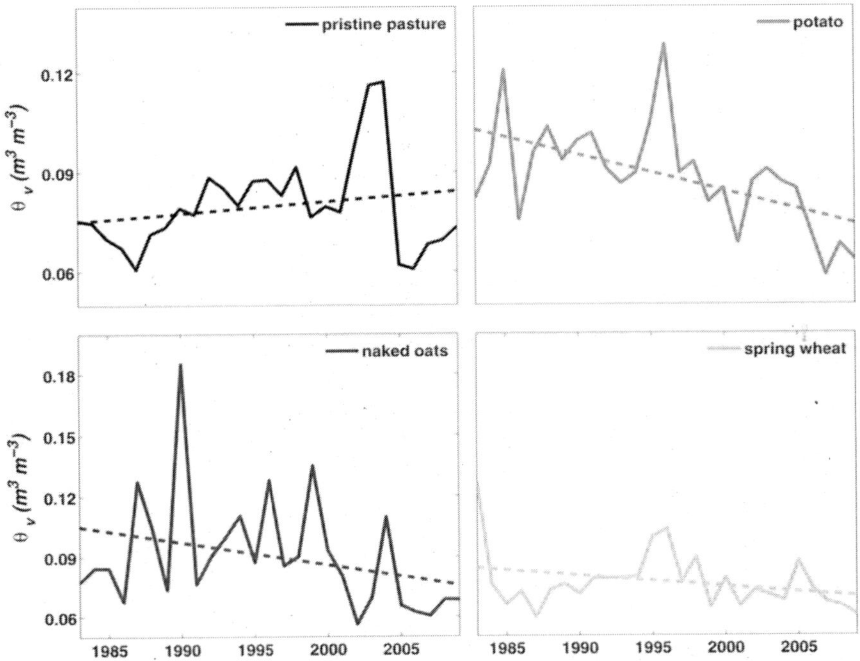

Figure 4: Variation of volumetric soil moisture (θ_v) in topsoil (0–50 cm) of pristine pasture and different crop fields in Wuchuan Agricultural Meteorology Observation Station during 1983–2009.

DISCUSSION

Food sufficiency has been achieved after nearly 50 years of sustained efforts in China37. Since the early 1980s, agriculture has intensified dramatically, and the food production system currently relies on application of chemical fertilizers and implementation of modern farming techniques38. The amount of fertilizer use in China accounts for 31.4% of the total global consumption39; this can be considered as a very inefficient system40, with rates of fertilizer application that are far greater than the needs of crops37,41,42,43. While the yield per hectare for all crops has increased from 1.21 $t\ ha^{-1}$ to 4.83 $t\ ha^{-1}$, the yield per unit chemical fertilizer use has decreased dramatically for the three main crops in China during 1961–1998: 164 to 10 $kg\ kg^{-1}$ for rice, 44 to 6 $kg\ kg^{-1}$for wheat and 93 to 9 $kg\ kg^{-1}$ for maize44.

Although the effects of excessive fertilizer use on agriculture and environment in China have been investigated in recent years37,42,43,45, only a few studies have drawn the attention to the potential effects of fertilizers on transpiration rates, thus on the hydrological cycle46,47. Excessive fertilizer use can increase water consumption and deplete soil moisture as revealed by our results (Fig 2a), while having little impact on crop yield37. In addition, fertilizer use and other agricultural practices may aggravate soil salinity and soil compaction, leading to decrease of infiltration rates and soil water holding capacity12, which further limits the available soil water.

It is important to note that limitations of this study arise from the scarcity of data on regional agricultural practices and the inability to capture all the interactive factors involved within a statistical framework. The increase in fertilizer use and the corresponding decline in soil moisture should be interpreted more broadly; changes in fertilizer use may be accompanied by the incorporation of new plant varieties, modified phenotypes or changes in agricultural technology, and therefore reflect a broad scenario of agricultural intensification. The increase in the yield per unit area for each of the main crops (wheat, maize, soybean, potato and rapeseed, Fig. 3c) is also an integrated effect of wide range of agricultural practices. New crop varieties have been extensively introduced in NC to enhance yields and improve crop quality in recent years48. These new varieties have different water and fertilizer use48, and thus potentially affect soil moisture. Similarly, irrigation area has drastically expanded20,49 and a range of farming techniques (e.g., reduced tillage, mulching) have been widely implemented in recent years across NC50. However, lack of historical information about cultivation of the large number of varieties – often exceeding 10,000 for individual crops48, irrigation management and implementation of farming techniques across different areas and over different eras make it infeasible to extract their impacts on regional soil moisture at present. Therefore, the results on fertilizer use reflect only the overall intensification of agriculture, whereas local factors that might have contributed to this intensification and to the decline in soil moisture (e.g. pesticide and herbicide application, changes in crop varieties, improvement in on-farm practices and implementation of modern technologies) could not be investigated at this stage.

Water scarcity has become increasingly prominent in NC as agriculture has intensified[51,52]. Current agricultural practices paired with inappropriate nutrient and irrigation management lead to severe environmental degradation[43,45] and excessive exploitation of water resources[50], as well as low nutrient use efficiency and water use efficiency that lag behind global averages[51]. As a result, alternative agriculture practices that could help meet food demand without compromising future water resources need to be pursued for sustainable agriculture. Recent years have seen progress in research on agriculture management[8,37,50,51,52], contributing to addressing the conflict between aggravating resources shortage and increasing food demand. The implementation of integrated soil-crop system management (e.g. synchronization of nutrient supply and crop demand)[8,52], the dissemination of water-saving technologies (e.g., mulching, drip irrigation), agronomic measures (e.g., reduced till, recycling of organic manures) and agro-climatic decision tools (using crop-climatic relations or data analytics), and the enhancement in crop varieties with resistance to drought and heat seem critical for sustainable agriculture production in NC.

METHODS

We analyse the soil moisture trends over NC using long-term measurements from 40 agricultural meteorological stations during 1983–2012. We support this analysis with a study of the discharge (Q) trends from the three outflow river basins in the region. To isolate the long-term impact of precipitation trends on the soil moisture changes, we examine how meteorological drought has evolved across NC using meteorological variables such as precipitation and air temperature from 307 weather stations. We use a Mann-Kendall test to evaluate the significance of the trends. Multiple linear regression is used to assess the correlated effects for changes in soil moisture and Q. Details of observed data and ancillary data used in this study are provided in the Supplementary Information.

Since the locations of the 40 stations, where the soil moisture measurements are conducted, are different from those of the weather stations, the meteorological data used for each site in the regression is interpolated from observations at four nearest weather stations using the inverse distance weighting method[53]. Historical records of (chemical) fertilizer use at these 40 stations are not available. To cope with this site-level data scarcity we use county-average rates[54]. Data at these 40 stations is combined into one multiple regression model.

In the regression for the soil moisture changes, we consider the impacts of meteorological variables such as air temperature (T, ^{o}C), precipitation (P, mm yr^{-1}), wind speed (u, m s^{-1}), relative humidity (RH) and radiation hours (R, hr), and agricultural activities as reflected by fertilizer use rate (F, kg ha^{-1} yr^{-1}, calculated as total weight of fertilizer use divided by cropland area) and crop types (c), as well as "county effect" representing differences in soil moisture level across different counties (C_m, the overall effects of county level environmental variables such as topography, soil texture and microclimate). We jointly consider nitrogen, phosphorus, and potassium fertilizers due to the unavailability of historical records of their separate use. These fertilizers we considered are active ingredients as nitrogen, phosphorus pentoxide, and potassium oxide, respectively. Based on multiple experiments on the crop water consumption that have been conducted in NC region30,31,32,33,34,35, we categorize the main crop types in NC into three groups: (i) soybean and potato ($c1$); (ii) wheat ($c2$); and (iii) maize and rapeseed ($c3$).

We use the ANCOVA model55 to investigate the correlated effects between the dependent and explanatory variables. The model can be summarized as:

$$Y_m = \beta_{m,0} + \beta_{m,1}X_1 + \cdots + \beta_{m,i}X_i + \sum_{j=2}^{3}\beta_{c,m,j}I\,(c = j) + C_m + \varepsilon_m \tag{1}$$

where Y_m is the mean growing season θ_v for each year during 1983–2010. β is the intercept term, $C_m = \sum_{j=2}^{k}\beta_{County,m,j}I\,(County = j)$ represents the county effect, $X_1 \sim X_i$ (i.e., T, P, RH, R, u and F) represent the *ith* explanatory variable and $\beta_{m,1} \sim \beta_{m,1}$ are the corresponding slopes, $I(\cdot)$ is the indicator function, and ε_m is a normally distributed error term. Since we have classified the crops into three groups, we use two parameters ($\beta_{c,m,2}$ for group 2 and $\beta_{c,m,3}$ for group 3) to represent the effects of crop groups. The factor terms do not contain the first level in their expression since we choose the first level of both county effect and crop effects as baselines. Note that the choice of baseline does not affect the regression results. Analysis of variance method56 is used to quantify the contribution of each variable to the total variation in the model.

We build separate models for the three river basins to investigate the effects of meteorological and agricultural variables on changes in Q. We use discharge data at the outlets of each river basin as dependent variable. Explanatory variables involve meteorology (i.e., T, P, RH, R and u) and agricultural activities (i.e., F, crop area of each major crop type) aggregated from all counties within each basin. The full model with all explanatory variables is given by:

$$Y_d = \beta_{d,0} + \beta_{d,1}X_1 + \cdots + \beta_{d,i}X_i + \sum_{k=1}^{5}\beta_{d,k}A_k + \varepsilon_d \quad (2)$$

where Y_d stands for the growing season mean Q for each year during 1983–2012 ($km^3\ yr^{-1}$), $\beta_{d,0}$ is the intercept term, $X_1 \sim X_i$ (i.e., *T, P, RH, R, u* and *F*) represent the *ith* explanatory variable and $\beta_{d,1} \sim \beta_{d,1}$ are corresponding slopes, A_k stands for the planting area of *kth* crop (i.e., wheat, maize, soybean, potato, rapeseed) and $\beta_{d,k}$ is the corresponding slope, and ε_d is a normally distributed error term. Besides this full model, a few reduced models that assign some of the coefficients equal to zero are also investigated. This is important because of the possible impact of the "Simpson Paradox" on parameter estimates as well as on the significance of effects36.

When we calculate the Palmer Drought Severity Index (*PDSI*)23 and the Standardized Precipitation Evapotranspiration Index (SPEI)25 we use the FAO56 Penman-Monteith equation57 instead of the Thornthwaite equation to estimate potential evapotranspiration (PET)3.

ACKNOWLEDGEMENTS

This research is supported by the National Basic Research Program of China (No. 2012CB956204), the National Natural Science Foundation of China (Grant Nos. 41371232 and 41271110), the National Science and Technology Support Program of China (No. 2012BAD09B02), the National Non-profit Research Foundation for Agriculture of China (201103039), the NASA Land Use and Land Cover Change program (NASA- NNX09AI26G, NN-H-04-Z-YS-005-N, and NNX09AM55G to Q.Z.), the Department of Energy (DE-FG02-08ER64599 to Q.Z.), the National Science Foundation (NSF-1028291 and NSF- 0919331 to Q.Z., NSF-CAREER AGS-0847472 to D.N.), the NSF Carbon and Water in the Earth Program (NSF-0630319 to Q.Z.). We acknowledge Chinese Meteorological Administration, Ministry of Water Resources of China and National Bureau of Statistics of China for provision of data used in this study. D.G.M. acknowledges financial support from Netherlands Organization for Scientific Research (NWO) through grant 863.14.004, and the Belgian Science Policy Office (BELSPO) in the frame of the STEREO III programme, project SAT-EX (SR/00/306). This research was primarily conducted when the lead author Yaling Liu worked at China Agricultural University and Purdue University and was revised during her ongoing affiliation with Pacific Northwest National Laboratory.

REFERENCES

1. Wang, A., Lettenmaier, D. P. & Sheffield, J. Soil moisture drought in China, 1950–2006. *J Climate* **24**, 3257–3271 (2011).
2. Pandey, S. & Bhandari, H. S. [Chapter 1 Introduction] *Economic Costs of Drought and Rice Farmers' Coping Mechanisms: A Cross-Country Comparative Analysis*, [Pandey, S., Bhandari, H. S., Hardy, B. (ed.)] [1–10] (International Rice Research Institute, Los Baños, Philippines, 2007).
3. Sheffield, J., Wood, E. F. & Roderick, M. L. Little change in global drought over the past 60 years. *Nature* **491**, 435–438 (2012).
4. Zou, X., Zhai, P. & Zhang, Q. Variations in droughts over China: 1951–2003. *Geophys Res Lett* **32**, doi:10.1029/2004GL021853 (2005).
5. Ma, Z. & Fu, C. Some evidence of drying trend over northern China from 1951 to 2004. *Chinese Science Bulletin* **51**, 2913–2925 (2006).
6. Piao, S. *et al.* The impacts of climate change on water resources and agriculture in China. *Nature* **467**, 43–51 (2010).
7. Population census Office of China. *Main Statistics on Sixth National Population Census in China.* (China Statistics Press, Beijing, 2011).
8. Chen, X. *et al.* Integrated soil-crop system management for food security. *Proc Natl Acad Sci* **108**, 6399–6404 (2011).
9. National Bureau of Statistics of China. *National data*, < http://data.stats.gov.cn/english/easyquery.htm?cn=E0103> (2014) Date of access: 01/12/2014
10. Trenberth, K. E. *et al.* Global warming and changes in drought. *Nat Clim Chang* **4**, 17–22 (2014).
11. Elliott, J. *et al.* Constraints and potentials of future irrigation water availability on agricultural production under climate change. *Proc Natl Acad Sci* **111**, 3239–3244 (2014).
12. Shaxson, F. & Barber, R. *Optimizing Soil Moisture for Plant Production. The Significance of Soil Porosity. FAO Soils Bulletin* **79**(FAO, Rome, 2003).
13. Fu, B., Wang, J., Chen, L. & Qiu, Y. The effects of land use on soil moisture variation in the Danangou catchment of the Loess Plateau, China. *Catena* **54**, 197–213 (2003).
14. Niyogi, D., Kishtawal, C., Tripathi, S. & Govindaraju, R. S.Observational evidence that agricultural intensification and land use change may be reducing the Indian summer monsoon rainfall. *Water Resour Res* **46**, doi:10.1029/2008WR007082 (2010).
15. Rosenzweig, C. *et al.* Attributing physical and biological impacts to anthropogenic climate change. *Nature* **453**, 353–357 (2008).
16. Dorigo, W. *et al.* Evaluating global trends (1988–2010) in harmonized multi-satellite surface soil moisture. *Geophys Res Lett*39, doi:10.1029/2012GL052988 (2012).
17. Nazareno, A. G., & Laurance, W. F. Brazil's drought: beware deforestation. *Science* **347**, 1427 (2015).
18. Liu, J. *et al.* The land use and land cover change database and its relative studies in China. *J Geogra Sci* **12**, 275–282 (2002).
19. Sakumura, C., Bettadpur, S. & Bruinsma, S. Ensemble prediction and intercomparison analysis of GRACE time variable gravity field models. *Geophys Res Lett* **41**, 1389–1397 (2014).
20. Fu, G., Chen, S., Liu, C. & Shepard, D. Hydro-climatic trends of the Yellow River basin for the last 50 years. *Climatic Change* **65**, 149–178 (2004).
21. Chen, J., He, D. & Cui, S. The response of river water quality and quantity to the development of irrigated agriculture in the last 4 decades in the Yellow River Basin, China. *Water Resour Res* **39**, doi:10.1029/2001WR001234 (2003).
22. Seneviratne, S. I. *et al.* Investigating soil moisture-climate interactions in a changing climate: A review. *Earth-Sci Rev* **99**, 125–161 (2010).
23. Palmer, W. C. *Meteorological drought.* (US Department of Commerce, Weather Bureau: Washington, DC, USA, , 1965).

24. McKee, T. B., Doesken, N. J. & Kleist, J. *The relationship of drought frequency and duration to time scales*: 8th Conference on Applied Climatology, Anaheim, California, USA. Boston, Massachusetts: American Meteorological Society (1993, January17-22).
25. Vicente-Serrano, S. M., Beguería, S. & López-Moreno, J. I. A multiscalar drought index sensitive to global warming: the standardized precipitation evapotranspiration index. *J Climate* **23**, 1696–1718 (2010).
26. Tao, F., Yokozawa, M., Hayashi, Y. & Lin, E. Changes in agricultural water demands and soil moisture in China over the last half-century and their effects on agricultural production. *Agr Forest Meteorol* **118**, 251–261 (2003).
27. Jalota, S. *et al.* Soil texture, climate and management effects on plant growth, grain yield and water use by rainfed maize-wheat cropping system: Field and simulation study. *Agr Water Manage* **97**, 83–90 (2010).
28. Liu, R. *et al.* Distribution pattern of winter wheat root system. *Chinese Journal of Ecology* **37**, 2024–2027 (2008) (in Chinese).
29. Liu, J. *et al.* Temporal variation and spatial distribution of the root system of corn in a soil profile. *Chinese Journal of Eco-Agriculture* **17**, 517–521 (2009) (in Chinese).
30. Xie, X. *Crop Water Demand, Consumption and Water Use Efficiency in the Main Agriculture Ecosystem of China.* (China Meteorological Press, Beijing, 1996) (in Chinese).
31. Wu, K., Chen, J. & Xie, X. On the water consumption characteristics of winter wheat and water-saving in agriculture. *Acta Geographica Sinica* **52**, 455–460 (1997) (in Chinese).
32. Ren, H. & Luo, Y. The Experimental Research on the Water-consumption of Winter Wheat and Summer Maize in the Northwest Plain of Shandong Province. *Journal of Irrigation and Drainage* **23**, 37–39 (2004) (in Chinese).
33. Liu, Z., Xiao, J. & Yu, X. Effects of different soil moisture treatments on morphological index, water consumption and yield of potatoes. *China Rural Water and Hydropower* **8**, 1–7 (2010) (in Chinese).
34. Zhang, J. *et al.* Effects of different film mulching modes on growth, yield and water use efficiency of winter rapeseed in dryland in Tianshui. *Agricultural Research in the Arid Areas* **31**, 79–80 (2013) (in Chinese).
35. He, J., Bian, X., Fu, Y. & Qin, Y. Research on Water Consumption and Its Law of Main Crops in West Liaohe River Plain. *Water-saving Irrigation* **11**, 1–4 (2012) (In Chinese).
36. Simpson, E. H. The interpretation of interaction in contingency tables. *J Roy Stat Soc. B Met* **13**, 238–241 (1951).
37. Ju, X. *et al.* Reducing environmental risk by improving N management in intensive Chinese agricultural systems. *P Natl Acad Sci* **106**, 3041–3046 (2009).
38. Guo, J. *et al.* Significant acidification in major Chinese croplands. *Science* **327**, 1008–1010 (2010).
39. Xin, L., Li, X. & Tan, M. Temporal and regional variations of China's fertilizer consumption by crops during 1998–2008. *J Geogra Sci* **22**, 643–652 (2012).
40. Raun, W. R. & Johnson, G. V. Improving nitrogen use efficiency for cereal production. *Agronomy J* **91**, 357–363 (1999).
41. Vitousek, P. M. *et al.* Nutrient imbalances in agricultural development. *Science* **324**, 1519–1520 (2009).
42. Gong, P., Liang, L. & Zhang, Q. China must reduce fertilizer use too. *Nature* **473**, 284–285 (2011).
43. Huang, Y., Sass, R., Sun, W., Zhang, W. & Yu, Y. *Reducing nitrogen fertilizer use to mitigate negative environmental impact in China* (James A. Baker III institute for public policy, Houston, TX, 2010).
44. Tong, C., Hall, C. A. & Wang, H. Land use change in rice, wheat and maize production in China (1961–1998). *Agr Ecosyst Environ* **95**, 523–536 (2003).
45. Kahrl, F. *et al.* Toward Sustainable Use of Nitrogen Fertilizers in China. *ARE Update* **14**, 5–7 (2010).

46. Huang, M., Dang, T., Gallichand, J. & Goulet, M. Effect of increased fertilizer applications to wheat crop on soil-water depletion in the Loess Plateau, China. *Agr Water Manage* **58**, 267–278 (2003).
47. Fan, T. *et al.* Long-term fertilizer and water availability effects on cereal yield and soil chemical properties in northwest China. *Soil Sci Soc Am J* **69**, 842–855 (2005).
48. Papademetriou, M. K. *Crop Diversification in the Asia-Pacific Region.* (FAO Regional Office for Asia and the Pacific, 2001).
49. Zhu, X., Li, Y., Li, M., Pan, Y. & Shi, P. Agricultural irrigation in China. *J Soil Water Conserv* **68**, 147A–154A (2013).
50. Wang, F. *et al.* Wheat cropping systems and technologies in China. *Field crop res* **111**, 181–188 (2009).
51. Fan, M. *et al.* Improving crop productivity and resource use efficiency to ensure food security and environmental quality in China. *J exp bot*, 1–12, doi:10.1093/jxb/err248 (2011)
52. Fan, M. *et al.* Integrated nutrient management for improving crop yields and nutrient utilization efficiencies in China. *J Soil Water Conserv* **63**, 126A–128A (2008).
53. Shepard, D. *Two-dimensional interpolation function for irregularly-spaced data*: 23rd ACM national conference, Las Vegas, Nevada, USA. New York: Association for Computing Machinery Press, doi: 10.1145/800186.810616 (1968, August 27-29).
54. Muramatsu, N. County-level income inequality and depression among older Americans. *Health Serv Res* **38**, 1863–1884 (2003).
55. Kutner, M. H., Nachtsheim, C. J., Neter, J. & Li, W. [Chapter 22 Analysis of Covariance] *Applied Linear Statistical Models* , Fifth edition, [917–950] (McGraw-Hill Irwin, New York, 2005).
56. Faraway, J. J. *Extending the Linear Model With R: Generalized Linear, Mixed Effects and Nonparametric Regression Models.* (CRC press, New York, 2005).
57. Allen, R. G., Pereira, L. S., Raes, D. & Smith, M. *Crop Evapotranspiration-Guidelines for Computing Crop Water Requirements-FAO Irrigation and Drainage Paper 56.* (FAO, Rome, 1998).

CHAPTER 6

A Global Analysis of Temperature Effects on Populus Plantation Production Potential

Natalya K. Kutsokon[1,2], Shibu Jose[1], Eric Holzmueller[3]

[1]*The Center for Agroforestry, University of Missouri, Columbia, MO, USA*

[2]*Institute of Cell Biology and Genetic Engineering NAS of Ukraine, Kiev, Ukraine*

[3]*Department of Forestry, Southern Illinois University, Carbondale, IL, USA*

ABSTRACT

The genus *Populus* contains some of the most economically important tree species and hybrids in the world. We compared productivity of short and long-rotation poplar plantations using published data from 23 countries to determine if climate, particularly temperature, had any effect on the observed patterns of productivity. We discovered that climate factors (related to temperature) and clone origin (pure species or hybrids) slightly influenced productivity of long rotation forests more than short rotation plantations. While long rotation plantation productivity exhibited positive correlations with increasing temperature during winter and decreasing heat during summer, short rotation plantations showed weak positive relationship among productivity and increasing yearly temperature and the number of hot days. It was apparent that short rotation plantations productivity was less dependent on regional climatic variables or origin of clone. However, it appears that overall, regardless of the system, *Populus* species are generally adapted to a range of climatic conditions where they are planted.

Keywords: Poplar, *Populus* sp., Hybrids, Fast-Growing Plantations, Intensive Forestry, Short and Long Rotation Plantations, Stem Volume, Climate

1. INTRODUCTION

The genus *Populus* is one of the most economically important woody plant genera in many parts of the world. Providing approximately 20 million m^3 of wood and fiber annually, Populus spp. are desirable because of their fast growth and versatile use of the wood. In addition to the many products from the wood, *Populus* spp. also provide many important environmental functions such as the rehabilitation of degraded lands, protection of soils and water, conservation of biological diversity, providing shelter and shade, and sequestering carbon in temperate and subtropical regions of the world [1].

Populus is a genus native to the Northern hemisphere, most commonly growing in mixed forests in North America and Eurasia. Globally, most (91%) of poplars are found in natural forests [2] , however, planted forests represent a small, but rapidly expanding component, of poplar forests. There are nearly 5 million ha of poplar plantations worldwide and another 1.5 million ha in agroforestry systems. Most of these plantations (73% of global total) are planted in China [1] and the most commonly planted poplars in the world are *Populus × euramericana* and *P. deltoides cultivars*.

In the United States, the first reported poplar artificial regeneration program began in 1893 when the Willamette Pulp and Paper Company planted 400 hectares of *P. trichocarpa* in the Oregon [3]. Additional plantations were soon established elsewhere in the region and due to the success of these initial plantings, the forest industry has looked to poplar production as one way to satisfy a continuously increasing demand for woody fiber.

Poplars are globally cultivated typically using two primary strategies: short rotation coppices (SRC) with very dense planting and short harvest rotation, 2 - 5 years, and long rotation plantations (LRP) with a wider initial spacing and longer rotation, generally 15 - 30 years. The main purpose of planting trees in SRC is generally production of biofuel wood or pulp, frequently used for pelleting or chopping, whereas in LRP production industrial roundwood is the main goal. Thus, those schemes are determined by the final products and cost-effec- tiveness [4]. Differences in sensitivity to biotic and abiotic stressors may also be a factor in making decisions on rotation length.

One of the most desirable characteristics of poplars is their fast growth. In the United States, individual studies indicate a wide range of production rates from 5.4 to 30 Mg/ha/yr and results are dependent on a number of factors [5] - [8]. In terms of biomass production, Holzmueller and Jose [9] reported these values were comparable to the production of annual grain-based crops such as maize (7 - 9.7 Mg/ha/yr) and sorghum (4.5 Mg/ha/yr) [10] [11]. Such fast growth, however, is linked to short life span, with some *Populus* species reaching maturity at the age of 40 - 50 years.

The other primary reason *Populus* is so desirable for plantation production is that it is a very adaptable tree genus. It includes many species and hybrids that are able to grow under various conditions and have a wide spectrum of adaptations to biotic and abiotic stressors [12]. For example, De Woody et al. [13] demonstrated that *Populus* nigra leaf size was reduced and branching architecture changed across a north-south gradient in Europe. *Villar et al.* [14] suggest that some poplar species have genetic and ecophysiological characteristics to cope with the ongoing climate change that may allow us to use them successfully in the future.

The huge diversity of species, clones and hybrids, and the easy ability for interspecies hybridization for some species may be of great significance for adaptations. It has been suggested that crosses between European, American and Asian poplars, some of which occur naturally, are more tolerant to biotic and abiotic stressors, have better growth and productivity, and competitive ability. In a study of poplar hybrids in the western United States Stettler et al. [15] attributed these characteristics primarily to heterosis effects. In another example, crosses between European and American poplars showed superior performance in germination, survival rate, and shoot growth under elevated temperature, suggesting they may be more beneficial than pure species for plantation purposes [16]. It is possible that further performance increases may be achieved through genetic engineering [17] - [22]. However, the easiness of spontaneous hybridization between introduced and wild relatives may pose risks to the genetic integrity of native species and should bring some caution before introduction [23].

While poplar species are planted across the globe, there has been little analysis to compare production rates among plantations in different regions. Given the range of temperature within which poplars are planted and the different types of plantings, the objective of this study was to analyze patterns in productivity across a temperature gradient of two different types of *Populus* plantings, short rotation coppices (SRC) and long rotation plantations (LRP) for both pure species and hybrids.

2. MATERIALS AND METHODS

Approximately 50 papers from 23 countries (North America, Eurasia, Australia) with more than 500 records (for short and long rotation plantations) were analyzed. While the list was not exhaustive we included as many studies from different regions and from different countries as possible. Only papers with available data about diameter and stem height at a given age, clearly described Populus species used, and type of plantation, SRC or LRP, were included. When possible, data about DBH were taken, but in a few studies, especially related to 1 year old SRC, only data on diameter at ground level were available. Annual stem volume indexes (SVI) were calculated based on DBH, height and age of trees by the formula: $SVI = (DBH^2 * height)/age$. Correlation analysis was applied for estimation of productivity for pure species and hybrids in two different systems of planting, SRC and LRP. Comparisons in the differences between regression lines were determined with the use of t-crite- ria ($P < 0.01$) [24].

As data about temperature conditions were presented in a small amount of publications, we collected parameters by obtaining them from an open-source database (http://www.weatherbase.com). Parameters were chosen based on importance for plant vegetation as well as accessibility from base data for most localities. Those parameters were: average annual temperature, average annual high temperature, average annual low temperature, highest recorded annual temperature, lowest recorded annual temperature, average number of days above 32°C, average number of days above 0°C. If the place signed in paper was not present in the base data, closest geographical place were chosen. Pearson correlation tests were used to quantify relationships between stem volume indexes and temperature data, relationships were accepted statistically significant under the level of $P < 0.01$.

Data about local soil conditions were included in our database but unfortunately were not used for analysis due to difficulties related to the wide range of soil classification. Data about irrigation, fertilization, chemicals used, altitude, longitude and elevation were not available in most publications, and therefore weren't applicable for analysis either.

3. RESULTS

Ten different Populus species and 21 hybrids were used for analysis (Table 1). Both Populus spp. and hybrids were planted in LRP and SRC. Most plantations were established using either 15 - 30 cm length dormant hardwood cuttings (SRC and LRP) or 1-year-old rooted cuttings (preferably

LRP). Spacing generally ranged from 0.5 × 0.5 m to 1.5 × 1.5 m for SRC and 2 × 2 m to 6 × 6 m for LRP. Calculated by stem volume index the average annual productivity per tree was 0.074 m^3/yr for LRP and 0.005 m^3/yr for SRC.

Comparison of growth parameters (DBH and height) for species and hybrids demonstrated that in LRP, hybrids are generally more productive than pure species, and the linear slope coefficient for hybrids was more than two times greater than those for non-hybrid Populus species (Figure 1). Tree height and DBH differences between slopes of regression lines were highly significant ($t = 10.80$, $t = 11.24$, respectively, $P < 0.01$). For SRC, tree height and DBH differences for hybrid and non-hybrid poplars were smaller and didn't differ significantly ($t = 1.84$ for DBH, $t = 0.03$ for tree height, $P > 0.05$; Figure 2).

Results of statistical analysis for estimation of the effects of temperature parameters on poplar tree productivity did not show any strong correlations, but there were some significantly weak correlations (within $0.20 < r < 0.32$, $P < 0.01$) (Table 2). For LRP, wood productivity slowly increased with increasing lowest recorded annual temperature and the numbers of days above 0°C per year and decreased with increasing number of hot days above 32°C. For SRC both average annual high temperatures and the number of hot days above 32°C have significant weak positive influence on SVI (Table 2).

4. DISCUSSION

Hybrid poplars are typically selected for high productivity and it is not surprising that on average hybrid poplars outperformed pure species planted as LRP. We expected to see similar results in SRC, but this was not the case. However if you were to compare the size of the best hybrids to the best species in SRC, the hybrids showed better growth (Figure 1 and Figure 2). Decreased mean size in the SRC hybrids could be the result of the testing of experimental hybrids that would not typically be planted on a wide spread basis.

The lack of a strong relationship between temperature parameters and poplar growth could be explained by several factors. The first could be that while data from over 23 countries were analyzed, most of these plantations were in temperate environments, and the range in temperature parameters used in the analysis was not wide enough. In addition, poplar species are adapted to a wide range of conditions and commonly have extensive ranges. For example, the range of *Populus deltoides* (eastern cottonwood) extends from the southernmost United States to northern Canada. Therefore, given

Table 1. Populus species analyzed in current research (listed by countries).

Populus sp.	Plantation	Country	Source
P. deltoides, P. × euroamericana	LRP	Australia	[25]
P. balsamifera × P. deltoides, P. balsamifera × P. maximowiczii, P. balsamifera × P. trichocarpa, P. deltoides × P. × petrowskyana, P. × euroamericana, P. maximowiczii × P. nigra, P. maximowiczii × P. trichocarpa	LRP, SRC	Canada	[26]-[29]
P. deltoides, P. × euroamericana × P. nigra, P. × interamericana, P. nigra × P. trichocarpa, P. tomentosa, P. tomentosa hybrids	LRP, SRC	China	[30]-[33]
P. angulata, P. deltoides, P. nigra, P. × euroamericana, P. maximowiczii × P. trichocarpa	LRP, SRC	Czech Republic	[34] [35]
P. tremula, P. tremula × P. tremuloides	LRP	Finland	[36] [37]
P. × euroamericana, P. × interamericana	SRC	France	[38]
P. trichocarpa, P × interamericana, P. maximowiczii × P. nigra, P. maximowiczii × P. trichocarpa	SRC	Germany	[39]
P. × euroamericana	SRC	Greece	[40]
P. alba, P. deltoides, P. alba × P. grandidentata, P. deltoides × P. × euroamericana, P. × interamericana	LRP	Hungary	[41]-[43]
P. deltoides	LRP	India	[44]-[48]
P. deltoides	LRP	Iran	[49]
P. deltoides × P. × euroamericana, P. × euroamericana, P. × generosa × P. nigra	SRC	Italy	[50]
P. maximowiczii	SRC	Japan	[51]
P. × euroamericana	LRP	New Zealand	[52]
P. alba, P. deltoides, P. nigra, P. trichocarpa, P. × euroamericana, P. × interamericana	LRP, SRC	Romania	[53]
P. balsamifera, P. tremula, P. tremuloides, P. × euroamericana	LRP	Russia	[54]-[56]
P. deltoides, P.deltoides × P. × euroamericana, P. × euroamericana	LRP, SRC	Serbia	[57]-[59]
P. nigra, P. angulata × P. nigra, P. deltoides, P. × euroamericana, P. × interamericana, P. maximowiczii × P. nigra	LRP	Slovak Republic	[4] [60]
P. × euroamericana, P. × interamericana, P. × interamericana × P. nigra	SRC	Spain	[61]
P. balsamifera, P. trichocarpa, P. × interamericana, P. maximowiszii × P. trichocarpa, P. deltoides × P. maximowiczii, P. tremula × P. tremuloides	LRP	Sweden	[62] [63]
P. deltoides, P. alba × P. deltoides, P. × euroamericana	LRP	Turkey	[64] [65]
P. nigra, P. balsamifera × P.trichocarpa, P. × euroamericana, P. laurifolia × P. nigra	LRP, SRC	Ukraine	[66] [67]
P. deltoides, P. nigra, P. trichocarpa, P. balsamifera x (P. laurifolia × P. nigra), P. deltoides × P. maximowiczii, P. × euroamericana, P. × interamericana, P. maximowiczii × P. nigra, P. nigra × P. trichocarpa	LRP, SRC	USA	[68]-[73]

Table 2. Pearson's correlation coefficients relationships between yearly stem volume indexes (SVI) and temperature data.

Climate data	Range for long rotation plantations	Range for short rotation plantations	r (SVI), long rotation plantations	r (SVI), short rotation plantations
Average annual temperature	1.2 to 18.0°C	1.2 to 15.1°C	0.155*	0.109
Average annual high temperature	6.0 to 24.0°C	7.0 to 20.3°C	0.113	<u>0.273*</u>
Average annual low temperature	−4.7 to 13.0°C	−4.7 to 11.0°C	0.167*	−0.062
Highest recorded annual temperature	28.0 to 45.4°C	33.0 to 43.2°C	−0.060	−0.027
Lowest recorded annual temperature	−48.3 to −3.9°C	−43.9 to −6.0°C	<u>0.206*</u>	0.189
Average number of days above 32°C per year	0 to 90	0 to 63	<u>−0.228*</u>	<u>0.314*</u>
Average number of days above 0°C per year	158 to 362	158 to 343	<u>0.300*</u>	0.058

*correlation is significant under $P < 0.01$; weak correlations (within $0.20 < r < 0.32$) are underlined.

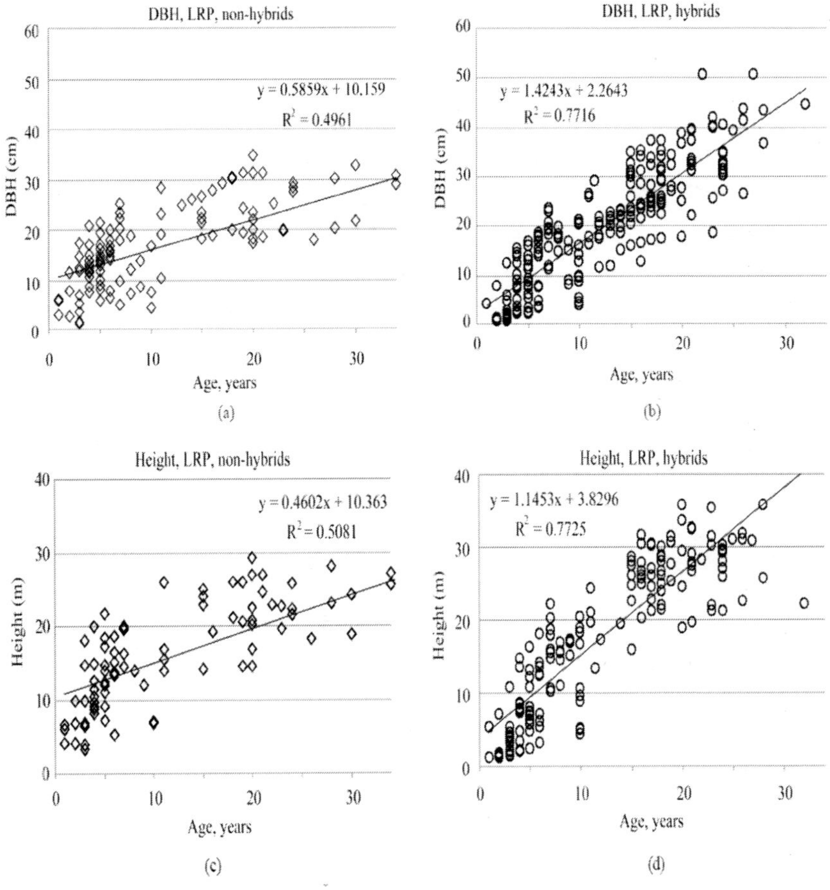

Figure 1. Age dynamics of DBH (a) (b) and height (c) (d) in long rotation plantations (LRP) for non-hybrid (a) (c) and hybrid (b) (d) poplar trees.

the wide latitude in which poplar species are naturally found differences in temperature may not influence poplar growth in most instances.

However, when analyzing productivity and temperature conditions, an interesting phenomenon was discovered. Indian plantations demonstrated sharp divergence from normal distributions and generally have higher productivity by stem volume indexes than plantations in other countries. We speculated that this effect was due to year-round favorable weather conditions as this was the only region that had zero days with temperature

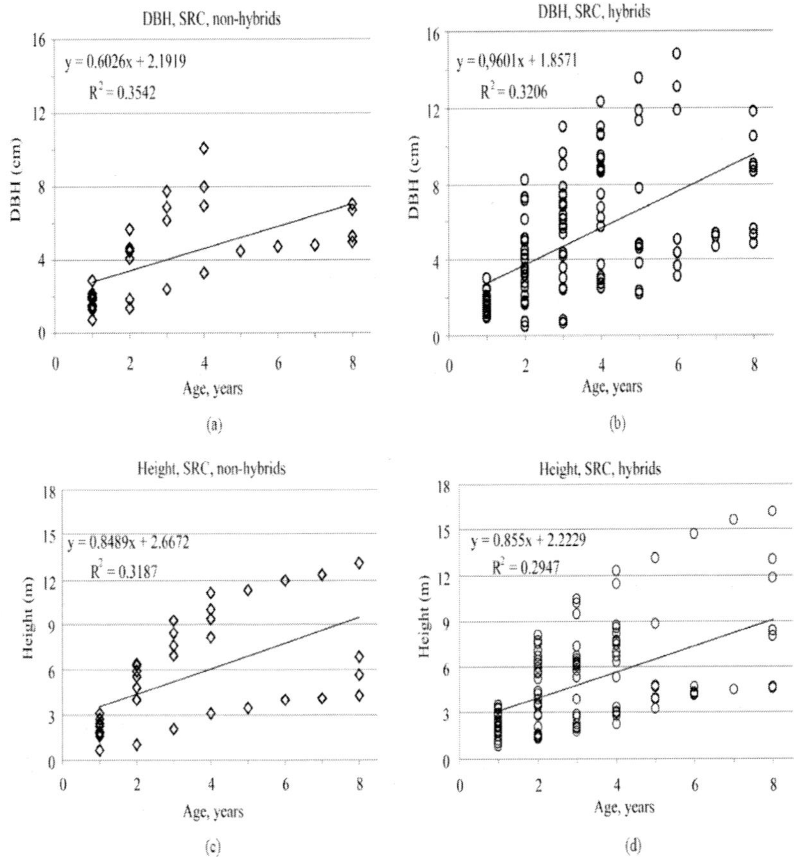

Figure 2. Age dynamics of DBH (a) (b) and height (c) (d) in short rotation coppices (SRC) for non-hybrid (a) (c) and hybrid (b) (d) poplar trees.

below 0°C. Therefore, trees could grow without (or very short) dormant periods. From the first look it was unexpected as there was a weak negative relationship between poplar growth and number of hot days above 32°C. But probably year-round favorable weather gives more benefits for poplar growth. Additional studies of poplars in warmer environments are needed to further test this hypothesis.

Other studies have suggested that water availability plays a larger role than temperature in poplar productivity [74]. Poplars are susceptible to drought and are commonly found in riparian regions, particularly in more arid environments. Jules et al. [75] observed that higher precipitation and cooler summers were positively correlated with plant growth. Conversely, some studies indicated that on a regional level, temperature can override the influence of precipitation levels. In a study of Populus tremuloides in British Columbia, positive correlations between tree height growth and mean annual temperature and mean annual summer temperature were observed, but no correlations were found for precipitation. Authors suggested that in this study growing season temperature was more limited than water availability [76].

In our study, LRP demonstrated some correlation between productivity and increasing temperature during winter and decreasing heat during summer; and for SRC, weak positive correlations between productivity and increasing yearly temperature and the number of hot days. Those findings may be partially based on the differences in range of parameters. While number of days above 32°C for LRP ranged 0 to 90 days, for SRC it ranged between 0 to 63 days (Table 2). We surmise that while too many hot days may be a limiting factor, it is probably reached after some critical threshold.

Climate parameters based on temperature, demonstrate that woody biomass production is possible under wide range of optimal temperature conditions. The number of hot summer days appears to decrease productivity in LRP; however, this may be offset by increased winter temperatures in these climates. High productivity for poplar plantations may be reached by using a broad genetic diversity of poplars and utilizing the highly adaptive potential of this genus. Overall, we observed an occurrence of a genotypic variability for productivity and some degree of sensitivity to environmental change in Populus, particularly for LRP. It is obvious that there is a wide range of adaptability for this genus, which is helpful to maintaining high levels of productivity even under predicted changes in future climate [75] [77] - [82].

5. CONCLUSION

We found that climate factors (related to temperature) and clone origin (pure species or hybrids) generally influenced production potential in long rotation plantations, but not in short rotation plantations. Opposite effects of number of hot days per year on LRP and SRC demonstrated that producing of wood in LRP was more vulnerable to the number of excessive heat days and an increase in number of excessive hot days. Our findings reveal that well managed SRC poplar plantations may have wider applications for regions with extreme temperature conditions, to which long rotation forests are more sensitive. It is most likely connected with more intensive early growth of poplar trees and may be offset with more active management of SRC, e.g. irrigation and fertilization. We conclude that SRC plantations appear to be less sensitive to environmental conditions than LRP irrespective of the geographic location. Overall, our analysis clearly indicates that growing Populus in SRC systems with proper cultural practices can decouple the effects of climate change on productivity.

ACKNOWLEDGEMENTS

Natalya K. Kutsokon, Shibu Jose, Eric Holzmueller This work was supported by the J. William Fulbright Foreign Scholarship Board and the Bureau of Educational and Cultural Affairs of the United States Department of State through a Fulbright Scholar Grant for 2012-2013 to Dr. N.K. Kutsokon, who is very thankful for the great opportunity to carry out the research.

REFERENCES

1. FAO (2005) International Poplar Commission Report of the 22nd Session of the Commission and of the 42nd Session of its Executive Committee. http://www.fao.org/forestry/9497-0be6293ee294ca6ff64cd8f5f570e1c1b.pdf.
2. Singh, A. and Kumar, A. (2012) Poplars and Willows for Increasing the Livelihood and Rural Development: A Review. Report on International Poplar Commission 24th Session, Dehradun, 29 October-2 November 2012, 155.
3. Berguson, B., Eaton, J. and Stanton, B. (2010) Development of Hybrid Poplar for Commercial Production in the United States: The Pacific Northwest and Minnesota Experience. In: Braun, R., Karlen, D. and Johnson E., Eds., Sustainable Alternative Fuel Feedstock Opportunities, Challenges and Roadmaps for Six U.S. Regions, 282-299.
4. Bartko, M. (2011) Analyza Biologickych, Produkcnych a Ekonomickych Aspektov Pestovania Rychlorastucich Drevin na Sovensku. Ph.D. Thesis, LF TU, Zvolen.
5. Geyer, W.A. (2006) Biomass Production in the Central Great Plains USA under Various Coppice Regimes. Biomass and Bioenergy, 30, 778-783. http://dx.doi.org/10.1016/j.biombioe.2005.08.002
6. Goerndt, M.E. and Mize, C.W. (2008) Short-Rotation Woody Biomass as a Crop on Marginal Lands in Iowa. Northern Journal of Applied Forestry, 25, 82-86.
7. Riemenschneider, D.E., Berguson, W.E., Dickmann, D.I., Hall, R.B., Isebrands, J.G., Mohn, C.A., Stanosz, G.R. and Tuskan, G.A. (2001) Poplar Breeding and Testing Strategies in the North-Central U.S.: Demonstration of Potential Yield and Consideration of Future Research Needs. Forestry Chronicle, 77, 245-253. http://dx.doi.org/10.5558/tfc77245-2
8. Tufekcioglu, A., Raich, J.W., Isenhart, T.M. and Schultz, R.C. (2003) Biomass, Carbon, and Nitrogen Dynamics of Multi-Species Riparian Buffers within an Agricultural Watershed in Iowa, USA. Agroforestry Systems, 57, 187-198. http://dx.doi.org/10.1023/A:1024898615284
9. Holzmueller, E.J. and Jose, S. (2012) Biomass Production for Biofuels Using Agroforestry: Potential for the North Central Region of the United States. Agroforestry Systems, 85, 305-314. http://dx.doi.org/10.1007/s10457-012-9502-z
10. Tollenaar, M. and Lee, E.A. (2002) Yield Potential, Yield Stability and Stress Tolerance in Maize. Field Crops Research, 75, 161-169. http://dx.doi.org/10.1016/S0378-4290(02)00024-2

11. USDA (U.S. Department of Agriculture) (2010) National Agricultural Statistics Service. http://www.nass.usda.gov/
12. Yaqoob, N., Karlsson, J., Riber, B.A., Solheim, H. and Fossdal, C.G. (2010) Constitutive and Induced Resistance between a Susceptible and Resistant Aspen Clone and Transcriptional Difference Towards Melampsora megnosiana. 5th International Poplar Symposium: Poplars and Willows: From Research Models to Multipurpose Trees for a Bio-Based Society, Orvieto, 20-25 September 2010, 204.
13. DeWoody, J., Trewin, H., Viger, M. and Taylor, G. (2010) Growing Large Leaves from a Small-Leaf Gene Pool: Evolutionary Trajectories in Populus nigra L. (Black Poplar) in Context of a Changing Climate. 5th International Poplar Symposium: Poplars and Willows: From Research Models to Multipurpose Trees for a Bio-Based Society, Orvieto, 20-25 September 2010, 19.
14. Villar, M., Chamaillard, S., Barbaroux, C., Bastien, C., Brignolas, F., Faivre Rampant, P., Fichot, R., Forestier, O., Jorge, V. and Rodrigues, S. (2010) Populus nigra as Keystone Species Able to Cope with the Ongoing Climate Change. 5th International Poplar Symposium: Poplars and Willows: From Research Models to Multipurpose Trees for a Bio- Based Society, Orvieto, 20-25 September 2010, 17.
15. Stettler, R.F., Fenn, R.C., Heilman, P.E. and Stanton, B.J. (1988) Populus trichocarpa × Populus deltoides Hybrids for Short Rotation Culture: Variation Patterns and 4-Year Field Performance. Canadian Journal of Forest Research, 18, 745-753. http://dx.doi.org/10.1139/x88-114
16. Pulkkinen, P., Vaario, L.M., Koivuranta, L. and Stenvall, N. (2013) Elevated Temperature Effects on Germination and Early Growth of European Aspen (Populus tremula), Hybrid Aspen (P. tremula × P. tremuloides) and Their F_2-Hybrids. European Journal of Forest Research, 132, 791-800. http://dx.doi.org/10.1007/s10342-013-0719-9
17. Confalonieri, M., Balestrazzi, A., Bisoffi, S. and Carbonera, D. (2003) In Vitro Culture and Genetic Engineering of Populus spp.: Forest Tree Improvement. Plant Cell, Tissue and Organ Culture, 72, 109-138. http://dx.doi.org/ 10.1023 /A:1022265504775
18. Giri, C.C., Shyamkumar, B. and Anjaneyulu, C. (2004) Progress in Tissue Culture, Genetic Transformation and Applications of Biotechnology to Trees: An Overview. Trees, 18, 115-135. http://dx.doi.org/10.1007/s00468-003-0287-6
19. Han, K.H., Meilan, R., Ma, C. and Strauss, S.H. (2000) An Agrobacterium tumefaciens Transformation Protocol Effective on a Variety of Cottonwood Hybrids (Genus Populus). Plant Cell Reports, 19, 315-320. http://dx.doi.org/ 10.1007/ s002990050019
20. Herschbach, C. and Kopriva, S. (2002) Transgenic Trees as Tools in Tree and Plant Physiology. Trees, 16, 250-261. http://dx.doi.org/10.1007/s00468-002-0178-2.
21. Kutsokon, N.K. (2011) Main Trends in the Genetic Transformation of Populus Species. Cytology and Genetics, 45, 352-361. http://dx.doi.org/ 10.3103/S009545271106003X
22. Yevtushenko, D.P. and Misra, S. (2010) Efficient Agrobacterium-Mediated Transformation of Commercial Hybrid Poplar Populus nigra L. × P. maximowiczii A. Henry. Plant Cell Reports, 29, 211-221. http://dx.doi.org/10.1007/s00299-009-0806-z.
23. Isabel, N., Talbot, P., Lamothe, M., Bousquet, J., Schroeder, W., Floate, K., Foottit, R.G., Ruiz, P., Cumming, S., McIntire, E. and Thompson, S.L. (2010) Contemporary Gene Flow between Poplars with Exotic Component and Natural Populations of Native Species across Canada. 5th International Poplar Symposium: Poplars and Willows: From Research Models to Multipurpose Trees for a Bio-Based Society, Orvieto, 20-25 September 2010, 21.
24. Fowler, J. and Cohen, L. (1990) Practical Statistics for Field Biology. Milton Keynes, Philadelphia.
25. Hopmans, P., Stewart, H.T.L., Flinn, D.W. and Hillman, T.J. (1990) Growth, Biomass Production and Nutrient Accumulation by Seven Tree Species Irrigated with

Municipal Effluent at Wodonga, Australia. Forest Ecology and Management, 30, 203-211. http://dx.doi.org/10.1016/0378-1127(90)90137-Z.

26. Benomar, L., DesRochers, A. and Larocque, G.R. (2012) The Effects of Spacing on Growth, Morphology and Biomass Production and Allocation in Two Hybrid Poplar Clones Growing in the Boreal Region of Canada. Trees, 26, 939-949. http://dx.doi.org/10.1007/s00468-011-0671-6.
27. Brown, K.R. and van den Driessche, R. (2002) Growth and Nutrition of Hybrid Poplars over 3 Years after Fertilization at Planting. Canadian Journal of Forest Research, 32, 226-232. http://dx.doi.org/10.1139/x01-191
28. Driessche, R., Thomas, B.R. and Kamelchuk, D.P. (2008) Effects of N, NP, and NPKS Fertilizers Applied to Four- Year Old Hybrid Poplar Plantations. New Forests, 35, 221-233. http://dx.doi.org/10.1007/s11056-007-9073-4
29. Labrecque, M. and Teodorescu, T.I. (2005) Field Performance and Biomass Production of 12 Willow and Poplar Clones in Short-Rotation Coppice in Southern Quebec (Canada). Biomass and Bioenergy, 29, 1-9. http://dx.doi.org /10.1016/j.biombioe.2004.12.004
30. Fang, S., Xie, B., Liu, D. and Liu, J. (2011) Effects of Mulching Materials on Nitrogen Mineralization, Nitrogen Availability and Poplar Growth on Degraded Agricultural Soil. New Forests, 41, 147-162. http://dx.doi.org/10.1007/s11056-010-9217-9
31. Fang, S., Xu, X., Lu, S. and Tang, L. (1999) Growth Dynamics and Biomass Production in Short-Rotation Poplar Plantations: 6-Year Results for Three Clones at Four Spacings. Biomass and Bioenergy, 17, 415-425. http://dx.doi.org/10.1016/S0961-9534(99)00060-4
32. Guo, X.Y. and Zhang, X.S. (2010) Performance of 14 Hybrid Poplar Clones Grown in Beijing, China. Biomass and Bioenergy, 34, 906-911. http://dx.doi.org/ 10.1016/j.biombioe.2010.01.036
33. Zhang, P., Wu, F. and Kang, X. (2012) Genotypic Variation in Wood Properties and Growth Traits of Triploid Hybrid Clones of Populus tomentosa at Three Clonal Trials. Tree Genetics & Genomes, 8, 1041-1050. http://dx.doi.org/10.1007/s11295-012-0484-9
34. Benetka, V., Vratny, V. and Salkova, I. (2007) Comparison of the Productivity of Populus nigra L. with an Interspecific Hybrid in a Short Rotation Coppice in Marginal Areas. Biomass and Bioenergy, 31, 367-374. http://dx.doi.org/ 10.1016/j.biombioe.2007.01.005
35. Cizkova, L., Cizek, V. and Bajajova, H. (2010) Growth of Hybrid Poplars in Silviculture at the Age of 6 Years. Journal of Forest Science, 56, 451-460.
36. Nikula, S., Manninen, S., Vapaavuori, E. and Pulkkinen, P. (2011) Growth, Leaf Traits and Litter Decomposition of Roadside Hybrid Aspen (Populus tremula L. × P. tremuloides Michx.) Clones. Environmental Pollution, 159, 1823-1831. http://dx.doi.org/10.1016/j.envpol.2011.03.022
37. Yu, Q. and Pulkkinen, P. (2003) Genotype-Environment Interaction and Stability in Growth of Aspen Hybrid Clones. Forest Ecology and Management, 173, 25-35. http://dx.doi.org/10.1016/S0378-1127(01)00819-2
38. Benbrahim, M., Gavaland, A. and Gauvin, J. (2000) Growth and Yield of Mixed Polyclonal Stands of Populus in Short-Rotation Coppice. Scandinavian Journal of Forest Research, 15, 605-610. http://dx.doi.org/10.1080/02827580050216851
39. Bungart, R. and Huttl, R.F. (2004) Growth Dynamics and Biomass Accumulation of 8-Year-Old Hybrid Poplar Clones in a Short-Rotation Plantation on a Clayey-Sandy Mining Substrate with Respect to Plant Nutrition and Water Budget. European Journal of Forest Research, 123, 105-115.
40. Spanos, K., Tsialtas, I. and Giakzidis, G. (2002) Biomass Production from a Short Rotation Coppice Experimental Planting of Ten Poplars Clones in N. Greece Study, Conservation and Utilization of Forest Resources. Proceedings of the 3rd Balkan Scientific Conference, Sofia, 2-6 October 2001, 326-335.
41. Barany, G. (2011) A Nemesnyar-Termesztes Fejlesztesenek Ujabb Eredmenyei. Ph.D. Thesis, Sopron, Budapest.

42. Redei, K., Keseru, Z. and Szulcsan, G. (2010) Early Evaluation of Promising White Poplar (Populus alba L.) Clones in Hungary. Acta Silvatica & Lingaria Hungarica, 6, 9-16.
43. Redei, K. and Keseru, Z.S. (2008) Promising White Poplar (Populus alba L.) Clones in Sandy Ridges between the Rivers Danube and Tisza in Hungary. International Journal of Horticultural Science, 14, 113-116
44. Chauhan, S.K., Brar, M.S. and Sharma, R. (2012) Performance of Poplar (Populus deltoides Bartr.) and Its Effect on Wheat Yield under Agroforestry System in Irrigated Agro-Ecosystem, India Caspian. Journal of Environmental Sciences, 10, 53-60.
45. Dhillon, G.P.S., Singh, A.P. and Sidhu, D.S. (2010) Field Evaluation of Populus deltoids Bartr. ex Marsh. at Two Sites in Indo-Gangetic Plains of India. Silvae Genetica, 59, 1-7.
46. Dhillon, G.P.S., Singh, A., Sidhu, D.S. and Brar, H.S. (2013) Variation among Poplar Clones for Growth and Crown Traits under Field Conditions at Two Sites of North-Western India. Journal of Forestry Research, 24, 61-67. http://dx.doi.org/10.1007/s11676-012-0269-x
47. Puri, S., Swamy, S.L. and Jaiswal, A.K. (2002) Evaluation of Populus deltoides Clones under Nursery, Field and Agrisilviculture System in Subhumid Tropics of Central India. New Forests, 23, 45-61. http://dx.doi.org/10.1023/A:1015620915654
48. Sidhu, D.S. and Dhillon, G.P.S. (2007) Field Performance of Ten Clones and Two Sizes of Planting Stock of Populus deltoides on the Indo-Gangetic Plains of India. New Forests, 34, 115-122. http://dx.doi.org/10.1007/s11056-007-9042-y
49. Salehi, A. and Maleki, M. (2012) Evaluation of Soil Physical and Chemical Properties in Poplar Plantations in North of Iran. Ecologia Balkanica, 4, 69-76.
50. Pannacci, E., Bartolini, S. and Covarelli, G. (2009) Evaluation of Four Poplar Clones in a Short Rotation Forestry in Central Italy. Italian Journal of Agronomy/Rivista di Agronomia, 4, 191-198.
51. Osawa, A. (1993) Effects of Mechanical Stresses and Photosynthetic Production on Stem Form Development of Populus maximowiczii. Annals of Botany, 71, 489-494. http://dx.doi.org/10.1006/anbo.1993.1065
52. McIvor, R., Douglas, G.B. and Benavides, R. (2009) Coarse Root Growth of Veronese Poplar Trees Varies with Position on an Erodible Slope in New Zealand. Agroforestry Systems, 76, 251-264. http://dx.doi.org/10.1007/s10457-009-9209-y
53. Filat, M. and Chira, D. (2004) Cercetari Pentru Introducerea in Cultura de Specii/Clone de Plop si Salcie cu Potential Silvoproductiv Superior si Rezistenta Sporita la Adversitat. Analele ICAS, 47, 83-99.
54. Sivolapov, V.A. (2012) Plantation Cultivation of Fast-Growing Species in the of Forest-Steppe Region with the Using in Vitro Biotechnology. Ph.D. Thesis, Voronezh.
55. Tsarev, A.P. (2012) Fodder Value of Populus euramericana. Green Biomass Journal of Agricultural Science and Technology, 2, 498-509.
56. Tsarev, V.A. and Tsarev, A.P. (2011) Breeding of Aspens in CCR. 6th Conference "Forests of Russia in 21st", St. Peterburg, March 2011, 186-191.
57. Andrasev, S., Roncevic, S. and Pekec, S. (2006) Karakteristike Rasta Nekih Klonova Crnih Topola (Growth Characteristics of Some Black Poplar Clones). Sumar inst Jastrebar, 41, 25-30.
58. Klasnja, B., Orlovic, S. and Galic, Z. (2012) Energy Potential of Poplar Plantations in Two Spacings and Two Rotations. Sumarski List, 136, 161-167.
59. Orlovic, S., Guzina, V., Krstic, B. and Merkulov, L. (1998) Genetic Variability in Anatomical, Physiological and Growth Characteristics of Hybrid Poplar (Populus × euramericana Dode (Guinier)) and Eastern Cottonwood (Populus deltoides Bartr.) Clones. Silvae Genetica, 47, 183-190.
60. Kohan, S. (2006) Valuation on Growth and Volume Production and Health Evaluation of Poplar Clones on Alluvial Soils of Laborec River in Eastern Slovakian Lowland. Forestry Journal, 52, 239-247.

61. Sixto, H., Salvia, H., Barrio, M., Ciria, M.P. and Canellas, I. (2011) Genetic Variation and Genotype-Environment Interactions in Short Rotation Populus Plantations in Southern Europe. New Forests, 42, 163-177. http://dx.doi.org/10.1007/s11056-010-9244-6
62. Christersson, L. (2010) Wood Production Potential in Poplar Plantations in Sweden. Biomass and Bioenergy, 34, 1289- 1299. http://dx.doi.org/10.1016/j.biombioe.2010.03.021.
63. Johansson, T. and Karacic, A. (2011) Increment and Biomass in Hybrid Poplar and Some Practical Implications. Biomass and Bioenergy, 35, 1925-1934. http://dx.doi.org/10.1016/j.biombioe.2011.01.040.
64. Gurses, M.K., Tufekci, S., Gulbaba, A.G., Ozkurt, N. and Ozkurt, A. (1999) The Results of Comparison Populeta Established in the Mediterranean Region. Teknik Bulten Orman Bakanlg Dogu Akdeniz Ormanclk Arastrma Enstitusu, 10, v+16.
65. Gurses, M.K., Gulbaba, A.G., Avcioglu, E., Ozkurt, N., Ozkurt, A. and Tufekci, S. (1997) Dogu Akdenizve Guneydogu Anadolu Bųlgeleri Oryantasyon Populetumlari Sonuclari. Dogu Akdeniz Ormancilik Arastirma Enstitusu, Teknik Bulten, Tarsus, 3, 18 p.
66. Fuchylo, Y.D., Sbytna, M.V., Fuchylo, Y.O. and Litvin, V.M. (2009) Experience and Perspectives of Cultivation of Poplar (Populus sp. L.) in Southern Steppe Region of Ukraine. Scientific Works of Ukrainian Forestry Academy, 7, 66-69.
67. Patlay, I.M. (Project Leader) (1991-1995) Breeding, Varieties Testing and Propagation of Important Native and Introduced Species of Forest Trees for Creation of Plantations for Different Purposes. Report of URIFFM Project No. 13.
68. Ceulemans, R., Scarascia-Mugnozza, G., Wiard, B.M., Braatne, J.H., Hinckley, T.M., Stettler, R.F., Isebrands, J.G. and Heilman, P.E. (1992) Production Physiology and Morphology of Populus Species and Their Hybrids Grown under Short Rotation. I. Clonal Comparisons of 4-Year Growth and Phenology. Canadian Journal of Forest Research, 22, 1937-1948. http://dx.doi.org/10.1139/x92-253
69. DeBell, D.S., Harrington, C.A., Clendenen, G.W. and Zasada, J.S. (1997) Tree Growth and Stand Development of Four Populus Clones in Large Monoclonal Plots. New Forests, 14, 1-18. http://dx.doi.org/10.1023/A:1006501229936
70. Kaczmarek, D.J., Coyle, D.R. and Coleman, M.D. (2013) Survival and Growth of a Range of Populus Clones in Central South Carolina USA through Age Ten: Do Early Assessments Reflect Longer-Term Survival and Growth Trends? Biomass and Bioenergy, 49, 260-272. http://dx.doi.org/10.1016/j.biombioe.2012.12.005
71. Netzer, D.A., Tolsted, D.N., Ostry, M.E., Isebrands, J.G., Riemenschneider, D.E. and Ward, K.T. (2002) Growth, Yield, and Disease Resistance of 7- to 12-Year-Old Poplar Clones in the North Central United States. Gen. Tech. Rep. NC-229, St. Paul.
72. O'Neill, M.K., Shock, C.C., Lombard, K.A., Heyduck, R.F., Feibert, B.G., Smeal, D. and Arnold, R.N. (2010) Hybrid Poplar (Populus ssp.) Selections for Arid and Semi-Arid Intermountain Regions of the Western United States. Agroforestry Systems, 79, 409-418. http://dx.doi.org/10.1007/s10457-010-9286-y
73. Pearson, C.H., Halvorson, A.D., Moench, R.D. and Hammon, R.W. (2010) Production of Hybrid Poplar under Short- Term, Intensive Culture in Western Colorado. Industrial Crops and Products, 31, 492-498. http://dx.doi.org/10.1016/j.indcrop.2010.01.011
74. Zsuffa, L., Giordano, E., Pryor, L.D. and Stettler, R.F. (1996). Trends in Poplar Culture: Some Global and Regional Perspectives. In: Stettler, R.F., Bradshaw Jr., H.D., Heilman, P.E. and Hinckley, T.M., Eds., Biology of Populus and Its Implications for Management and Conservation, Part II, Chapter 19, NRC Research Press, National Research Council of Canada, Ottawa, 515-539.
75. Jules, E.K., Carroll, A.L. and Kauffman, M.J. (2010) Relationship of Climate and Growth of Quaking Aspen (Populus tremuloides) in Yellowstone National Park. Final Report for RWO 81 Rocky Mountain Ungulates.

76. Messaoud, Y. and Chen, H.Y. (2011) The Influence of Recent Climate Change on Tree Height Growth Differs with Species and Spatial Environment. PLOS One, 16, 1-8. [Citation Time(s):1]
77. Chhin, S. (2010) Influence of Climate on the Growth of Hybrid Poplar in Michigan. Forests, 1, 209-229.
78. Henderson, D.E. and Jose, S. (2010) Biomass Production Potential of Three Short Rotation Woody Crop Species under Varying Nitrogen and Water Availability. Agroforestry Systems, 80, 259-273. http://dx.doi.org/10.1007/s10457-010-9283-1
79. Monclus, R., Villar, M., Barbaroux, C., Bastien, C., Fichot, R., Delmotte, F.M., Delay, D., Petit, J.M., Bruchet, C., Dreyer, E. and Brignolas, F. (2009) Productivity, Water-Use Efficiency and Tolerance to Moderate Water Deficit Correlate in 33 Poplar Genotypes from a Populus deltoids × Populus trichocarpa F_1 Progeny. Tree Physiology, 29, 1329- 1339. http://dx.doi.org/10.1093/treephys/tpp075
80. Morelli, T.N. and Carr, S.C. (2011) Review of the Potential Effects of Climate Change on Quaking Aspen (Populus tremuloides) in the Western United States and a New Tool for Surveying Aspen Decline. Report USDA Forest Service Pacific Southwest Research Station, General Technical Report PSW-GTR-235.
81. Rohde, A., Bastien, C. and Boerjan, W. (2011) Temperature Signals Contribute to the Timing of Photoperiodic Growth Cessation and Bud Set in Poplar. Tree Physiology, 31, 472-482. http://dx.doi.org/10.1093/treephys/tpr038
82. Zalesny, R.S., Hall, R.B., Zalesny, J.A., McMahon, B.J., Berguson, W.E. and Stanosz, G.R. (2009) Biomass and Genotype × Environment Interactions of Populus Energy Crops in the Midwestern United States. Bioenergy Research, 2, 106-122. http://dx.doi.org/10.1007/s12155-009-9039-9

CHAPTER 7

Effects of Forests, Roads and Mistletoe on Bird Diversity in Monoculture Rubber Plantations

Rachakonda Sreekar[1,2], Guohualing Huang[1], Mika Yasuda[1,3], Rui-Chang Quan[1], Eben Goodale[4], Richard T. Corlett[1] & Kyle W. Tomlinson[1]

[1]*Center for Integrative Conservation, Xishuangbanna Tropical Botanical Garden, Chinese Academy of Sciences, Menglun, Yunnan, China*

[2]*School of Biological Sciences, University of Adelaide, Australia*

Products Research Institute, Tsukuba, Ibaraki, Japan

[3]*College of Forestry, Guangxi University, Nanning, Guangxi, China*

ABSTRACT

Rising global demand for natural rubber is expanding monoculture rubber (*Hevea brasilensis*) at the expense of natural forests in the Old World tropics. Conversion of forests into rubber plantations has a devastating impact on biodiversity and we have yet to identify management strategies that can mitigate this. We determined the life-history traits that best predict bird species occurrence in rubber plantations in SW China and investigated the effects of surrounding forest cover and distance to roads on bird diversity. Mistletoes provide nectar and fruit resources in rubber so we examined mistletoe densities and the relationship with forest cover and rubber tree diameter. In rubber plantations, we recorded less than half of all bird species extant in the surrounding area. Birds with wider habitat breadths and low conservation value had a higher probability of occurrence. Species richness and diversity increased logarithmically with surrounding forest cover, but roads had little effect. Mistletoe density increased exponentially with rubber tree diameters, but was unrelated to forest cover. To maximize bird diversity in rubber-dominated landscapes it is therefore necessary to preserve as much forest as possible, construct roads through plantations and not forest, and retain some large rubber trees with mistletoes during crop rotations.

INTRODUCTION

Human-driven habitat loss due to agricultural intensification is currently expanding faster than any time in the past 50 years, with a conservative estimate of c. 120 million hectares of additional land, mostly in developing nations, required to meet the projected demand by 2050[1]. Croplands particularly threaten the tropical forests that hold the majority of the world's species[2]. The future of global biodiversity will therefore increasingly be placed in the hands of agricultural policy makers[3]. Assessments of the impacts of croplands on native ecosystems are an urgent priority for conservation, especially in tropical East Asia where both species richness and conversion of forests to monoculture plantations are among the highest globally[4,5]. Rubber is the most rapidly expanding tree crop in mainland SE Asia (with 84% of the global total rubber area), and an estimated 4.3–8.5 million hectares of additional rubber plantations are required to meet global demand by 2024[6]. This threat is imminent: for example, the Cambodian government has recently allocated 346,000 hectares inside 23 protected areas to rubber companies[7].

Conversion of natural forests into monoculture rubber plantations reduces species richness and changes the composition of animal assemblages[8,9,10,11,12,13], but we are yet to identify conservation strategies that could improve biodiversity within monoculture rubber landscapes. Studies in oil palm plantations have shown that nearby natural forests can enhance biodiversity in palm plantations, but not enough to maintain original assemblages[14,15,16,17]. However, riparian reserves within oil palm plantations were important to maintain riparian ants, dung beetles and freshwater fish communities[5,18]. Research is still needed to understand how neighbouring forests support biodiversity in rubber[6].

Millions of kilometers of new roads are anticipated by 2050, partly to facilitate movements of resources in and out of these plantations[19]. However, roads often open up tropical forest regions to colonization and exploitation[20]. Roads escalate forest degradation, spread invasive species, act as a barrier for many sensitive forest species, and provide easy access for overexploitation[20]. However, roads within plantations may potentially increase animal diversity by creating bordering plant communities distinct from the monoculture plantations[21]. Research is still needed to determine if roads can have a positive or negative effect on the biodiversity in rubber.

Birds are often used as a proxy for the biodiversity value of human-dominated landscapes because they 1) are easy to survey, 2) are diverse, with better known life history attributes than any other group[22], 3) have diet guilds that respond differently to different threats[22,23,24,25,26], and 4) provide multiple ecosystem functions, including seed dispersal and predation, pollination, scavenging, nutrient deposition, and pest predation[22,23]. Though birds are the most studied vertebrate group, studies on the effects of habitat change on their life history traits are still scarce[27,28] and research is needed to determine the life-history traits that best predict species occurrence in rubber. Small insects pollinate rubber trees (*Hevea brasiliensis*) and seeds are dispersed explosively, but ten families of nectarivorous and frugivorous birds were recorded in monoculture rubber in SE Asia[11,12,13]. We suspected that mistletoes, a diverse group of parasitic plants that are found on rubber trees, provide key resources in rubber plantations through provision of abundant fruits and nectar[29].

In this paper, 1) we compiled a comprehensive checklist of resident birds that occur in the entire landscape (forests, villages, rubber and banana plantations) and compared it with a checklist of resident birds that occur in rubber plantations to determine life history traits that predict occurrence probabilities of birds in rubber plantations. 2) We conducted 52 bird point counts (repeated four times) across a distance gradient from roads and primary forests within monoculture rubber to determine the effects of forest and roads on bird diversity and composition (Fig. 1). We modeled bird richness at each point as a function of forest cover and distance to roads. It has been shown that species richness increases non-linearly with increasing forest cover or distance to roads30·31 and that thresholds can be detected in some cases[32,33], so we also determined the shape of model fits and searched for thresholds of forest cover and road distance, if any. 3) Finally, we determined the effects of forest cover and rubber diameter at breast height (DBH) on mistletoe densities in rubber plantations, and made focal point observations to determine if frugivorous and nectarivorous birds feed on them.

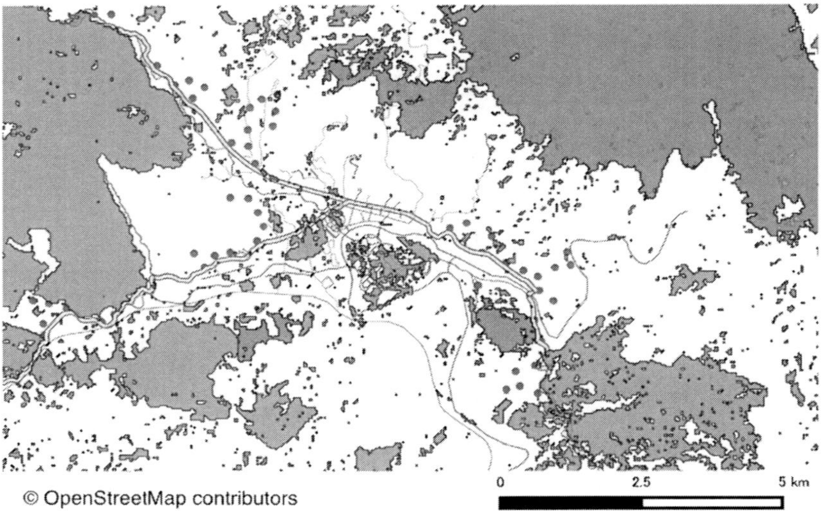

Figure 1: Map of Menglun landscape, Xishuangbanna Prefecture, Yunnan, China, in 2010. In the map, green is forest cover, white is rubber plantations and other human land-use, red circles are points sampled for birds in rubber, thick pale green lines are the paved roads along which sampling was conducted, thin light blue lines are unpaved roads, and maroon lines are a river and streams. We produced the forest-cover map for the study landscape by calculating the Normalized Difference Vegetation Index (NDVI) of multispectral data from Landsat-2 Multi Spectral Scanner (MSS) and Landsat-5 Thematic Mapper (TM) images (30 m spatial resolution) acquired in February 2010 using the free and open source software QGIS 1.8.0 (http://www.qgis.org) and GRASS 6.4.3 (http://grass.osgeo.org). The Landsat images were downloaded from the open database at the Earth Resources Observation and Science centre (EROS), U.S. Geological Survey (USGS; http://eros.usgs.gov), and road data from the open database OpenStreetMap (http://www.openstreetmap.org). The map tiles of road data are available under the Creative Commons Attribution-ShareAlike 2.0 licence (CC BY-SA; http://www.openstreetmap.org/copyright).

RESULTS

Trait determinants of bird occurrence

Among the 156 extant resident diurnal bird species in the Menglun landscape between 2011 and 2015, we recorded 58 species (37.1%) in rubber (Table S1). Considering forest birds only, we recorded 33.7% (33 of 98 recorded in forests) at least once in rubber (Table S1). Bird species occurrence probability increased with habitat breadth ($X^2_{(1,147)} = 10.45$, $P = 0.001$; Fig. 2), but was not influenced by forest preference ($X^2_{(1,147)} = 0.09$, $P = 0.75$), body size ($X^2_{(1,147)} = 0.78$, $P = 0.37$) or diet type ($X^2_{(4,147)} = 3.1$, $P = 0.54$).

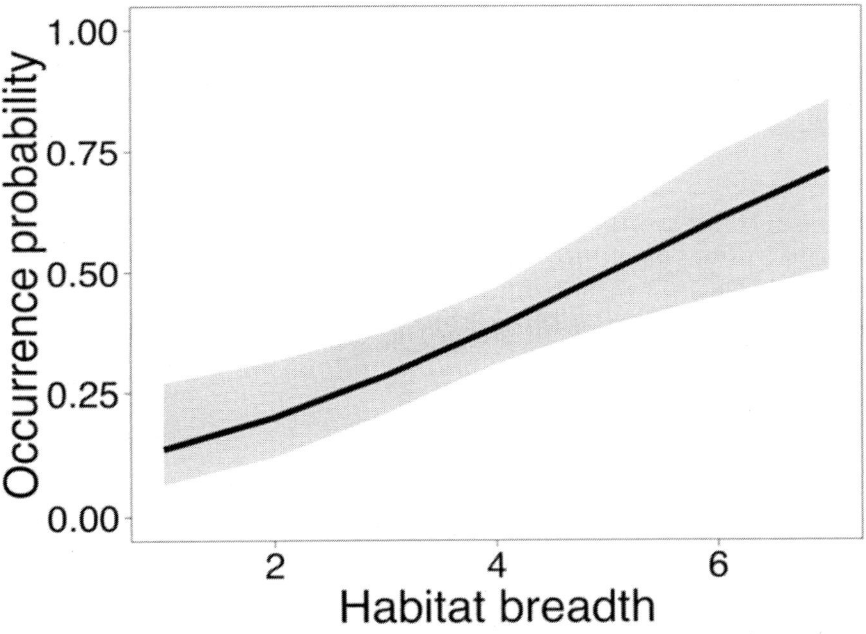

Figure 2: Occurrence probability of Menglun birds in monoculture rubber plantations as a function of habitat breadth. The black line is the prediction of the model fitted to the data and the shaded grey area is its 95% confidence interval.

Environmental determinants of bird diversity

We recorded 3140 individuals of 45 resident and migrant species in rubber during point counts (Table S2). We only considered 32 species (71%; 3076 individuals) in the analysis, because we recorded the other 13 species in fewer than four points (Table S2; including the excluded species did not change results). We used original species richness instead of extrapolated richness estimators because the former had reached an asymptote (Fig. S1). Additionally, the extrapolated richness at each sample point of all species ($r = 0.84$, $P < 0.001$), frugivores ($r = 0.98$, $P < 0.001$) and insectivores ($r = 0.89$, $P < 0.001$) was highly correlated with original species richness.

Bird species richness and diversity in monoculture rubber plantations increased logarithmically with forest cover, as did frugivore and insectivore richness and diversity (Table 1; Fig. 3). The piecewise linear regressions estimated threshold values of forest cover as 4.01 ha (95% CIs: 1.1–6.91), 5.92 ha (95% CIs: 1.2–10.63) and 1.86 ha (95% CIs: 0.69–3.04) within the radius of 50 m from the point (78 ha) for the richness of all species, frugivores and insectivores respectively, which increased their species richness by c. 4, 3 and 3 species per point respectively. Similarly, the piecewise linear regressions estimated threshold values of forest cover as 2.76 ha (95% CIs: 0.97–4.55), 4.63 ha (95% CIs: 1.89–7.36) and 1.84 ha (95% CIs: 0.69–2.98) for the diversity of all species, frugivores and insectivores respectively (Fig. 3).

Table 1: Model-averaged estimates and covariate weights of environmental variables (forest cover (in hectares) =forest cover within 500 m radius from the sampling points; road distance (in meters) =closest distance to the nearest non-rubber road) for bird species richness, Shannon diversity and individual species occurrence probability in Menglun, Xishuangbanna, Yunnan, China.

Response	Covariate	Model averaged estimate ± SE	Covariate weight	Deviance explained
Species richness				
All species	Forest cover	1.75 ± 0.29	1.00	0.43
Frugivore	Forest cover	0.74 ± 0.15	1.00	0.35
Insectivore	Forest cover	1.02 ± 0.26	1.00	0.23
Shannon diversity				
All species	Forest cover	0.11 ± 0.02	1.00	0.32
Frugivore	Forest cover	0.18 ± 0.04	1.00	0.33
Insectivore	Forest cover	0.12 ± 0.04	0.98	0.16
Occurrence probability				
Ashy Drongo	Forest cover	0.78 ± 0.33	0.90	0.12
Black-crested Bulbul	Forest cover	1.86 ± 0.76	1.00	0.23
Blue-throated Barbet	Forest cover	0.77 ± 0.30	0.93	0.13
Brown-cheeked Fulvetta	Forest cover	0.86 ± 0.38	0.88	0.13
Crimson Sunbird	Road distance	− 0.65 ± 0.27	0.89	0.09
Plaintive Cuckoo	Forest cover	1.94 ± 0.67	1.00	0.38
Yellow-vented Flowerpecker	Forest cover	0.91 ± 0.49	0.75	0.13
Great Barbet	Forest cover	1.32 ± 0.59	0.94	0.21
Grey-headed Canary Flycatcher	Road distance	0.59 ± 0.28	0.82	0.07
Indian Cuckoo	Forest Cover	0.92 ± 0.39	0.90	0.17
Magpie Robin	Road distance	− 0.68 ± 0.30	0.85	0.11
Puff-throated Babbler	Forest cover	0.56 ± 0.27	0.79	0.07
Red-whiskered Bulbul	Forest cover	1.50 ± 0.46	1.00	0.30
Rufous-capped Babbler	Forest cover	1.25 ± 0.55	0.94	0.07

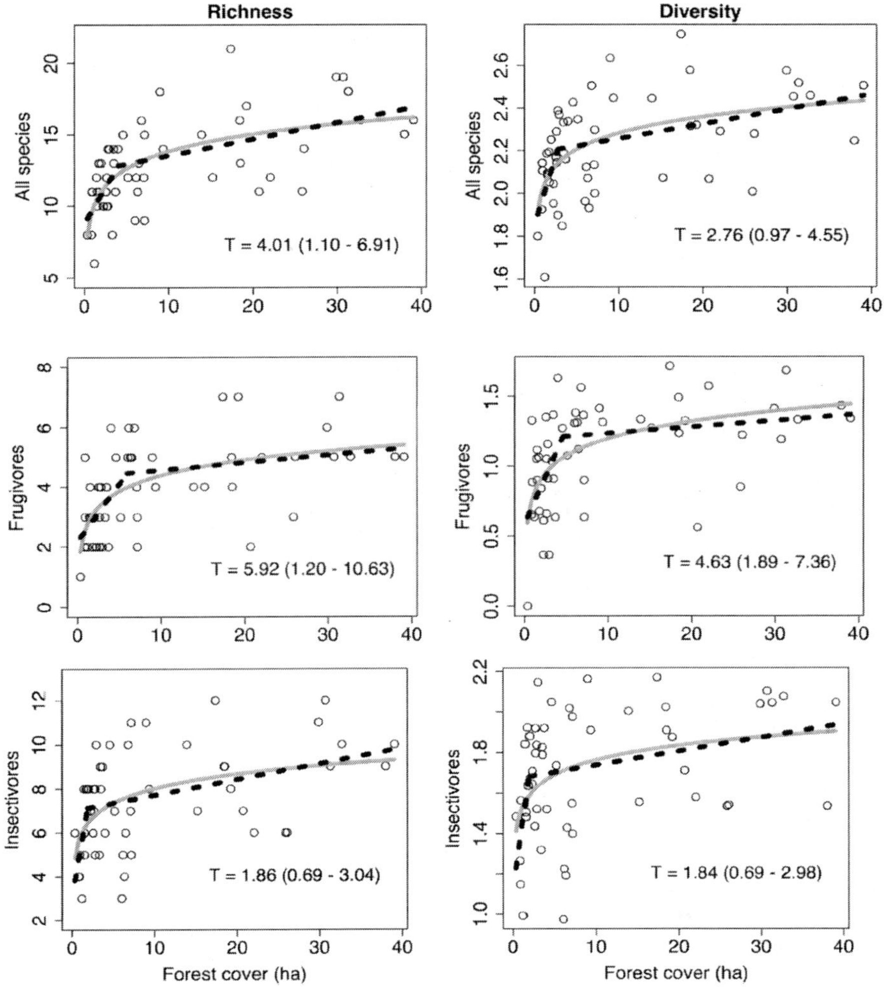

Figure 3: All, frugivorous and insectivorous bird richness and Shannon diversity as a function of forest cover (in hectares) within 500 m from a sampling point. The grey lines are predictions of the general linear models fitted to the data and the dashed black lines are predictions of piecewise linear regressions. The 'T' in each plot is the threshold value of forest cover with 95% confidence intervals in parenthesis.

Forest cover also influenced bird species composition ($P = 0.001$; Fig. 4). Among the 32 species used in the analyses, the occurrence probability (OP) of 11 species increased with forest cover (Table 1). Very few responded to distance to roads. The OP of Grey-headed Canary Flycatcher (*Culicicapa ceylonensis*) increased with distance to the road, and the OP of Crimson Sunbird (*Aethopyga siparaja*) and Magpie Robin (*Copsychus saularis*) decreased with distance to the road (Table 1).

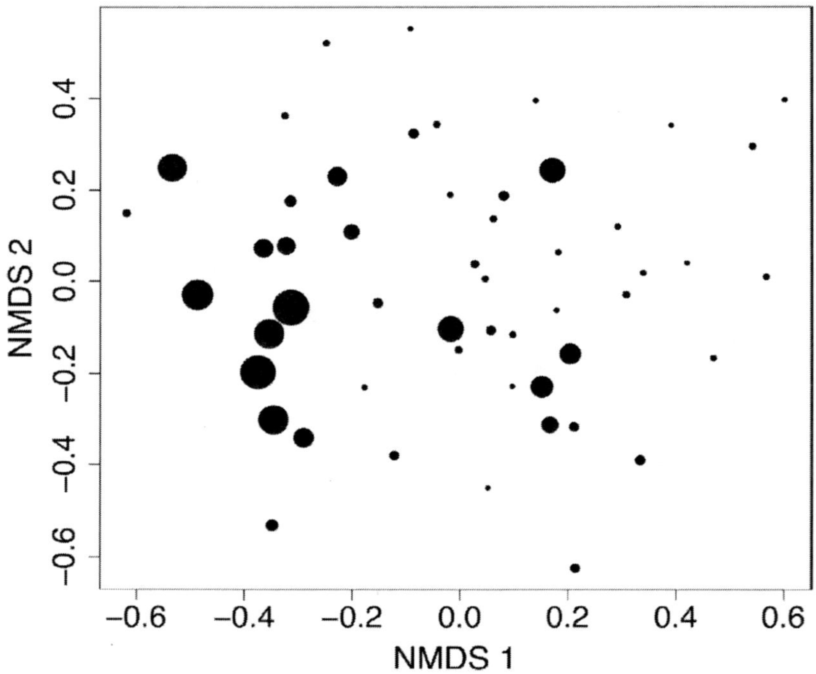

Figure 4: A non-metric dimensional scaling (NMDS) based on the Bray-Curtis distance with a two-dimensional solution of the bird assemblages in rubber plantations of Menglun, Xishuangbanna, Yunnan, China. The size of the circle is proportional to the forest area (in hectares) within 500 m radius of a sampling point. The ordination diagram is for visualization only. We conducted all tests using multivariate generalized linear models (see text for statistical analyses).

Determinants of mistletoe densities

We recorded 95 mistletoe individuals after scanning 800 rubber trees in 20 different rubber plots. The number of mistletoes in a plot increased exponentially with the DBH of the trees ($\beta \pm SE = 0.034 \pm 0.005$,

$P < 0.001$; Fig. 5), but not with forest cover ($\beta \pm SE = 0.000002 \pm 0.000003$, $P = 0.456$). We observed 16 species of resident frugivores and nectarivores in rubber between 2011 and 2015, of which we recorded 13 during point counts. Frugivores and nectarivores comprised 47% of total bird abundance (1439 of 3076) in rubber. We observed 68 individuals belonging to five species (four families) feeding on flowers and/or fruits of mistletoes during our mistletoe observations (Table S1).

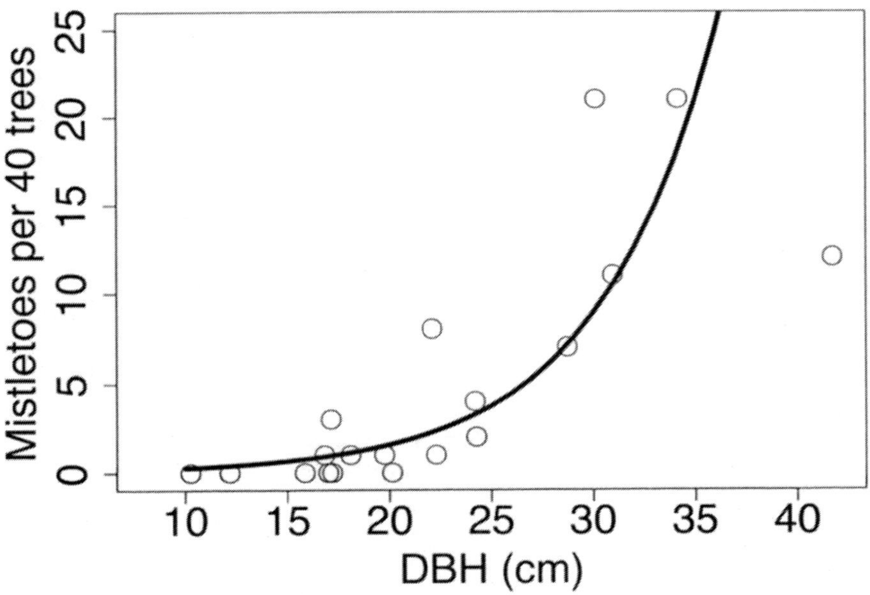

Figure 5: Density of mistletoes (per forty trees) in monoculture rubber plantations as a function of rubber tree diameter at breast height (DBH in cms). The black line is the prediction of the model fitted to the data.

DISCUSSION

Our study shows that natural forests enhance bird diversity within nearby monoculture rubber plantations. However, less than half of all the bird species in the study area were found in rubber even when forests where in close proximity. Habitat breadth was the most important trait that increased bird occurrence. The most parsimonious model reveals that the difference in occurrence probability between birds with habitat breadth one and seven was 57% (Fig. 2), suggesting that the birds that use multiple habitats, and are therefore typically of low conservation concern, tend to occur in rubber. Therefore, retention of the remaining forest patches not only improves bird diversity in the neighboring rubber plantations, but also protects the 66% (65 of 98) of forest bird species that do not use plantations. These results are in line with studies conducted in other monoculture rubber plantations, which also showed that less than half of the bird species occur in monoculture rubber[6,34]. However, studies conducted in jungle rubber (low intensity multi-cropping systems that contain natural colonizing vegetation) showed that such practices could retain more than half of the bird species[6,35].

Forest cover alone explained 43% of the species richness variation in rubber (Table 1). The most parsimonious model reveals that an increase in forest cover within 500 m radius of a sampling point from 0.35 to 39 ha increased the average bird richness by eight species (Fig. 3). Forest cover had similar influences on frugivore and insectivore species within rubber (Fig. 3), but based on the estimated threshold values, frugivorous bird species that occur in rubber required more forest area (5.92 ha) than insectivores (1.84 ha; Fig. 3). Typically frugivores have lower fragmentation sensitivity and travel longer distances between food resources[27], however frugivores in rubber may require higher surrounding forest cover due to the lack of adequate resources in rubber. These results are in line with previous studies, which showed that the insectivorous bird composition in rubber was more similar to primary forests than it was for frugivorous species (see Mang & Brodie[34]). This suggests that rubber plantations are better for insectivores than frugivores. In contrast, studies showed that cocoa plantations are better for frugivores than insectivores[34], whereas in oil palm plantations both frugivores and insectivores are equally sensitive[36].

Road proximity had little effect on most birds in rubber plantations because species that occupied them were habitat generalists (Fig. 2). Grey-headed Canary Flycatcher (*Culicicapa ceylonensis*) avoided roads, but Crimson Sunbird (*Aethopyga siparaja*) and Magpie Robin (*Copsychus saularis*) preferred them. Previous studies from around the world also show that roads and their marginal vegetation can enhance foraging and roosting opportunities, and minimize predation pressure for a few bird species, especially in unnatural anthropogenic habitats like rubber[21]. Therefore, building roads through plantations will affect biodiversity less than if they are built through forest[20].

Maintenance of large rubber trees with mistletoes might further enhance frugivore and nectarivore diversity. Around half of the total bird abundance in rubber plantations consisted of frugivores and nectarivores, although rubber produces neither fleshy fruits nor accessible nectar. Luo *et al.*[37] showed that the Plain Flowerpecker (*Dicaeum concolor*) and Red-whiskered Bulbul (*Pycnonotus jocosus*) were the most important dispersers of *Dendrophthoe pentandra,* the most common mistletoe in our study area. These two-bird species alone account for 37% (536 of 1439) of all frugivores and nectarivores recorded in rubber. Additionally, Scarlet-backed Flowerpeckers (*Dicaeum cruentatum*) and Japanese White-eyes (*Zosterops japonicus*) were also commonly observed feeding on mistletoes in rubber and all four species together account for 74% (1070 of 1439) of the recorded frugivores and nectarivores. Although a 40 year rotation is optimal for rubber[38], farmers in Xishuangbanna tend to cut down their plantations at 15–30 years in order to replant with new, higher yielding, clones (G. Huang *pers. observ.*). Such short rotations limit the build-up of mistletoe populations. Obiang and Salle[39] have shown that mistletoe parasitism on rubber trees in Africa did not affect yield, so retaining a few old trees with mistletoes while replacing the plantation with new high yielding clones could provide a low-cost benefit to biodiversity.

Although ten families of frugivorous and nectarivorous birds were recorded in monoculture rubber in tropical East Asia[10,11,12,13], only four were recorded in Xishuangbanna (Table S1). Among the others, parrots (Psittaculidae) and mynas (Sturnidae) were common in rubber plantations in Malaysia and Indonesia[11,12]. Indeed, Wee[40] recorded parrots feeding on fruits of *Dendrophthoe pentandra,* the most common mistletoe in our study area[37]. The absence of these birds is almost certainly due to hunting. Parrots and forest mynas used to inhabit forests around the study area (Menglun) but have been locally extirpated or exist in very small populations (<10 individuals; possibly vagrants or escaped cage birds) due to hunting and the pet trade[24,41]. Hunting has been reported as a serious threat in the study area since 1985[42]. Sreekar *et al.*[41] recently showed that birds with body sizes greater than 21 cm had a 52% extirpation probability in this area, primarily due to hunting, and that hunting alone increased frugivore extirpation probability in Menglun by around two-fold[24]. Therefore, hunting might have extirpated or reduced populations of a few larger frugivore species that might have otherwise occurred in rubber plantations.

Some limitations of this study need discussion, which also can guide future research. First, our study placed sampling points near large forest patches (>1000 ha; Fig. 1), but a question for future research is the role of degraded smaller forest fragments (<20 ha) in retaining rubber birds. Previous studies have already shown such fragments can conserve birds and trees[43,44]. Second, the result that a third of the bird species persist in rubber represents a best-case estimate, as we do not know if they can complete their entire life cycle there. It is possible that the individuals in rubber represent sink populations. Therefore, future studies should use spatio-temporal counts, tracking techniques and nest surveys to understand the life cycles of these species in rubber. Third, we also recommend investigating a larger range of organisms to understand how to effectively manage and conserve biodiversity in landscape mosaics with monoculture rubber plantations. Finally, we suggest conducting mistletoe exclusion experiments to better understand the importance of mistletoes in rubber for frugivorous and nectarivorous birds.

Our results suggest that natural forests are irreplaceable for biodiversity conservation as more than half of the local bird species do not occupy monoculture rubber even when in close proximity to natural forests (<100 m). The birds that occur in rubber are common species that tend to use multiple habitats and are of low conservation concern. In order to maximize biodiversity in rubber-dominated landscapes, local smallholders, rubber companies and governments should work together. First, based on the evidence summarized here, there is an obvious need to preserve as much of the remaining natural forest as possible. Second, new roads should be placed through rubber plantations where they appear to have little effect on bird diversity, and outside the remaining forests, where they are known to have negative impacts on diversity[20]. Third, farmers should be encouraged to retain at least a few large rubber trees (>25 cm DBH) with mistletoes within plantations during rotations.

METHODS

Study area

We carried out our study in monoculture rubber plantations within 10 km radius of Xishuangbanna Tropical Botanical Garden (XTBG) at Menglun, Xishuangbanna Prefecture, Yunnan Province, tropical SW China (21°55′N, 101°15′E). In this area, a change in land tenure regulations resulted in a massive expansion of smallholder, monoculture rubber plantations that resulted in around 65% forest loss between 1976 and 2010[24]. Hunting is also common in the area and increased bird extirpations by at least 1.3 to 1.6 fold over deforestation alone[24]. Currently, four large primary forest patches (each >1000 ha) contribute >90% of remaining forest cover in the study area[24] (Fig. 1).

Trait determinants of bird occurrence

We compiled a list of all resident birds that occurred in the region by conducting exhaustive bird surveys in the study area (10 km radius from XTBG) between 2011 and 2015. The first author and three other experienced birders, who lived in the study area for four years, conducted all bird sampling by repeatedly walking through forests, villages and rubber plantations. We also shared observations and sightings with other amateur and professional ornithologists who visit the study area in large numbers (c. 400 birdwatchers per year). We used a mixture of methodologically constrained and unconstrained bird sampling, which is an efficient way to obtain high coverage of birds in an area[45]. We included only the resident diurnal terrestrial species in the rubber bird-list because we did not conduct systematic surveys for nocturnal and migratory species. Initial surveys did not include migratory species and our migratory bird list does not appear to be complete; therefore, we excluded them prior to analysis. Furthermore, a list of resident bird species that occurred in rubber plantations was also prepared to compare with the overall list and determine how best to explain which species occur in rubber. Our overall and rubber species accumulation curves had reached an asymptote at the end of our four-year (>2500 man-hours) bird survey of the study area (see Sreekar *et al.*[24]). Thus, we are confident that our inventories are nearly complete and the bird species that could not be found in rubber can be reasonably classified as absent.

We collected data on four life-history traits of birds using Yang *et al.*[46] and Robson[47] to investigate the correlates of occurrence in rubber plantations among bird species. These were forest specialization (specialist or non-specialist), body size (bill to tail length), habitat breadth (observed number of habitats the species occurs in; range: 1–8) and primary diet type (frugivore, insectivore, carnivore, granivore, nectarivore). We used a generalized linear model (GLM) with binomial error structure to model the effects of bird traits on bird occurrence probability in rubber, and an ANOVA (type III sum of squares) to determine the variance explained by each variable.

Environmental determinants of bird diversity

We produced forest-cover maps for the study landscape by calculating the Normalized Difference Vegetation Index (NDVI) of multispectral data from Landsat-2 Multi Spectral Scanner (MSS) and Landsat-5 Thematic Mapper (TM) images (30 m spatial resolution) acquired in February 2010. We made sure that the selected images had the least cloud cover possible. Rubber trees are briefly leafless at this time, enhancing the contrast with the largely evergreen rain forest. NDVI values >0.1 were converted into 1 and everything else into 0. We layered roads upon the forest cover map using images downloaded from OpenStreetMap and ground-truthed road data (see Fig. 1). Other human altered land types in the study area include villages, river and banana plantations. We converted the resulting raster to a vector shape file as a forest-road map to calculate the distance to nearest paved roads, distance to nearest forest (>1000 ha), forest cover within 250 m, 500 m, 750 m and 1000 m from the sampling point (Table S3). Forest cover (in ha) within 500 m radius of a sampling point was highly correlated with distance to nearest forest ($r = -0.77$; $P < 0.001$), forest cover at 250 m ($r = 0.79$; $P < 0.001$), 750 m ($r = 0.90$; $P < 0.001$) and 1000 m ($r = 0.81$; $P < 0.001$), so we only used forest cover within 500 m radius (hereafter 'forest cover') in the model.

Using GIS and ground-truthing, we placed 52 point counts in rubber at different distances from paved road (13–1272 m) and forest patches (21–1927 m) in order to maximize the variation between these factors, and minimize their correlation (Fig. 1, Table S3). The minimum distance between each point was 300 m, we sampled each four times in the winter and spring seasons (December 2014–March 2015). All point counts were conducted between 550 and 650 m elevation in rubber plantations with little or no undergrowth. Younger plantations (<10 cm DBH) were avoided as they had relatively open canopies and were structurally different. We recorded all bird detections (sighting and aural) using ten-minute fixed radius (50 m) point counts, conducted between 8:00 and 10:00 AM (Beijing Standard Time). We used a Marantz solid-state recorder PMD661 MK-II (Marantz America, Mahwah, New Jersey, USA) with a Seinheiser microphone (Seinheiser, Wedemark, Germany) to record calls during the point count and identified unrecognized calls in the lab. We recorded both migrant and resident birds during point counts and used a species accumulation curve to assess the effectiveness of sampling effort.

To evaluate the effectiveness of sampling effort, we transformed the original species richness into estimated richness using Chao's non-parametric estimator[48]. We used Pearson's correlation test to determine the correlation between original and Chao's estimated richness. We removed species that were detected fewer than four times from the data to reduce the influence of accidental occurrences (including the excluded species did not change results)[26,41,49].

We used general linear models to determine the effects of distance to roads from the sampling point (13–1272 m) and forest cover within 500 m radius of the sampling point radiuses (0.35–39 ha) on species richness and diversity of all birds, frugivores only and insectivores only, respectively. Road distance (in meters) and forest cover (in hectares) were $\log_e$ transformed prior to analysis to improve their spread so that they form good linear fits with the response variable. We used the Shannon diversity index as a measure of species diversity because, unlike Simpson's diversity index, it does not give more weight to common or dominant species[50]. We checked for spatial autocorrelation among model residuals using Moran's I and found no evidence of autocorrelation ($P > 0.1$). To determine the effects of distance to road and forest cover on the occurrence probability (presence-absence) of individual species, we used generalized linear models with binomial error structures.

We generated full model, null model and models with all valid combinations of the explanatory variables (road distance and forest cover). There were no interaction effects between forest cover and distance to road in the models, so it was not included in the combinations of explanatory variables. We compared and ranked models using the Akaike Information Criterion (AIC_c)[51]. We also calculated Akaike weights (wAIC), which provide relative weights for any particular model in relation to the entire model set, which varies from 0 (no support) to 152 (complete support). We summed up the wAIC of all the models containing a particular covariate (covariate weight) within the subset to identify the covariates that had the strongest influence41. We present model-averaged estimates and their unconditional standard errors for covariates with weights (w) ≥0.75. We used McFadden's pseudo-R^2 to calculate the deviance explained by the generalized linear models[53]. After determining the best fitting covariates ($w \geq 0.75$) for species richness and diversity, we additionally fitted piecewise models to determine the existence and estimates of thresholds across the chosen environmental predictors[54].

We used a multivariate generalized linear model with negative-binomial error structure to determine the effects of distance to road and forest cover on bird community composition in rubber55. We assessed the significance of explanatory variables using 999 permutations of a Monte-Carlo test. We used non-metric multi-dimensional scaling based on the Bray-Curtis distance with a two-dimensional solution only to visualize changes in bird composition across rubber sites.

Determinants of mistletoe densities

The rubber trees in Xishuangbanna briefly shed their leaves in February and produce new leaves and flowers in March[56]. We surveyed mistletoes, which were evergreen and thus conspicuous, on leafless rubber trees in a subset of 20 plots during a c. 15-day period in February 2015. We scanned forty trees around the sampling point for mistletoes and measured the diameter at breast height (DBH) for 10 rubber trees around the sampling point. At the same time, we also recorded birds we saw eating mistletoe fruits or visiting flowers. We used a GLM with negative binomial error structure to model the effects of rubber DBH and forest cover on mistletoe density in rubber.

All analyses were conducted in the programming and statistical language R 3.1.1[57]. The csv files and R codes are available on request from the corresponding author.

ACKNOWLEDGEMENTS

We would like to thank Alison Wee, Ling Zhang, Alice Hughes, Jiang-Bo Zhao, Ximin Wang, Bonifacio Pasion, Rhett Harrison and Salindra Dayananda for material support and suggestions.

REFERENCES

1. Laurance, W. F., Sayer, J. & Cassman, K. G. Agricultural expansion and its impacts on tropical nature. *Trends Ecol. Evol.* **29**, 107–116 (2014).
2. Gibson, L. *et al.* Primary forests are irreplaceable for sustaining tropical 3.
3. Gilroy, J. J., Edwards, F. A., Uribe, C. A. M., Haugaasen, T. & Edwards, D. P. Surrounding habitats mediate the trade-off between land-sharing and land-sparing agriculture in the tropics. *J. Appl. Ecol.* **51**, 1337–1346 (2014).
4. Wilcove, D. S., Giam, X., Edwards, D. P., Fisher, B. & Koh, L. P.Navjot's nightmare revisited: logging, agriculture, and biodiversity in Southeast Asia. *Trends Ecol. Evol.* **28**, 531–540 (2013).
5. Giam, X. *et al.* Mitigating the impact of oil palm monoculture on freshwater fishes in Southeast Asia. *Conserv. Biol.* **29**, 1357–1367 (2015).
6. Warren-Thomas, E., Dolman, P. M. & Edwards, D. P. Increasing demands for natural rubber necessities a robust sustainability initiative to mitigate impacts on tropical diversity. *Conserv. Lett.* **8**, 230–241 (2015).
7. Watson, J. E. M., Dudley, N., Segan, D. B. & Hockings, M. The performance and potential of protected areas. *Nature* **515**, 67–73 (2014).
8. Aratrakorn, S., Thunhikorn, S. & Donald, P. F. Changes in bird communities following the conversion of lowland forest to oil palm and rubber plantations in Southern Thailand. *Bird Conserv. Int.* **16**, 71–82 (2006).
9. Meng, L.-Z., Martin, K., Weigel, A. & Liu, J.-X. Impact of rubber plantation on carabid beetle communities and species distribution in a changing tropical landscape (southern Yunnan, China). *J. Insect Conserv.* **16**, 423–432 (2011).
10. Phommexay, P., Satasook, C., Bates, P., Pearch, M. & Bumrungsri, S. The impact of rubber plantations on the diversity and activity of understorey insectivorous bats in southern Thailand. *Biodivers. Conserv.* **20**, 1441–1456 (2011).
11. Peh, K. *et al.* Conservation value of degraded habitats for forest birds in southern Peninsular Malaysia. *Divers. Distrib.* **12**, 572–581 (2006).
12. Beukema, H., Danielsen, F., Vincent, G., Hardiwinoto, S. & Andel, J.Plant and bird diversity in agroforests in the lowlands of Sumatra, Indonesia. *Agrofor. Syst.* **70**, 217–242 (2007).
13. Li, S., Zou, F., Zhang, Q. & Sheldon, F. H. Species richness and guild composition in rubber plantations compared to secondary forest on Hainan Island, China. *Agrofor. Syst.* **87**, 1117–1128 (2013).

14. Koh, L. P. Can oil palm be made more hospitable for forest butterflies and birds? *J. Appl. Ecol.* **45**, 1002–1009 (2008).
15. Edwards, D. P. *et al.* Wildlife-Friendly Oil Palm Plantations Fail to Protect Biodiversity Effectively. *Conserv. Lett.* **3**, 236–242 (2010).
16. Edwards, F. A. *et al.* Does logging and forest conversion to oil palm agriculture alter functional diversity in a biodiversity hotspot? *Anim. Conserv.* **17**, 163–173 (2014).
17. Yue, S., Brodie, J. F., Zipkin, E. F. & Bernard, H. Oil palm plantations fail to support mammal diversity. *Ecol. Appl.* doi: 10.1890/14-1928.1 (in press).
18. Gray, C. L., Lewis, O. T., Chung, A. Y. C. & Fayle, T. M. Riparian reserves within oil palm plantations conserve logged forest leaf litter ant communities and maintain associated scavenging rates. *J. Appl. Ecol.* **52**, 31–40 (2015).
19. Laurance, W. F. *et al.* A global strategy for road building. *Nature***513**, 229–232 (2014).
20. Laurance, W. F., Goosem, M. & Laurance, S. G. W. Impacts of roads and linear clearings on tropical forests. *Trends Ecol. Evol.***24**, 659–669 (2009).
21. Morelli, M., Beim, M., Jerzak, L., Jones, D. & Tryjanowski, P. Can roads railways and related structures have positive effects on birds? –A review. *Transportation Res. D* **30**, 21–31 (2014).
22. Sekercioglu, C. H. Increasing awareness of avian ecological function. *Trends Ecol. Evol.* **21**, 464–471 (2006).
23. Whelan, C. J., Sekrcioglu, C. H. & Wenny, D. G. Why birds matter: from economic ornithology to ecosystem services. *J. Ornithol.* doi: 10.1007/s10336-015-1229-y (in press).
24. Sreekar, R. *et al.* The use of species-area relationships to partition the effects of hunting and deforestation on bird extirpations in a fragmented landscape. *Divers. Distrib.* **21**, 441–450 (2015).
25. Sreekar, R. *et al.* The effect of land-use on the diversity and the mass-abundance relationships of understory avian insectivores in Sri Lanka and southern India. *Sci. Rep.* **5**, 11569 (2015).
26. Sreekar, R., Mohan, A., Das, S., Agarwal, P. & Vivek, R. Natural windbreaks sustain bird diversity in a tea-dominated landscape. *PLOS ONE* **8**, e70379 (2013).
27. Lees, A. C. & Peres, C. A. Avian life history determinants of local extinction risk in a hyper-fragmented neotropical forest landscape. *Anim. Conserv.* **11**, 128–137 (2008).
28. Vargas, R. R., Fonturbel, F. E., Bonacorso, E. & Simonetti, J. A.Variation in reproductive life-history traits of birds in fragmented habitats: a review and meta-analysis. *Bird Conserv. Int.* **22**, 462–467 (2012).
29. Watson, D. M. Mistletoe–A keystone resource in forests and woodlands worldwide. *Annu. Rev. Ecol. Syst.* **32**, 219–249 (2001).
30. Arrhenius, O. Distribution of the species over the area. *Meddelanden fran K. Vetenskapsakademiens Nobelinstitut* **4**, 1–6 (1920).
31. Mathews, T. J., Guilhaumon, F., Triantis, K. A., Borregaard, M. K. & Whittaker, R. J. On the form of species-area relationships in habitat islands and true islands. *Global Ecol. Biogeogr.* doi: 10.1111/geb.12269 (in press).
32. Mathews, T. J., Steinbauer, M. J., Tzirkalli, E., Triantis, K. A. & Whittaker, R. J. Thresholds and species-area relationship: a synthetic analysis of habitat island datasets. *J. Biogeogr.* **41**, 1018–1028 (2014).
33. Drinnan, I. N. The search for fragmentation thresholds in a Southern Sydney Suburb. *Biol. Conserv.* **124**, 339–349 (2005).
34. Mang, S. L. & Brodie, J. F. Impacts of non-oil tree plantations on biodiversity in Southeast Asia. *Biodivers. Conserv.* doi: 10.1007/s10531-015-1022-5 (in press).
35. Thiollay J. -M. The role of traditional agroforests in the conservation of rain forest bird diversity in Sumatra. *Conserv. Biol.* **9**, 335–353 (1995).
36. Senior, M. J. M. *et al.* Trait dependent declines of species following conversion of rainforest to oil palm plantations. *Biodivers. Conserv.* **22**, 253–268 (2013).

37. Luo, Y., Siu, Y., Gan, J. & Zhang, L. Host compatibility interacts with seed dispersal to determine small scale distribution of a mistletoe in Xishuangbanna, China. *J. Plant Ecol.* doi: 10.1093/jpe/rtv024 (in press).
38. Nizami, S. M., Yiping, Z., Liqing, S., Zhao, W. & Zhang, X. Managing carbon sinks in rubber (*Hevea brasilensis*) plantation by changing rotation length in SW China. *PLOS ONE* **9**, e115234 (2014).
39. Obiang, N. L. E. & Salle, G. Is there any point to eradicate *Phramanthera capitata* parasitizing African rubber trees? *Comptes Rendus Biologies* **329**, 185–195 (2006).
40. Wee, Y. C. *Long-tailed Parakeet and the Dendrophthoe pentandra mistletoe*. (2011) Available at: http://www.besgroup.org/. (Accessed: 31st March 2015).
41. Sreekar, R., Zhang, K., Xu, J. & Harrison, R. D. Yet another empty forest: Considering the conservation value of a recently established nature reserve. *PLOS ONE* **10**, e0117920 (2015).
42. Wang, Z. The structure of bird community in Xishuangbanna forest. *Zool. Res.* **12**, 169–174. [in Chinese] (1991)
43. Chang, X., Quan, R. & Wang, L. Bird conservation in extremely small tropical rainforest patches in southwest China. *Biol. Conserv.* **158**, 188–195 (2013).
44. Liu, J. -J. & Slik, J. W. F. Forest fragment spatial distribution matters for tropical tree conservation. *Biol. Conserv.* **171**, 99–106 (2014).
45. Bibby, C. J., Burgess, N. D., Hill, D. A. & Mustoe, S. H. *Bird census techniques* (Academic Press, 2000).
46. Yang, L. *et al. The avifauna of Yunnan, China*, **vol. 1 & 2** (Yunnan Science and Technology Press, [in Chinese] 1995).
47. Robson, C. *New Holland field guide to the birds of south-east Asia* (Tien Wah Press, 2000).
48. Chao, A. Non-parametric estimation of the number of classes in a population. *Scand. J. Stat.* **11**, 265–270 (1984).
49. Barlow, J., Gardner, T. A., Louzada, J. & Peres C. A. Measuring the conservation value of tropical primary forests: The effects of occasional species on estimates of biodiversity uniqueness. *PLOS ONE* **5**, e9609 (2010).
50. Magurran, A. E. *Measuring biological diversity* (Blackwell Publishing, 2004).
51. Hobbs, N. T. & Hilborn, R. Alternatives to statistical hypothesis testing in ecology: A guide to self teaching. *Ecol. Appl.* **16**, 5–19 (2006).
52. Burnham, K. P. & Anderson, D. R. *Model selection and inference: A practical information-theoretic approach* (Springer-Verlag, 1998).
53. Veall, M. R. & Zimmermann, K. F. Pseudo-R^2 measures for some common limited dependent variable models. *J. Econ. Surv.* **10**, 241–259 (1996).
54. Muggeo, V. M. R. Package 'segmented'. *Biometrika*, **58**, 525–534 (2010).
55. Wang, Y., Naumann, U., Wright, S. T. & Warton, D. I. mvabund – an R package for model based analysis of multivariate abundance data. *Methods Ecol. Evol.* **3**, 471–474 (2012).
56. Li, P., Zhang, J. & Feng, Z. Mapping rubber tree plantations using landsat-based phonological algorithm in Xishuangbanna, southwest China. *Remote Sens. Lett.* **6**, 49–58 (2015).
57. R Development Core Team. R*: A language and environment for statistical computing* (R foundation for statistical computing, 2014).

FOOTNOTES

[1]Center for Integrative Conservation, Xishuangbanna Tropical Botanical Garden, Chinese Academy of Sciences, Menglun, Yunnan, China.

[2]School of Biological Sciences, University of Adelaide, Australia.

[3]Products Research Institute, Tsukuba, Ibaraki, Japan.

[4]College of Forestry, Guangxi University, Nanning, Guangxi, China. Correspondence and requests for materials should be addressed to R.S. (email: rachakonda.sreekar@outlook.com)

CHAPTER 8

Identification of Potential Areas for Establishment of Plantations of *Agave lechuguilla* Torr. in Coahuila, Mexico

David Castillo-Quiroz*, Oscar Ulises Martínez-Burciaga, Diana Yemilet Ávila-Flores, Francisco Castillo-Reyes, Juan David Sánchez-Chaparro

National Institute of Forestry, Agriculture and Livestock Research (INIFAP), Experimental Station Saltillo, Saltillo, Mexico

ABSTRACT

Adverse weather conditions and intensive use for fiber production have caused decline of natural populations of Agave lechuguilla. By these reasons, it is important to develop a sustainable management of this specie to avoid this situation. One of the first steps is to establish forest plantations as an alternative to restore A. lechuguilla habitat and increase its range while at the same time indirectly reducing erosion rates. The aim of this study was to identify the potential areas for establishment of plantations of A. lechuguilla in the State of Coahuila, Mexico. Areas for planting establishment were selected according to multi-criteria analysis using Geographic Information System (GIS) Idrisi32 Ver 2.0 and 3.2 Arc.View. The analysis was based on comparison of climate and soil requirements in contrast to environmental conditions of the tested area. The evaluated variables were: soil type, texture, depth, average annual temperature, annual precipitation, elevation and slope. Results derive in potential area maps, geo-referenced and their distribution by Agricultural Development District. An area of 5 million 159,273 hectares was determined with potential for establishment of A. lechuguilla forest plantations in the State of Coahuila. It was concluded that, use of GIS is a reliable tool to determine potential areas for A. lechuguilla forest plantations, where this specie has higher probability of success in survival and biomass production. This study will support programs aimed to strengthen the economy of rural producers and also recover and preserve the arid state.

Keywords: *Agave lechuguilla*, forest plantations, natural-fiber, Geographical Information Systems

1. INTRODUCTION

Natural fibers of Lechuguilla (Agave lechuguilla Torr.) are one of the most produced fibers around the world. This specie is a native succulent plant, distributed in arid and semi-arid areas of the Southern part of United States of America and Northern part of Mexico (Castillo et al., 2011). In Mexico, A. lechuguilla is distributed in the Coahuila, Chihuahua, Nuevo Leon, Durango, San Luis Potosi, Tamaulipas and Zacatecas States (Nobel & Quero, 1986; Berlanga, 1991) and to lesser range in the center and south of the country (Marroquin et al., 1981). However, the main producing states of A. lechuguilla fiber are Coahuila, Tamaulipas, Nuevo Leon, Zacatecas and San Luis Potosi. The lechuguilla fiber with higher commercial quality and value is obtained from the growth bud, called "cogollo". This is formed by gentle leaves, grouped in the center, which contains lesser lignin than the lateral leaves (Castillo et al., 2005). For habitants of arid and semi-arid areas of Mexico, harvesting of lechuguilla fiber has great socioeconomic impact and for generations it has been a subsistence activity (Berlanga et al., 1992; Pando et al., 2004; Castillo et al., 2008). The 93% of the national production of lechuguilla fiber is exported, generating a source of foreign exchange for the country. Because of this, Mexico is the only exporting country. Between 2003 and 2007 sales for $350,000 dollars were recorded (SEMARNAT, 2010; Castillo et al., 2011). The fiber has application in the manufacture of brushes for various uses (Castillo et al., 2005).

In arid and semi-arid regions of the Coahuila State, agricultural soils have been used without previous studies where its potential uses and environment impact by different uses are determined. Use of land through agriculture, livestock and forestry is performed on an inadequate way, which has resulted in environmental degradation, an inefficient exploitation commercial and loss of natural resources, mainly soil and water. On the other hand, productive potential of crops such as corn and beans is low, mainly because of limited availability of water and soil degradation due to erosive action of natural factors and the human actions. Like many plant species from arid and semiarid regions of Mexico, lechuguilla has been subjected to excessive and irrational harvesting which has caused decline of natural populations, to the extent that, it is increasingly difficult to obtain raw material for fiber production. In addition, a decline in lechuguilla ecosystem has been observed which obligates to look for a sustainable production, involving methodologies from evaluation and management under natural conditions until domestication through techniques for establishment and agronomic management of commercial plantations (Velasco et al., 2009, Pando-Moreno et al., 2008, Castillo et al., 2008, Berlanga et al., 1991 and Berlanga et al., 1992). A key aspect that has been little considered and should be taken into account if results of this technology want to be maximized, is the agroecological aspect; the detailed factors of the physical environment which determine the species production potential, according to their physiological and morphological characteristics at the environment where it can be more productive. Therefore, these characteristics can be identified through technology applications for future plantings. This will allow high yields and a sustainable development. Based on the above, the aim of this study was to determine potential areas with high productive potential for establishment of commercial plantations of A. lechuguilla in the State of Coahuila, Mexico, based on attributes of each of the agro-weather factors that influence A. lechuguilla production.

2. MATERIALS AND METHODS

2.1. Study area

Arid and semiarid regions of Coahuila State were chosen to perform this study. This State is located in northeastern part of Mexico, between 24°32' and 29°53' North Latitude (LN) and 99°51' and 103°68' West Longitude (LW). Coahuila is the third largest state in the country with an area of 151,562 km^2 (INEGI, 2013), bordered in north by the United States of America, to the west by the State of Chihuahua, to the south with Zacatecas and San Luis Potosí, to the west and east with Durango and Nuevo León respectively (Figure 1).

Subdivision of the study area

Coahuila State is divided in 38 counties, which for this study purposes was subdivided into five regions based on the five regions of agricultural production, according to the regionalization by Secretary of Agriculture and Rural Development, Fisheries and Food (SAGARPA) which divides the State into five Rural Development Districts (**Table 1**).

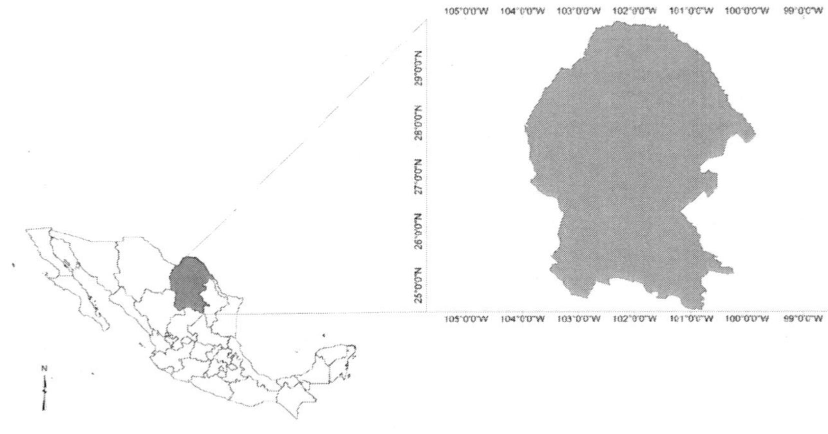

Figure 1. Geographical location of the study area in Coahuila, Mexico.

Table 1. Rural development districts and counties, according to the Ministry of Agriculture and Rural Development (SAGARPA).

District Rural Development	Municipalities
DDR 001	Acuña, Allende, Hidalgo, Jiménez, Nava, Piedras Negras, Morelos, Guerrero, Zaragoza and Villa Unión.
DDR 002	Juárez, Múzquiz, Progreso, Sabinas and San Juan de Sabinas.
DDR 003	Cuatro Ciénegas, Sierra Mojada, Ocampo, Candela, Monclova, Escobedo, Frontera, Abasolo, Castaños, Lamadrid, Nadadores, Sacramento and San Buenaventura.
DDR 004	Arteaga, General Cepeda, Parras de la Fuente, Ramos Arizpe and Saltillo.
DDR 005	Matamoros, Francisco I. Madero, San Pedro, Torreón and Viesca.

2.2. Selection Criteria for determining potential area for Agave lechuguilla Torr. establishment

The criteria considered relevant to A. lechuguilla growth involved climatic, soil and topography aspects. These criteria were obtained through literature reviewing, particularly the work done by Eguiarte & Silva 2001; Reyes et al., 2000; Gentry, 1985; Nobel & Quero, 1986; Sheldon, 1980 and the research results by the National Institute of Forestry, Agriculture and Livestock (INIFAP) (Martínez et al., 2011; Martínez et al., 2010; Castillo et al., 2011; Castillo et al., 2008; Velasco et al., 2009; Narcia et al., 2012; Martinez & Lara, 2003; Berlanga et al., 1991; Berlanga et al., 1992; Marroquín et al., 1981; Zapién, 1981). The minimum requirements established for determining the geographic areas with potential for planting and production of A. lechuguilla Torrare shown in Table 2.

2.3. Determination of the production potential

Weather parameters for identification of potential areas for the A. lechuguilla commercial plantations establishment consisted of a multi-criteria analysis conducted by the Geographic Information System (GIS) Idrisi 32 Ver 2.0 (Eastman, 1999). This analysis was based on comparison of climate and soil requirements of A. lechuguilla against environmental conditions of the study area according to the methodology proposed by Medina et al., (1997). The climatic parameters were obtained from the Climate Information System of the National Institute of Agricultural and Forestry Research (INIFAP), which come from a process of collection, management, analysis and interpretation of data on a daily base from maximum and minimum temperature, precipitation and evaporation, corresponding to meteorology station of ordinary type belonging to the network of stations of the National

Table 2. Environmental requirements considered to assess areas potentially suitable for establishment of plantations of Agave lechuguilla Torr. in Coahuila.

Altitude m. s. n. m	Precipitation mm	Temperature °C	Slope %	Soil Texture
950 - 1990	200 - 500	16 - 26	4 - 20	Media and course

Water Commission in Coahuila (CONAGUA). In addition, edaphic information, land use and topographic conditions were obtained from the cartography from National Institute of Statistics, Geography and Informatics (INEGI) at scale of 1:250,000. Identification and mapping of potential area derived from multi-criteria analysis was performed using Arc View GIS 3.2 software (ESRI, 1999). Starting from the thematic images produced by variable, was generated the cartography of each variable by exporting the images and converting them to vector by format "Shapefile", which were edited to obtain the general maps with potential areas and the number of hectares that they represented. Distribution of potential areas for forest plantations of A. lechuguilla in Coahuila was based on the regionalization made by Secretariat of Agriculture, Livestock and Rural Development, Fisheries and Food (SAGARPA) which divide the State into five Districts of Rural Development (**Table 1**).

2.4. Validation of potential areas

Field visits were performance to verify the specific points with potential for artificial establishment of the species under study, in order to check the proper ecological conditions for A. Lechuguilla growth and therefore, for establishment of plantations.

3. RESULTS AND DISCUSSION

Combination of physical environment factors n the GIS software allowed generating aps with eo-referenced potential areas. Based on previously developed databases of climate, soil and topographic factors, the study allowed differentiating areas with similar environments characteristics in which A. lechuguilla could have better growth. In a first step, each of these physical factors was analyzed independently to determine its spatial distribution within the ranges required by A. lechuguilla. From binary map were indicating areas that are or are not within these ranges of lechuguilla growth. In this step, the combining factors and overlaying different maps resulted in binary maps on which it relied to quantify the potential surface for each of the Districts of State Rural Development.

Area with potential for commercial planting of Agave lechuguilla Torr. in Coahuila

An area of 5159273.59 hectares with potential for plantations of A. lechuguilla was estimated. The district with the greatest potential area for plantation establishment was DDR 003 with a total superficial from 3029096.74 ha, followed by DDR 004 with 1074831.11 ha. At the county level, Ocampo stands out with a higher recording of surface potential with 1434878.11 ha, and the lowest record was the Frontera county with a surface with only 30.83 ha. Full results are shown in **Table 3** and their geographical distribution is presented by District Rural Development in **Figure 2.**

According to the obtained results, over a third part of the State surface has agroclimatic characteristics required to produce A. lechuguilla with high potential, mainly in the western, central and south regions of the State, which is related to agro-climatic conditions of arid and semiarid ecosystems. Martínez et al., 2010; Martínez & Lara (2003), and Nobel & Quero, (1986) mentions that A. lechuguilla inhabits sites with a range of rainfall between 200 - 500 mm, whose average value is not more than 300 mm, another important factor is altitude above sea level which should ranges between 950 and 2300 meters, however the optimal conditions for establishment of plantations of A. lechuguilla ranging from 950 to 1990. In additions, this plant requires over 200 mm rainfall and average annual temperature ranges from 17°C to 26°C with 30°C as maximum and 10°C as a minimum, grows in shallow soils, stony ridges or, well, flat stony soils with good drainage conditions without salinity problems. It can be said that, few factors restricting the potential area of A. lechuguilla, perhaps altitude above sea level is the most extensively discriminate though; problems of low slope and soil salinity also take into greatly limiting its effect. The obtained results agree with those reported by Gentry, 1985 who said that the natural distribution of A. lechuguilla is located in the western and southeastern parts of the State.

Table 3. Area of potential sites (ha) suitable for forest plantations of Agave lechuguilla Torr. in the Coahuila counties.

District Rural Development	Counties	Area (ha)
DDR 001	Acuña	130611.13
	Zaragoza	41244.59
DDR 002	Múzquiz	96636.10
	Progreso	1643.28
DDR 003	Abasolo	2254.44
	Candela	10148.21
	Castaños	116511.41
	Cuatro Ciénegas	536734.36
	Escobedo	497.96
	Frontera	30.83
	Lamadrid	36520.78
	Monclova	19235.632
	Nadadores	6735.56
	Ocampo	1434878.11
	Sacramento	6982.47
	San Buenaventura	245914.71
	Sierra Mojada	612652.27
DDR 004	Arteaga	4054.78
	General Cepeda	177159.26
	Parras de la Fuente	564744.56
	Ramos Arizpe	308099.46
	Saltillo	20773.05
DDR 005	Francisco I. Madero	134734.03
	Matamoros	23022.57
	San Pedro	274558.36
	Torreón	60364.93
	Viesca	292530.75
TOTAL		5159273.73

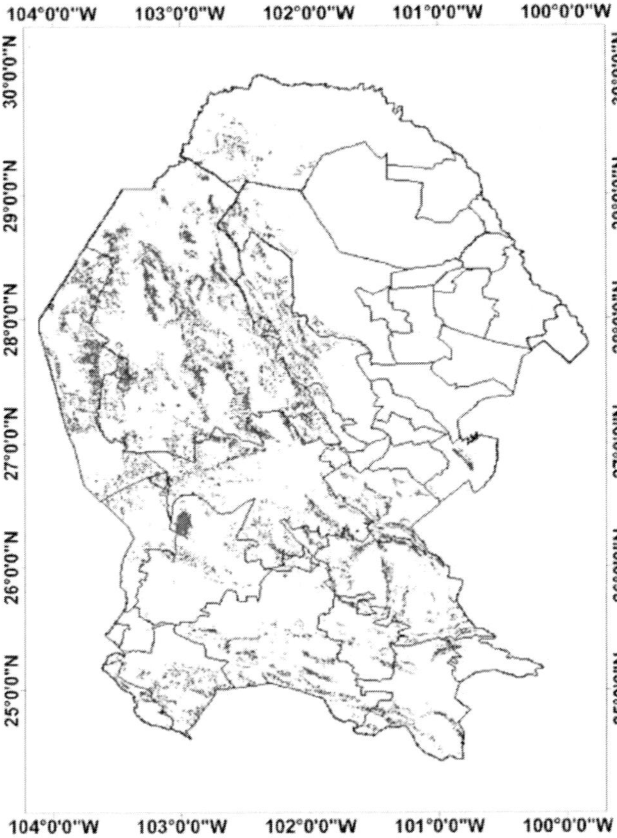

Figure 2. Areas with potential for establishment of forest plan- tations of lechuguilla (Agave lechuguilla Torr.) in Coahuila.

4. CONCLUSION

David Castillo-Quiroz, Oscar Ulises Martínez-Burciaga, Diana Yemilet Ávila-Flores, Francisco Castillo-Reyes, Juan David Sánchez-Chaparro We identified 5,159,273 hectares in the state of Coahuila that have favorable conditions for the establishment of A. lechuguilla, where the Rural Development District 003 has the largest area with potential for establishment of A. lechuguilla. This study will support programs aimed to strengthen the economy of rural producers in arid and semi-arid areas of the Coahuila State and also to recover and conserve natural populations and reduce erosion problems.

REFERENCES

1. Berlanga, R. C. A. (1991). Producción y recuperación de lechuguilla (Agave lechuguilla Torr.) en poblaciones naturales. In III Simposio Nacional Sobre Ecología, Manejo y Domesticación de Plantas Útiles del Desierto, INIFAP, Saltillo, Coah, México.
2. Berlanga, R. C. A., González, L. L. A., & Franco, L. H. (1992). Metodología para evaluación y manejo de lechuguilla en condiciones naturales. Folleto Técnico No. 1 SARH-INIFAP-CIRNE. Campo Experimental. "La Sauceda", Saltillo, Coa- huila, México, 22 p.
3. Castillo, Q. D., Berlanga, R. C. A., Pando, M. M., & Cano, P. A. (2008). Regeneración del cogollo de Agave lechuguilla Torr., de cinco procedencias bajo cultivo. Revista Ciencia Forestal en México, 33, 27-40.
4. Castillo, Q. D., Berlanga, R. C. A., & Cano, P. A. (2005). Recolección extracción y uso de la fibra de lechuguilla (Agave lechuguilla Torr.) en el estado de Coahuila. INIFAP-CIRNE. Campo Experimental Saltillo. Publicación Especial Núm. 6. Coahuila, México, 13 p.
5. Castillo, Q. D., Mares, A. O., & Villavicencio, G. E. E. (2011). Lechuguilla (Agave lechuguilla Torr.) planta suculenta de importancia económica y social de las zonas áridas y semiáridas de México. Vol. 8, No. 2. Boletín de la Sociedad Latinoamericana y del Caribe de Cactáceas y otras Suculentas, 6-9.
6. Eastman, J. R. (1999). IDRISI32 Ver 2.0: Guide to GIS and Image Processing. Clark Labs, Clark University. Worcester, MA 01610-1477, USA.
7. Eguiarte, L. E., & Silva, A. (2001). Ecología evolutiva de Agave lechuguilla: Variación geográfica ybiología reproductiva. XV Congreso Mexicano de Botánica. Los retos de la botánica mexicana en el presente siglo. Querétaro-Querétaro14-19.
8. ESRI (1999). ArcView GIS, Using ArcView GIS (340 p). Redlands, CA: Environmental Systems Research Institute Inc.
9. Gentry, S. H. (1985). Agaves of Continental North America. Tucson: The University of Arizona Press. 670 p.
10. Instituto Nacional de Estadística Geografía e Informática (INEGI) (2013). México en cifras. Información Nacional por Entidad Federativa (lastchecked 1 june 2014). http://www3.inegi.org.mx/sistemas/mexicocifras/
11. Martínez, B. O. U., & Lara, G. G. J. (2003). Potencial productivo de áreas de temporal en el estado de Coahuila. Una pro- puesta de conversión productiva. INIFAP-CIRNE. Campo Experimental Saltillo. Publicación Especial Núm. 1. Coahuila, 89 p.
12. Marroquín, J. S., Borja, L. G., Velásquez, C. R., & De la Cruz, J. A. (1981). Estudio ecológico dasonómico de las zonas áridas del norte de México. Publicación Especial Núm. 2. 2nd Edición, Instituto Nacional de Investigaciones Forestales. México, 166 p.
13. Medina, G., Ruiz C. G., Martínez, P. J. A., & Ortiz, V. M. (1997). Metodología para la determinación del potencial producti- vo de especies vegetales. Agricultura Técnica en México, 23, 69-90.
14. Narcia, V. M., Castillo, Q. D., Vázquez R. J. A., & Berlanga, R. C. A. (2012). Turno Técnico de la lechuguilla (Agave lechuguilla Torr.) en el noreste de México. Revista Mexicana de Ciencias Forestales, 3, 81-88.
15. Nobel, P. S., & Quero, E. (1986). Environmental Productivity Indices for a Chihuahuan Desert CAM Plant, Agave lecheguilla. Ecology, 67, 1-11. http://dx.doi.org/10.2307/1938497
16. Pando, M. M., Eufracio, O., Jurado, E., & Estrada, E. (2004). Post-Harvesting Growth of Lechuguilla (Agave lechuguilla Torr. Agavaceae) in Northeastern Mexico. Economic Botany, 58, 78-82.
17. Pando-Moreno, M., Pulido, R. R., Castillo, Q. D., Jurado, E., & Jiménez, J. (2008). Estimating Fiber for Lechuguilla (Agave lecheguilla Torr., Agavaceae), a Traditional

Non-Timber Forest Product in Mexico. Forest Ecology and Management, 255, 3686-3690.

18. Reyes A. J. A., Aguirre R. J., & Peña, V. C. B. (2000). Biología y aprovechamiento de Agave lechuguilla. Torrey. Boletin de la Sociedad Botanica de Mexico, 67, 75-88.
19. Secretaría del Medio Ambiente y Recursos Naturales (SEMARNAT) (2010). Fichas de información comercial de productos forestales. Comisión Nacional Forestal y Fondos Mixtos para la Investigación, el Desarrollo y la Innovación Tecnológica Forestal. México City, 13 p.
20. Sheldon, S. (1980). Ethnobotany of Agave lechuguilla and Yucca carnerosana in Mexico's Zona Ixtlera. Economic Botany, 34, 376-390. http://dx.doi.org/10.1007/BF02858314
21. Velasco, B. E., Arredondo, G. A., Zamora-Martínez, M. C., & Moreno, S. F. (2009). Modelos Predictivos para la Producción de Productos Forestales No Maderables: Lechuguilla. Manual Técnico Núm. 2. CENID-COMEF. INIFAP, México City, 56 p.
22. Zapién, B. M. (1981). Evaluación de la producción de ixtle de lechuguilla en cuatro sitios diferentes. Primera Reunión Re- gional sobre Ecología, Manejo y Domesticación de las Plantas Útiles del Desierto. Publicación Especial Núm. 31. Instituto Nacional de Investigaciones Forestales. SARH. México City, 385-389.

CHAPTER 9

Areas with Potential for Commercial Timber Plantations of *Enterolobium cyclocarpum* (Jacq.) Griseb. in Michoacán, México

Hipólito Jesús Muñoz-Flores[1], J. Trinidad Sáenz-Reyes[1], Agustín Rueda-Sánchez[1], David Castillo-Quiroz[2], Francisco Castillo-Reyes[2], Diana Yemilet Avila-Flores[2]

[1]*National Institute of Forestry, Agriculture and Livestock Research (INIFAP)-Experimental Station Uruapan, Uruapan, México*

[2]*National Institute of Forestry, Agriculture and Livestock Research (INIFAP)-Experimental Station Saltillo, Saltillo, México*

ABSTRACT

Michoacán has a deforested area of 525,260 ha, representing 52,526 ha per year, mainly caused by anthropogenic disturbances such as agricultural burning and forest fires (50%), change of use of land for extensive livestock farming (28%), agriculture (17%) and illegal logging (5%). The establishment of forest plantations is an alternative for reducing the pressure on natural forests and creating options for sustainable development and diversification of production and conversion of land for agricultural and livestock fragmented for forestry purposes. The aim of this study was to determine the potential areas for commercial forest plantations of Enterolobium cyclocarpum (Jacq.) Griseb. in the State of Michoacán, México, through the use of Geographic information systems (GIS). Areas were identified using IDRISI 32, and ArcView software.

Screening variables include annual temperature, annual precipitation, land use, soil type, elevation and slope. Products obtained were field verified. Two maps where potential areas for the establishment of commercial forest plantations of E. cyclocarpum shown were obtained. Potential areas for commercial forest plantation establishment are (0% - 15% slope) for commercial forest plantations mechanized was 57,227 ha and (15% - 30% slope) for manual plantations was 6273 ha. The total area in the state with potential for the establishment of the species in the study was 63,500 ha.

Keywords: Potential Areas, Michoacán, Enterolobium cyclocarpum, Commercial Forest Plantations, Geographic Information System

1. INTRODUCTION

Mexico ranks 10th place in the world as for the rich diversity in its natural forests. Mexican forests cover 141.7 million ha. Timberland in Mexico amounts to 56.7 million ha. Temperate forests occupy 30.4 million ha, and 26.4 million ha are covered by tropical forests (FAO, 2000) . Despite its considerable forestry potential, the country faces serious and complex problems such as deforestation and erosion of large land areas. Deforestation has been the direct result of illegal logging, forest fires, and changes in land use. In the 2000-2005 period, deforestation averaged a net annual loss of 314,000 ha (CONAFOR, 2005) .

The state of Michoacán is endowed with 59,864 km^2 of territory, which amounts to 3.04% of the national territory. Its natural wealth takes Michoacán to 5th place as for its national biodiversity. Timberlands represent nearly 2,602,727 ha. They contain 1,544,353 ha of temperate forests, 1,058,374 ha of tropical forests, and 247,846 ha of arid lands, wetlands and saline environments. Michoacán has 1,355,878 ha of degraded areas, even though it is a region with a tradition in forestry. Deforestation, illegal logging, over-exploitation, forest fires, pests and diseases are the forces behind the deforestation rate of 30,000 ha per year (COFOM, 2003) . This condition can only be reversed by reforestation with forest tree species, which can be adapted to the zone or limiting the areas of choice using the Geographic Information System (GIS) (Sáenz et al., 2000) . The GIS is a convenient tool that facilitates geographic features location and delimitation. Carmona and Monsalve (1999) explain that it is a system composed of hardware, software and procedures designed to handle spatially explicit objects and data processing such as input, administration, processing, analysis, modeling, and plotting. GIS helps solve complex problems in planning and management. The United Nations Food and Agriculture Organization (FAO, 2000) defines as GIS those computer systems used for data storage, summarization and retrieval. GIS data management, according to FAO, is carried out with specially designed programs that handle both geographic attributes and spatial information. Moreno and Moreno (1995) point out that the National Institute of Forestry, Agricultural and Livestock Research (INIFAP) have conducted research at national level using GIS since 1991. The study of the productive potential of crops is frequently carried out using GIS. This research has enriched and promoted the use of INIFAP's geographic data bank. The results have provided grounds for decision on the plans for land use. Michoacán forest plantations have become testing grounds for a great variety of conifers and hardwoods: Pinus michoacana Martínez and P. oocarpa (INIFAP, 1993) ; Abies religiosa (Kunth) Schltdl. & Cham, Pinus pseudostrobus Lindl. P. michoacana, P. montezumae, P. teocote, P. oocarpa, P. ayacahuite, P. lawsonii, P. herrerae, P. cembroides, P. radiata, P. greggii, P. engelmannii, P. arizonica P. durangensis and P. chihuahuana (Martínez & Prieto, 2011; Muñoz et al., 2011; Sáenz et al., 2010) ; and other shrub species of arid land of the state of Coahuila as Nolina cespitifera Trel. (Martínez & Castillo, 2007) Agave lechuguilla Torr. (Castillo et al., 2014) . The usual findings are characterized by poor survival and limited growth, due to various reasons such as: lack of a specific reason for planting, wrong selection of species and stocking and limited silvicultural tending (Sáenz et al., 2000) .

An alternative may be Enterolobium cyclocarpum (Jacq.) Griseb. Fabaceae (Tropicos, 2016) . E. cyclocarpum has been used in ecological restoration programs for degraded ecosystems, soil conservation and erosion control (Vázquez et al., 1999) . Studies on growth and volumetric efficiency in plantations of this species are reporting 15 to 30 m^3/ha/year, depending on the density and technical turning, could conservatively grow, in 20 years of age, to a total estimated volume of 130 m^3/ha (Manzanilla et al., 2001) . The wood obtained from this species has a wide variety of uses as in the manufacture of staves, trimmings, agricultural devices such as carts, carpentry and helps in the production of panels, fine furniture, interior building materials, kitchenettes, halls, mills, fittings, closets, and thermal insulation (Pennington & Sarukhán, 1998; Manzanilla et al., 2001) . The E. cyclocarpum species is native from America and is characteristic in places where the vegetation has been damaged. Usually grows in areas with a high degree of tampering (changes in land use, clearing for agriculture, etcetera), especially in humid tropics and sub-humid, low altitude in Mexico and Central America. This species is not related to any kind of primary vegetation; however, you can find it in forests, tropical deciduous forest galleries, and as secondary vegetation in the tropical evergreen forest (Vázquez et al., 1999; CONABIO, 2008) . The object of this paper was to locate and determine potential areas for commercial forest plantations of Enterolobium cyclocarpum (Jacq.) Griseb. in Michoacán, through the use of geographic information systems (GIS).

2. MATERIALS AND METHODS

2.1. Study Area

The state of Michoacán in Western Mexico was the study area. Coordinates for Michoacán are 100°04'45''W to 103°44'49'' W, and 20°23'43''N to 18°09'47''N. Michoacán territory expands 59,864 km^2, and it is Northwest of Guerrero, North of the Pacific Ocean, South from Guanajuato and Querétaro, and East of Colima and Jalisco. Michoacán is administratively divided into 113 municipalities (INEGI, 2012; COFOM, 2014) (Figure 1).

The climate in Michoacán is affected by one main geographic factor: the changes in elevation caused by rough topography. Michoacán is characterized by a series of mountain ranges parallel to the coastline. Climate patterns tend to be predominantly sub-humid with summer rains and a sharp winter dry season. Predominant soils in the state are the result of volcanic ash (Andosols). These soils are found at Eje Neovolcánico and Sierra Madre del Sur. In this edaphic province, Vertisols also appear in ranges and valleys. On hills and mountains, Leptosols, Luvisols, and Feozems might be present. Andosols dominate the MesetaTarasca. Soils from Sierra Mil Cumbres at high elevation sites are also Andosols. Acrisols and Luvisols are found at piedmonts, creeks and valleys. Feozems normally can be seen at the edge of crop lands. Low lands and flat ground places with non-irrigated and irrigated agriculture, or natural vegetation of subtropical woodlands and grasslands tend to be Vertisols. Leptosols are distributed in random patterns associated to steep terrain and rocky outcrops.

Subdivision of the Study Area

Michoacán state is divided in 113 counties, which for this study purposes was subdivided in ten regions and were defined according the regionalization of the Forestry Commission of state of Michoacán (COFOM, 2014) , which are shown in Table 1.

2.2. Natural Distribution and Environmental Requirements to Determine Potential Areas of *Enterolobium cyclocarpum* (Jacq.) Griseb

To obtain the natural distribution and environmental requirements from E. cyclocarpum, an extensive research of written material was conducted. The findings show the climate where E. cyclocarpum is located to be warm humid (SEDER, 1995) with an annual rainfall of 750 to 2500 mm, average annual temperature between 22°C - 32°C; at altitudes ranging from 0 to 1600 m and Regosol and Cambisol soil types, though it develops best in Vertisol soils (SEDER, 1995; CNIC, 2005; Manzanilla et al., 2001; Rueda et al., 2007). The information was gathered mostly considering the dry tropical conditions found in the state of Michoacán, and once the information was systemized, the environmental requirements for E. cyclocarpum was determining the potential areas for commercial forest plantations were determined, based on altitude, total annual precipitation, average annual temperature, soil type and slope. The research scheme for potential areas chose sites meeting the following environmental features for E. cyclocarpum: elevation, total annual rainfall, annual average temperature, soil type and slope (**Table 2**).

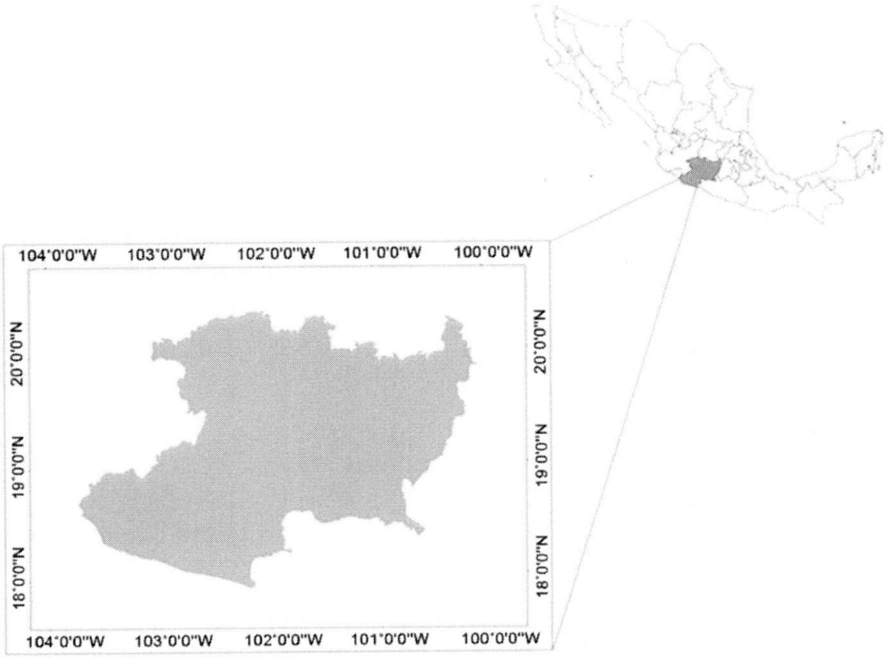

Figure 1. Geographic location of Michoacán, the study area.

Table 1.Municipalities and Michoacán forest regions, according to the forestry commission of the state of Michoacán (COFOM, 2014).

Region	Municipalities
I. Lerma-Chapala	Briseñas, Chavinda, Cojumatlán, Ixtlán, Jacona, Jiquilpan, Marcos Castellanos, Pajacuarán, Purépero, Sahuayo, Tangamandapio, Tangancícuaro, Tlazazalca, Venustiano Carranza, Villamar, Vista Hermosa, Zamora.
II. Bajío	Angamacutiro, Coeneo, Churintzio, Ecuandureo, Huaniqueo, Jiménez, José Sixto Verduzco, Morelos, Numarán, Penjamillo, La Piedad, Panindícuaro, Puruándiro, Tanhuato, Yurécuaro, Zináparo, Zacapu.
III. Cuitzeo	Acuitzio, Álvaro Obregón, Copándaro, Cuitzeo, Charo, Chucándiro, Huandacareo, Indaparapeo, Morelia, Queréndaro, Santa Ana Maya, Tarímbaro, Zinapécuaro.
IV. Oriente	Angangeo, Aporo, Contepec, Epitacio Huerta, Hidalgo, Irimbo, Juárez, Jungapeo, Maravatío, Ocampo, Senguio, Susupuato, Tlalpujahua, Tuxpan, Tuzantla, Tiquicheo, Tzitzio, Zitácuaro.
V. Tepalcatepec	Aguililla, Apatzingán, Buenavista, Cotija, Tepalcatepec, Tingüindín, Tocumbo, Parácuaro, Peribán, Los Reyes.
VI. Meseta Tarasca	Charapan, Cherán, Chilchota, Nahuatzen, Nuevo Parangaricutiro, Paracho, Tancítaro, Taretan, Tingambato, Uruapan, Ziracuaretiro.
VII. Pátzcuaro-Zirahuén	Erongarícuaro, Huiramba, Lagunillas, Pátzcuaro, Quiroga, Salvador Escalante, Tzintzuntzan.
VIII. Tierra Caliente	Parácuaro, Huetamo, Madero, Nocupétaro, San Lucas, Tacámbaro, Turicato.
IX. Costa	Aquila, Arteaga, Coahuayana, Coalcomán, Chinicuila, Lázaro Cárdenas, Tumbiscatío.
X. Infiernillo	Ario de Rosales, Churumuco, La Huacana, Gabriel Zamora, Múgica, Nuevo Urecho.

Table 2. Environmental requirements considered to assess sites potentially suitable for commercial forest plantations of Enterolobiumcyclocarpum (Jacq.) Griseb. in Michoacán, México.

Environmental requirement	Optima requirement
Elevation (msnm)	0 - 800
Annual rainfall (mm)	800 - 1200
Annual average temperature (°C)	20 - 32
Soil type	Vertisol
Slope	0% - 15% and 15% - 30%

2.3. Determining Areas of Potential Maps

IDRISI 32 (2.0) (Eastman, 1999) software was used to produce maps and estimates of area with potential for commercial forest plantations of Enterolobium cyclocarpum (Jacq.) Griseb. IDRISI is one of many commercial software products that handle GIS databases, as well as charts, imagery and climate digital information from National Institute of Forestry, Agriculture and Livestock Research were used. The Digital Elevation Model (DEM) employed had a 360 m × 360 m resolution, equivalent to 12.96 ha, that is, each pixel (cell) covers this area; all variables were measured within this spatial frame. Screening sorted sites using Reclass command. Excludes pixels that do not meet stated species habitat criteria. A hierarchical screening sequence was designed for this purpose, according to the following categories or fitness classes: Unsuitable = 0 Sites lacking required ecological features for growth and development of the studied taxon.

Suitable = 1 Sites that offer the required ecological features for a certain species.

GIS thematic layers displayed each variable and its range of values spread throughout the geographic distribution range of E. cyclocarpum. These layers were, then joined through Overlay command. The resulting product represents the potential areas in Raster format. Command Area produced a quantitative estimate of potential area. IDRISI command Reformat was used to convert to vectorial format, and then exporting it to Shapefile for further work on ARCVIEW 3.2. GIS software for rendering 1:100,000 and 1:250,000 scale outputs.

2.4. Validation of Potential Areas

Selected sites within the potential zones of the studied specie were visited in the field to assess the existence of suitable ecologic conditions to favor for the establishment and development of E. cyclocarpum plantations, and therefore considered suitable for commercial forest plantations, for it were chosen randomly and were georeferenced data accurately in selected areas.

3. RESULTS AND DISCUSSION

According to the methodology used and the agro-ecological requirements for E. cyclocarpum growth, the required characteristics are: elevation, total annual precipitation, average annual temperature, slope, land use, and soil type, two maps were produced which show the sites and potential areas for plantations and developing of E. cyclocarpum. The first map shows a slope ranging between 0% to 15% and is indicate for mechanized plantations, and the other map shows a slope with an inclination of 15% to 30%, where manual planting is best in these conditions.

Maps potential areas for Enterolobium cyclocarpum (Jacq.) Griseb.

Two maps were drawn from the GIS analysis. This maps show the potential areas for the establishment of commercial forest plantations of E. cyclocarpum according within a two slope range proposal. First slope range (0% - 15%) where mechanized plantations have greater potential for planting E. cyclocarpum, equivalent to 57,227 ha and are located in the IV (Oriente), V (Tepalcatepec), VII (Tierra Caliente), IX (Sierra Costa) and X (Infiernillo) regions (**Figure 2**).

The other map obtained shows potential areas for establishing new commercial forestry plantations non-mechanized or manual plantations of E. cyclocarpum on a slope range of 15% - 30% which is a 6273 ha area and are located in the IV Oriente, V Tepalcatepec, VII Tierra Caliente, IX Sierra Costa y X Infiernillo regions (**Figure 3**).

The total surface considering the range of both slopes (0% - 15% and 15% - 30%) adds up to 63,500 ha and showed potential for establishing new commercial forestry plantations with E. cyclocarpum in the state of Michoacán, and are mostly distributed in the V, VII, IX, X regions (Tepalcatepec, Tierra Caliente, Sierra and Costa Infiernillo, respectively), with 43,967 ha. The lower height range was taken in reference to sea level, which is the same as for SEDER (1995) and the upper limit, 800 m, which is slightly less than that found by SEDER (1995) and other authors for 1200 m.

The results are similar to those reported by Rueda et al. (2007) , who used the same variables for his work, to generate maps of potential areas for forest plantations of 11 species of pine and six tropical species in the state of Jalisco. It also matches those of Sáenz et al. (2000) , who, using the same variables, determined the potential areas in the eastern region of Michoacán, for plantations with Abies religiosa, Pinus pseudostrobus, P. michoacana, P. montezumae, P. teocote, P. oocarpa, P. ayacahuite, P. lawsonii and P. herrerae.

Similarly, and consistent with Sáenz et al. (2007, 2011) , is their finding of potential areas for commercial forest plantations of Pinus pseudostrobus, P. michoacana, P. lawsonii and P. montezumae combined with forage species in the basin of Lake Cuitzeo, in Michoacán. Furthermore, and consistent with Meza (2003) to locate potential areas for the establishment of plantations of damiana (Turnera diffusa Willd), and as indicated by the National Institute of Forestry, Agriculture and Livestock (INIFAP) in their assessment to determine potential areas in the state of Michoacán, considering the same variables (altitude, precipitation, mean annual temperature, slope, uses and soil type, and using the same GIS).

The potential areas determined by GIS maps and verified against the ecological requirements show results with a high level of accuracy (100%) between map sites and ecological sites, so the sampled sites match in soil type and altitude requirements for commercial plantations of E. cyclocarpum (**Table 3**).

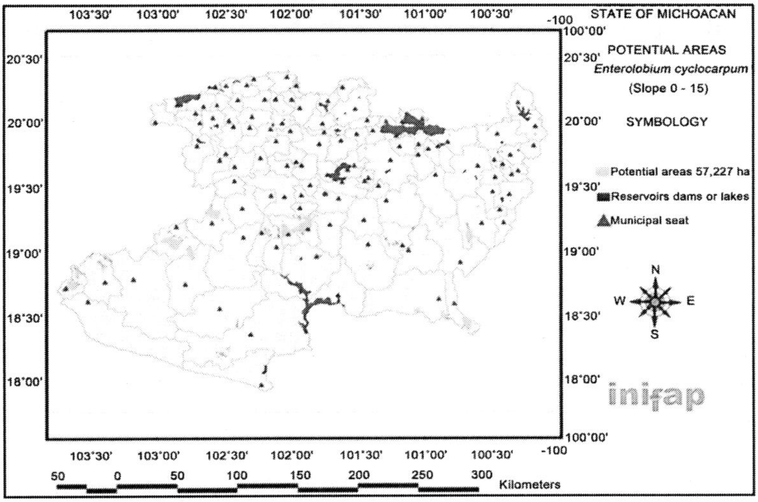

Figure 2. Areas with potential for mechanized establishment of commercial plantations of Enterolobium cyclocarpum (Jacq.) Griseb. in Michoacán, México(slope 0% to 15%).

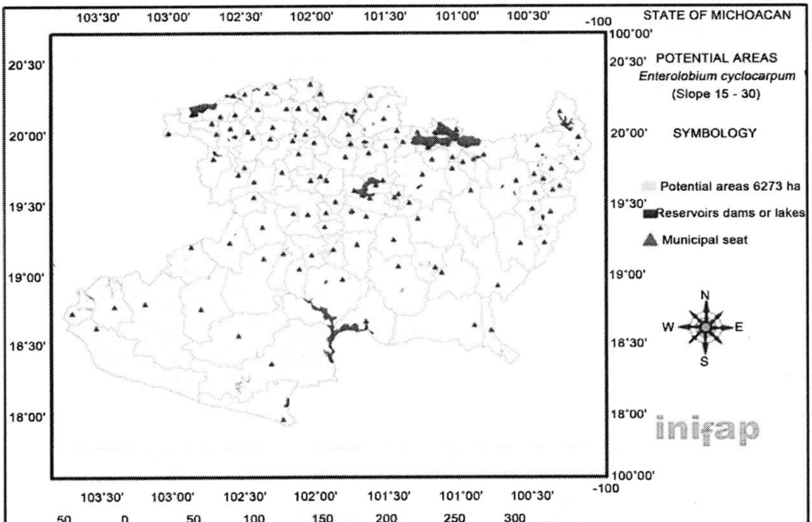

Figure 3. Sites with potential for non-mechanized establishment of commercial forest plantations of Enterolobium cyclocarpum (Jacq.) Griseb. in Michoacán, México (slopes 15% to 30%).

Table 3. Altitude, slope and soil type showing in validation sites for potential areas for commercial forest plantations Enterolobium cyclocarpum (Jacq.) Griseb. in the state of Michoacán.

Sample site	Municipality	Elevation (msnm)	Slope (%)	Soil type
1	San Lucas	440	0 - 15	Vertisol
2	Huetamo	433	0 - 15	Vertisol
3	Nuevo Urecho	625	15 - 30	Vertisol
4	Nuevo Urecho	514	0 - 15	Vertisol

4. CONCLUSION

The ecologic conditions in Michoacán are frequently favorable for successful establishment and development of commercial forest plantations of Enterolobium cyclocarpum (Jacq.) Griseb. where there are a totals area suitable for mechanized establishment of commercial plantations of E. cyclocarpum near of 57,227 ha low slope less than 15% and 6273 ha more in slopes upper 15%. Both superficies added 63,500 ha totals and V region (Tepalcatepec) presents 19,883 ha for new commercial forest plantations.

Geographic information systems are dependable tools in assessing the potential sites for growing forest species. Results have provided support to decisions on land use planning and establishment of commercial forest plantations of E. cyclocarpum in Michoacán.

REFERENCES

1. Carmona, A. J., & Monsalve, R. J. J. (1999). Geographic Information Systems (p. 44). (In Spanish) http://dds.cepal.org/infancia/guia-para-estimar-la-pobreza-infantil/bibliografia/capitulo IV/Carmona%20Alvaro%20y%20Monsalve%20Jhon%20(1999)%20Sistemas%20de%20informacion%20geografica.pdf
2. Castillo, Q. D., Martínez, B. O. U., ávila, F. D. Y., Castillo, R. F., & Sánchez, Ch. J. D. (2014). Identification of Potential Areas for Establishment of Plantations of Agave lechuguilla Torr. in Coahuila, Mexico. Open Journal of Forestry, 4, 520-526. http://www.scirp.org/journal/ojf http://dx.doi.org/10.4236/ojf.2014.45056
3. Eastman, J. R. (1999). IDRISI32 Ver 2.0: Guide to GIS and Image Processing. Worcester, MA: Clark Labs, Clark University.
4. Food and Agriculture Organization of the United Nations (FAO) (2000). Status of Forestry in Mexico. Forestry Commission 20th Session North America. Informative Note. Canada. http://www.fao.org/docrep/meeting/X4702S.htm
5. Forestry Commission of State of Michoacán (COFOM) (2003). Forest Plantation Program of Michoacán. Revista Bosques y Selvas de Michoacán, 1, 6-7. (In Spanish)

6. Forestry Commission of State of Michoacán (COFOM) (2014). State Forest and Soil Inventory of Michoacán de Ocampo. Vol. 1. No.1. Morelia. Michoacán. p. 376. (In Spanish)
7. Manzanilla, B. H., Martínez, D. M., & Martínez, M. A. (2001). Monographs of Promising Native Species for the Establishment for Commercial Timber Plantations in Jalisco, Cedar and Purple Rose. Documento Técnico. No. 31. FIPRODEFO. Guadalajara, Jal. México. pp. 51-73. (In Spanish)
8. Martínez, B. O. U., & Castillo, Q. D. (2007). Identification of Sites for Plantations Cortadillo (Nolina cespitifera Trel.) in Southeast Coahuila. FolletoTécnicoNúm. 34. CIRNE-INIFAP. Saltillo, Coah. p. 38. (In Spanish)
9. Martínez, M. S., & Prieto, R. J. A. (2011). Establishment for Commercial Timber Plantations in the Northern Region of México. SAGARPA-CIRPAC-INIFAP. CIR NORTE CENTRO. Folleto técnico Núm. 47. Durango, Dgo. p. 27. (In Spanish)
10. Moreno, S. R., & F. Moreno, S. (1995). Geographic Information Systems in Natural Resource Management: Experiences Recommendations INIFAP. Revista de Ciencias Forestales en México, 20, 93-109. (In Spanish)
11. Muñoz, F. H. J., Sáenz, J. T. R., García, J. J. S., Hernández, E. M., & Anguiano, J. C. (2011). Areas with Potential for Commercial Timber of Plantations of Pinus pseudostrobus. Revista Mexicana de Ciencias Forestales, 2, 29-44. (In Spanish)
12. National Center of Coffee Research (CNIC) (2005). Tree Associated to Coffee Plantations (149 p). (In Spanish) https://www.yumpu.com/es/ document/ view/15450777/arboles-encontrados-en-zonas-cafetaleras-eco-index
13. National Commission for Knowledge and Use of Biodiversity (CONABIO) (2008). Forestry Biological Diversity from Mexico. (In Spanish) www.Conabio .gob.mx/institución/cooperación_internacional/doctos/dbf_méxico.html
14. National Forestry Commission (CONAFOR) (2005). Coordination General of Conservation and Restoration. México. http://app1.semarnat.gob. mx/dgeia/ informe_resumen/pdf/2_info_resumen.pdf
15. National Institute of Forestry, Agriculture and Livestock Research (INIFAP) (1993). Areas with Potential for Commercial Timber Plantations of Plant Species in Mexico from Jalisco State. In: Memorias. VI Reunión Científica y Técnica Forestal y Agropecuaria (pp. 7-11). Guadalajara: INIFAP. (In Spanish)
16. National Institute of Statistics and Geography (Mexico) (INEGI) (2012). Statistical Perspectives of Michoacán de Ocampo (p. 97). Aguascalientes: INEGI. (In Spanish)
17. Pennington, T. D., & Sarukhán, K. J. (1998). Tropical treefrom Mexico. Fondo de Cultura Económica. Segunda edición. UNAM-Fondo de Cultura Económica. México.pp. 138-139. (In Spanish)
18. Rueda, S. A., Ruiz, J. A., Benavides, S. J. de D., & Flores, G. J. G. (2007). Definition of Favorable Areas for Six Tropical Forest Species in the State of Jalisco (p.165). Libro Técnico. Núm. 5. Tepatitlán de Morelos, Jalisco: SAGARPA-CIRPAC-INIFAP. (In Spanish)
19. Rural Development Secretary (SEDER) (1995). Main Features of Some Forest Species Susceptible to Commercial Use in the State of Jalisco (p. 32). Guadalajara: Integral Forestry Development Program of Jalisco State. (In Spanish)
20. Sáenz, R. J. T., Anguiano, C. J., Muñoz, F. H. J., Villaseñor, R. F. J., Rueda, S. A., & García, M. J. J. (2007). Potential Areas for Commercial Forest Plantations in the Lake Cuitzeo Basin, Michoacán (p. 29). Morelia: Actions and Outcomes Symposium for Sustainable Development of Lake Cuitzeo Basin, Michoacán. (In Spanish)
21. Sáenz, R. J. T., Chávez, G. L., García, J. M., & Anguiano, J. C. (2000). Regionalization of Potential Areas for Forest Plantations in Eastern Michoacán. In: Primer Congreso Nacional de Reforestación (6 p). México, D.F.: PRONARE-SEMARNAT. (In Spanish)
22. Sáenz, R. J. T., Jiménez, O. J., Rueda, S. A., Ibáñez, R. C., & García, M. J. J. (2011). Regionalization of Potential Areas for Silvopastoral Systems in Temperate Forest of Michoacán (p. 42). Folleto Técnico Núm. 23. SAGARPA-INIFAP-CIRPAC. Uruapan: Campo Experimental Uruapan. (In Spanish)

23. Sáenz, R. J. T., Muñoz, F. H. J., Rueda, S. A., Villaseñor, R. F. J., Anguiano, C. J., & García, M. J. J. (2010). Potential Areas for Commercial Forest Plantations in Michoacán. In: 2nd National Congress of Environmental Damage Mitigation in the Agricultural Sector of Mexico (pp. 2000-2009). Morelia: INIFAP-COECYT Michoacán-UMSNH-IT Celaya-SEDRU-IPN-UACH-COFUPRO-SEMARNAT-SAGARPA-Gobierno del estado de Michoacán. (In Spanish)

24. Tropicos (2016). Missouri Botanical Garden. http://www.tropicos.org/

25. Vázquez, Y. C., Batís, M. A., Alcocer, S. S., Gual, D. M., & Sánchez, D. C. (1999). Potentially Valuable Trees and Shrubs Native for Ecological Restoration and Reforestation. Instituto de Ecología, Universidad Autónoma de México. (In Spanish) http://www.conabio.gob.mx/conocimiento/info_especies/arboles/doctos/inicio.pdf

CHAPTER 10

Characterization of Annur and Bedakam Ecotypes of Coconut from Kerala State, India, Using Microsatellite Markers

M. K. Rajesh,[1] K. Samsudeen,[1] P. Rejusha,[1] C. Manjula,[1,2] Shafeeq Rahman,[1] and Anitha Karun[1]

[1]*Division of Crop Improvement, Central Plantation Crops Research Institute, Kasaragod Kerala, 671124, India*
[2]*Nehru Arts and Science College, Kanhangad Kerala, 671314, India*

ABSTRACT

The coconut palm is versatile in its adaptability to a wide range of soil and climatic conditions. A long history of its cultivation has resulted in development of many ecotypes, which are adapted to various agro-eco factors prevalent in a particular region. These ecotypes usually are known by the location where they are grown. It is important to explore such adaptation in the coconut population for better utilization of these ecotypes in coconut breeding programs. The aim of the present study was to identify the genetic diversity of the Bedakam and Annur ecotypes of coconut and compare these ecotypes with predominant West Coast Tall (WCT) populations, from which they are presumed to have been derived, using microsatellite markers. All the 17 microsatellite markers used in the study revealed 100% polymorphism. The clustering analysis showed that Annur and Bedakam ecotypes were two separate and distinct populations compared to WCT. It was also evident from the clustering that Annur ecotype was closer to WCT than Bedakam ecotype.

1. INTRODUCTION

Coconut (*Cocos nucifera L.*), a monotypic species, is one of the major perennial oil crops of the tropics. The palm forms the basis in many developing countries for food products as well as serving industrial purposes [1]. It is often referred to as "*Kalpavriksa,*" the tree which provides all the necessities of life. The coconut palm is versatile in its adaptability to a wide range of soil and climatic conditions. In India, it grows well in coastal alluvium of both the West and East coasts, river alluvium of the deltaic regions, and the literate and red loam soil of the inland areas. It is estimated that about 70% of the cropped area is under sandy loam soil. Sandy loam soil with good cation-exchange capacity and soil water level at about 4 m depth is considered as the best soil type for coconut [2]. Coconut cultivars are generally classified into tall and dwarf types. The tall types are primarily outcrossing, while the dwarf types are predominantly self-pollinated [3]. In a cultivar, certain members have been reported to differ from one another in a single or a constellation of characters [4].

Coconut is a perennial crop with indeterminate flowering and the productive features of the palm are considerably influenced by environmental variables; weather factors like sunshine hours, light intensity, ambient temperature, humidity, and rainfall have been reported to play a significant role in fluctuations of coconut yield [5, 6]. The influence of seasonal variations on gas exchange characteristics and biochemical constituents of coconut palms in a particular area have also been reported in earlier studies [7–10]. Weather factors are known to influence crop production especially under rainfall condition and of all the climatic factors, rainfall has maximum influence on the seasonal variation in yield [11]. Variations in environmental factors usually cause genotypes to respond differently from one environment to another resulting in genotype environmental interactions.

The understanding of the distribution of a particular species is also dependent on knowledge of the interrelationship of particular species to their environment. There are wide variations in the crop and cultivation method in different regions.

Ecotypes are cultivars which are grown for a long period to be superior to other local cultivars by the farmers. Wide morphological variability has been observed in these native populations of coconut in various coconut growing countries of the world. Studies have indicated high genetic variability and diversity for whole nut weight, husked nuts weight, kernel weight, copra weight, copra and oil yield of the palms [12]. In India, coconut is generally considered to be a crop of coastal region, even though it is found to be grown well in inlands like Assam State of India [13].

Genetic diversity is thought of as the amount of genetic variability among individuals of a variety, or population of a species [14]. Assessment of genetic diversity of a population provides valuable information that can decide its proper utilization, sustainable management, and design of conservation strategies. There are different techniques available to assess genetic diversity. Among these, molecular markers have the potential to significantly increase the efficiency of coconut genetic improvement, specifically in the areas of germplasm management, genotype identification, and marker-assisted selection of economically important traits [15]. Microsatellites or simple sequence repeats (SSRs) are tandemly repeated motifs of 1–6 bases found in all prokaryotic and eukaryotic genomes analyzed to date. They are present in both coding and noncoding regions and are usually characterized by a high degree of length polymorphism. The origin of such polymorphism is still under debate though it appears most likely to be due to slippage events during DNA replication [16]. The increased number of SSR markers has greatly improved the knowledge about the genetic diversity/relationships between coconut varieties/populations [17–20].

One of the major coconut growing states in India is Kerala, where the most popular variety grown by the farmers is the West Coast Tall (WCT), which occupies over 95 percent of the area under coconut. The WCT palm grows well in all types of soil and is relatively tolerant to moisture stress. It is recommended for large scale cultivation in the coastal regions of Kerala and Karnataka states of India [21]. WCT cultivar has spread inwards from the coastal regions and is now found cultivated even in high ranges. In the process, the cultivar diverged into different ecotypes known by the location where they are cultivated, some of them being Annur, Bedakam, Kuttiyadi, Attingal, and Kanjirappally. These diverse coconut ecotypes have been reported to exhibit morphological or physiological phenotypic differences. The objective of this study was to decipher the genetic diversity among Annur (ANR) and Bedakam (BDK) ecotypes of coconut from Kerala State, India, and compare them with WCT populations, from which they are presumed to have been derived, using molecular markers, which might throw light on the crop's evolutionary diversification.

2. MATERIALS AND METHODS

2.1. Plant Materials and DNA Extraction

The plant material for this study consisted of 50 leaf samples collected from the two different ecotypes, namely, ANR (17 palms) from Kannur district and BDK (16 palms) from Kasaragod district, both in Kerala State of India, and WCT cultivar (17 palms) from CPCRI, Kasaragod, Kerala State. Genomic DNA was extracted from spindle leaves of palms following the standardized protocol [22]. The DNA was quantified spectrophotometrically and the bands were checked on 0.8% agarose gel electrophoresis.

2.2. Molecular Analysis

A total of 17 SSR primer pairs specific to coconut were used in the present study to assess the genetic diversity of the respective coconut collections (Table 1). The PCR reactions were carried out in 20 µL volume with standardized components: 20 ng genomic DNA, 0.2 µM each of forward and reverse primers, 10 µM dNTPs (M/s Bangalore Genei Pvt. Ltd.), 10x buffer (10 mM Tris-Hcl (pH 8.3), 50 mM KCl, 1.5 mM $MgCl_2$), and 3 units of Taq DNA polymerase (M/s Bangalore Genei Pvt. Ltd.). After amplification, a volume of 3 µL of loading buffer was added to each of the amplified products. The amplified products were run on 3.0% high resolution agarose gel, stained with ethidium bromide following the protocol of Sambrook et al. [23], and were visualized in a gel documentation system.

2.3. Data Analysis

Data analysis was done by scoring of bands. The alleles were scored individually based on comparison in the molecular ladder. The size of the amplicons was compared using a 100 bp ladder. Each band generated by SSR primers was considered as an independent locus. Clearly resolved unambiguous bands were scored visually for their presence or absence with each primer. The scores were obtained in the form of a matrix with "1" and "0" indicating the presence and absence of bands, respectively. Based on the number of polymorphic bands, percentage polymorphism was calculated for each primer.

Table 1: SSR primers, their sequences, banding patterns, and percent polymorphism.

Sl. no.	Primer	Primer sequence (5′–3′)	Number of bands	Number of polymorphic bands	Polymorphism (%)
1	CAC2	AGCTTTTTCATTGCTGGAAT CCCCTCCAATACATTTTTCC	4	4	100
2	CAC3	GGCTCTCCAGCAGAGGCTTAC GGGACACCAGAAAAAGCC	3	3	100
3	CAC4	CCCCTATGCATCAAAACAAG CTCAGTGTCCGTCTTTGTCC	4	4	100
4	CAC6	TGTACATGTTTTTTGCCCAA CGATGTAGCTACCTTCCCC	3	3	100
5	CAC8	ATCACCCCAATACAAGGACA AATTCTATGGTCCACCCACA	3	3	100
6	CAC10	GGAACCTCTTTTGGGTCATT GATGGAAGGTGGTAATGCTG	3	3	100
7	CAC13	GGGTTTTTTAGATCTTCGGC CTCAACAATCTGAAGCATCG	4	4	100
8	CNZ1	ATGATGATCTCTGGTTAGGCT AAATGAGGGTTTGGAAGGATT	4	4	100
9	CNZ2	CTCTTCCCATCATATACCAGC ACTGGGGGGATCTTATCTCTG	4	4	100
10	CNZ3	CATCTTTCATCATTTAGCTCT AAACCAAAAGCAAGGAGAAGT	4	4	100
11	CNZ4	TATATGGGATGCTTTAGTGGA CAAATCGACAGACATCCTAAA	4	4	100
12	CNZ5	CTTATCCAAATCGTCACAGAG AGGAGAAGCCAGGAAAGATTT	4	4	100
13	CNZ6	ATACTCATCATCATACGACGC CTCCCACAAAATCATGTTATT	4	4	100
14	CNZ10	CCTATTGCACCTAAGCAATTA AATGATTTTCGAAGAGAGGTC	4	4	100
15	CnCir56	AACCAGAACTTAAATGTCG TTTGAACTCTTCTATTGGG	4	4	100
16	CnCirH9	CACAATCCTTACATCAAA TCTCAAGTTCTTACAGCAGT	3	3	100
17	CnCirG4	AGTATAGTCACGCCAGAAAA AAACCCATAACCAGCAAG	4	4	100
	Total		63	63	
	Average		3.70	3.70	

The genetic associations between varieties were evaluated by calculating Dice's similarity coefficient for pairwise comparisons based on the proportions of shared bands produced by the primers [24]. Similarity matrix was generated using the NTSYS-PC software, version 2.0 [25]. The similarity coefficients were used for cluster analysis and dendrogram was constructed by the Unweighted Pair-Group method (UPGMA) [26]. Shannon's Information Index, expected and observed heterozygosity, unbiased expected heterozygosity, fixation index, principal component analysis (PCA), and analysis of molecular variance (AMOVA) were worked out for the coconut populations using the software GenAlEx 6.5 [27]. For PCA, genetic distance was calculated from the allele data and the genetic distance was plotted as PCA using GenAlEx.

3. RESULTS

In the present study, phylogenetic analysis was carried out using 50 palms that belong to three distinct coconut populations, one a cultivar (WCT) and two ecotypes, namely, ANR and BDK. The 17 SSR primers employed for the study gave clear, unambiguous bands and the data derived were used for further analysis. The details of banding patterns produced in the accessions are given in Table 1. A total of 63 bands were produced by 17 primers, with an average of 3.70 bands/primer. Polymorphism was calculated for all the 17 primers after scoring the bands. All the 17 primers showed 100% polymorphism.

The similarity index, based on Dice's coefficient, was obtained after pair wise comparison of the three coconut populations. The percentage similarity varied between the palms. Maximum similarity was seen between ANR6 and ANR5 (0.97) palms and minimum similarity was seen in BDK16 and ANR1 (0.30) (data not shown). Cluster analysis, based on UPGMA, was performed using NTSYS software in order to obtain a dendrogram. On clustering, it was found that all the three populations formed two major clusters—one containing all WCT and a few BDK palms, while the second cluster exclusively contained BDK palms (Figure 1). In general, palms of ANR, BDK, and WCT grouped together according to the population; however, a few palms did show intergroup affinity. Two palms (ANR1 and WCT8) stood outside their respective clusters. A group of five palms of Annur ecotype clustered separately and this group showed more affinity towards WCT. The palms of BDK ecotype were all clustered together except for one palm that aligned with Annur group. WCT palms were mostly clustered together but one palm aligned with BDK and three palms with ANR ecotype. Among palms studied from BDK, four palms showed affinity towards ANR and WCT palms. Remaining palms of BDK were clustered in a different group. However, one WCT palm (WCT 8) showed variation from all other WCT palms and grouped with BDK ecotype. Other WCT palms were mostly clustered together showing affinity towards ANR palms. The intergroup affinity might be due to sharing of alleles between the three coconut populations studied. The clustering shows that ANR and BDK ecotypes were two separate populations with clear distinction, with just one palm of BDK (BDK 1) showing affinity towards ANR. It is also evident from the clustering that ANR ecotype is closer to WCT than BDK ecotype. When the populations were considered individually, WCT palms clustered at 61% similarity, ANR palms at 70% similarity, and BDK palms at 58% similarity.

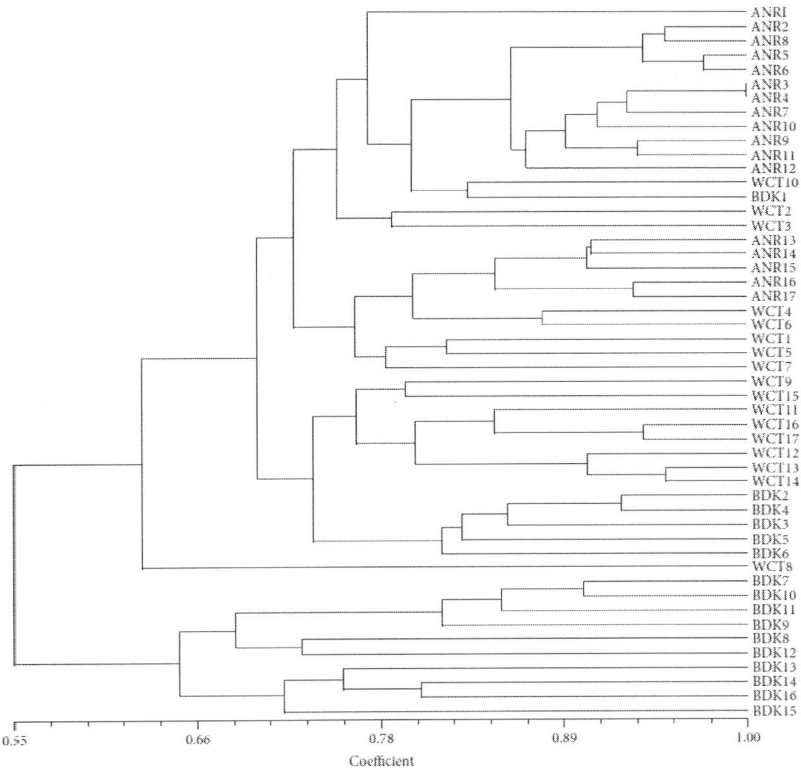

Figure 1: UPGMA phylogenetic tree constructed based on Dice's coefficient showing the genetic relationships among the three coconut populations.

Population-wise mean Shannon's Information Index ranged from 0.513 (ANR) to 0.712 (BDK) and the mean observed heterozygosity from 0.257 (ANR) to 0.339 (BDK) (Table 2). The mean unbiased expected heterozygosity (uHe) ranged from 0.325 (ANR) to 0.442 (BDK). Mean fixation index () ranged from 0.205 to 0.372 (Table 2). Pair-wise population matrix of Nei's genetic identity calculated using GenAlEx program showed a higher average identity between ANR and WCT (0.92) than between WCT and BDK (0.869) and ANR and BDK (0.783) (Table 3). The AMOVA estimation based on 99 permutations using GenAlEx showed a significant () within population variation (79%) compared to among population variation (21%) (Table 4).

Table 2: Shannon's Information Index, expected and observed heterozygosity, unbiased expected heterozygosity, and fixation index for the three populations.

Population	I	H_o	H_e	uH_e	F_{ST}
ANR					
Mean	0.513	0.257	0.315	0.325	0.205
SE	0.066	0.060	0.041	0.043	0.110
WCT					
Mean	0.708	0.339	0.427	0.440	0.244
SE	0.061	0.052	0.037	0.038	0.092
BDK					
Mean	0.712	0.284	0.428	0.442	0.372
SE	0.062	0.057	0.035	0.036	0.105

I = Shannon's Information Index = −1 * Sum (pi * Ln (pi)).
H_o = observed heterozygosity = Number of Hets/N.
H_e = expected heterozygosity = 1 − Sum pi^2.
uH_e = unbiased expected heterozygosity = $(2N/(2N-1)) * H_e$.
F = fixation index = $(H_e - H_o)/H_e = 1 - (H_o/H_e)$,
where pi is the frequency of the ith allele for the population and Sum pi^2 is the sum of the squared population allele frequencies.

Table 3: Pairwise population matrix of Nei's genetic identity.

ANR	WCT	BDK	
1.000			ANR
0.920	1.000		WCT
0.783	0.869	1.000	BDK

Table 4: Analysis of molecular variance in the coconut populations.

ANR	WCT	BDK	
1.000			ANR
0.920	1.000		WCT
0.783	0.869	1.000	BDK

For a deeper understanding of the clustering pattern of the coconut populations, we also carried out genetic distance-based principal component analysis (PCA) using GenAlEx. The results showed clear segregation of all the three populations into different quadrates of the PCA (Figure 2). The first and second axes accounted for 85.13 and 14.87% of the total inertia, respectively. As shown in Figure 2, the first axis separated BDK from the other two populations, while the second axis separated WCT from the other two populations.

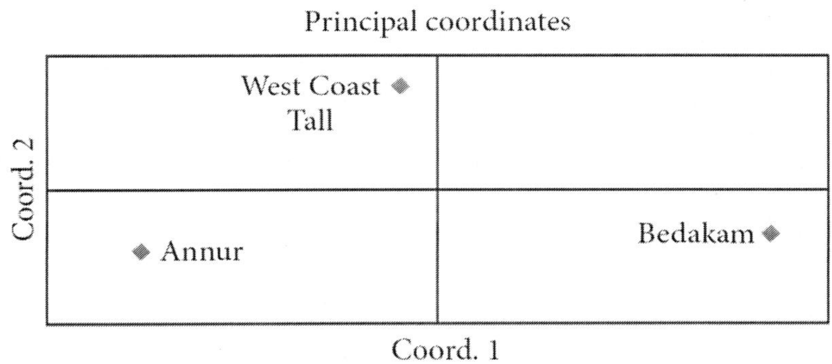

Figure 2: Diagram of PCA based on Nei's genetic distance.

4. DISCUSSION

Ecotypes are groups of similar populations within the same plant species that are adapted to certain climatic and edaphic conditions. Coconut has been cultivated for centuries in India and the long history of its cultivation along the length and breadth of the country has resulted in development of many ecotypes, which are generally named or known by the location where they are grown. These ecotypes are likely to have developed certain adaptation to the particular environment conditions in which they have been growing for a long time and possess a rich source of valuable genes for coconut breeding. It is important to identify and conserve these ecotypes, which could withstand the vagaries of nature and still perform better in terms of growth and yield.

West Coast Tall (WCT) coconut population developed on the West coast of India and came to be known by the region where it was cultivated. Though the origin of WCT is not traceable to any area from where it spread, it is obvious that sea journey by nuts was involved in its spread. The cultivar over the years moved from coastal region to the interior areas, which also resulted in further adaptation and diversity in the cultivar. Such adapted

WCT populations in certain localities are designated with local names by farmers to differentiate it from other WCT populations. Annur, Bedakam, Kuttiyadi, Attingal, and Kanjirappally are some of these. The variability shown by WCT and ecotypes derived from it might be genetic or due to physiological factors such as climatic change, pH of the soil, annual rainfall, or any other environmental factors or even human involvement. These coconut ecotypes may possess higher variability which might be an important source for plant breeding, in comparison to the present day cultivars, which have been specifically chosen for their growth performance under certain specific environment.

Molecular markers play an important role in conservation and use of plant genetic resources. Molecular markers are relatively independent of environment. DNA-based markers are a way of exploring the genetic relations between populations and these markers have acted as versatile tools in various fields. Among the various molecular markers currently available, microsatellite markers are the most popular because they are reproducible, enabling their parallel analysis in different laboratories and exchange of the resulting data [28]. Also, microsatellites form an ideal marker system creating complex banding patterns by simultaneously detecting multiple DNA loci. They have been used successfully by many researchers to characterize the genetic diversity of the coconut population [18, 21].

With the objective of deciphering the diversity among and within WCT populations from Kerala region, in the present study, Annur and ecotypes were compared with WCT populations using SSRs. On clustering, it was found that the two ecotypes and WCT formed two major clusters. Annur, Bedakam, and WCT grouped separately, in general, with a few palms showing intergroup affinity. However, most of the Bedakam palms were grouped in separate cluster proving that Annur and Bedakam ecotypes were two separate populations. It is also evident from the clustering that Annur ecotype was closer to WCT than to Bedakam ecotype. Pair-wise population matrix of Nei's genetic identity also revealed a higher average identity between ANR and WCT than between WCT and BDK and ANR and BDK.

The comparison of average observed and expected heterozygosity values did not show great differences between the three coconut populations studied. All the three populations displayed smaller observed than expected heterozygosities-inbreeding may be a factor contributing to this. Among the three populations, palms of ANR ecotype displayed more genetic similarity amongst themselves, with lower observed heterozygosity and fixation index () compared to the other two populations.

The values of heterozygosities obtained in this study confirm that the coconut ecotypes represent an important reservoir of genetic diversity. The three populations showed significant genetic differentiation, as indicated by values, indicating that a high level of differentiation for the alleles has occurred in members of the subpopulation compared to the total population, and therefore members of the subpopulation tend to carry unique alleles compared to the total population. The among population variation was more than within population variation, based on AMOVA calculations, suggesting very rare genetic exchange between the populations, at least in recent history.

Diversity in the original WCT population allowed it to spread in to different ecoregions resulting in the evolution of ecotypes suitable for the local environment and its subsequent adaptation. The result of molecular characterization revealed significant differences in midland and coastal growing palms. The WCT and ecotypes might have differed due to genetic factors or adaptation resulting from environmental variations like climatic change, pH of the soil, annual rainfall, or any other environmental factors or even human interventions.

One of the important prerequisites for evolutionary change is genetic variation as in its absence populations lack the capacity to evolve. Within a single species, natural selection, coupled with heterogeneity in its habitat, might result in multiple ecotypes, which are genetically distinct [29]. The total genetic variation of a species is likely to be distributed among populations as the impact and direction of natural selection varies from one to another, due to environmental variation and genetic drift [30]. Therefore, with germplasm conservation programmes, it is imperative to accurately measure the amount of genetic diversity and its distribution within and between populations. To these ends, molecular markers provide an efficient and unbiased estimate of these statistics, free of environment effects. Molecular characterization of genetic diversity provides base information which could be utilized in selection of a promising range of accessions for different breeding programs. The microsatellites used in this study appeared to possess a significant potential in this respect. This is one of the first studies to probe the diversity of coconut ecotypes using molecular markers. The results of this study may be used in developing a strategy for conservation of these ecotypes and their utilization in future coconut breeding programmes.

ACKNOWLEDGMENTS

The authors thank Director, CPCRI, Kasaragod, for his guidance and facilities provided. The authors also thank the farmers and extension personnel from Annur and Bedakam for their cooperation.

REFERENCES

1. L. H. Jones, "Perennial vegetable oil crops," in Agricultural Biotechnology: Opportunities for International Development, pp. 213–224, CAB International, Wallingford, UK, 1990.
2. J. G. Ohler, "Coconut, a tree of life," Plant Production and Protection Paper, 1984.
3. D. V. Liyanage, "Preliminary studies on the floral biology of the coconut palm," Tropical Agriculture, vol. 105, pp. 171–175, 1949.
4. K. Satyabalan, "Yield variation in west coast tall coconut palms: yield attributes which cause variation in annual yield of nuts in the palms of different yield groups," Indian Coconut Journal, vol. 1, pp. 5–8, 1993
5. P. Coomans, "Influence of climate factors on seasonal fluctuations of coconut production," Oleagineux, vol. 30, no. 4, pp. 153–157, 1975.
6. D. B. Murray, "Coconut palm," in Ecophysiology of Tropical Crops, P. T. Alvin and T. T. Kozlowski, Eds., pp. 24–27, Academic Press, New York, NY, USA, 1977.
7. K. V. Kasturi Bai and A. Ramadasan, "Changes in the levels of carbohydrates as a function of environmental variables in hybrids and tall coconut palm," in Coconut Research and Development, N. M. Nair, Ed., pp. 203–209, Wiley Eastern, New Delhi, India, 1983.
8. V. Rajagopal, A. Ramadasan, K. V. Kasturi Bai, and D. Balasimha, "Influence of irrigation on leaf water relations and dry matter production in coconut palms," Irrigation Science, vol. 10, no. 1, pp. 73–81, 1989.
9. K. V. Kasturi Bai, V. Rajagopal, C. D. Prabha, M. J. Ratnambal, and M. V. George, "Evaluation of coconut cutivars and hybrids for dry matter production," Journal of Plantation Crops, vol. 24, pp. 23–28, 1996.
10. C. Jayasekara, N. P. A. D. Nainanayake, and K. S. Jayasekara, "Photosythetic characteristic and productivity of coconut palm," Journal of Plantation Crops, vol. 24, pp. 538–547, 1996.
11. K. N. Krishna Kumar, Coconut phenology and yield response to climate variability and change [Ph.D. thesis], Cochin University of Science and Technology, Kerala, India, 2011.
12. K. Ganesamurthy, C. Natarajan, and M. Jayaramachandran, "Genetic diversity and its exploitation in coconut improvement," in Proceedings of the National Conference on Agro-Biodiversity, Conducted by National Biodiversity Authority Chennai, February 2006.
13. U. Parthasarathy, A comparative study of coconut cultivation in coastal and inland River Plain Ecosystem of Kasaragod District of Kerala and Kamrup District of Assam [Ph.D. thesis], Gauhati University, Assam, India, 2004.
14. H. D. Brown, "Isozymes, plant population genetic structure and genetic conservation," Theoretical and Applied Genetics, vol. 52, no. 4, pp. 145–157, 1978.
15. G. R. Ashburner, W. K. Thompson, G. M. Halloran, and M. A. Foale, "Fruit component analysis of south pacific coconut palm populations," Genetic Resources and Crop Evolution, vol. 44, no. 4, pp. 327–335, 1997.

16. C. Schlotterer and D. Tautz, "Slippage synthesis of simple sequence DNA," Nucleic Acids Research, vol. 20, no. 2, pp. 211–215, 1992.
17. R. Rivera, K. J. Edwards, J. H. A. Barker et al., "Isolation and characterization of polymorphic microsatellites in Cocos nucifera L," Genome, vol. 42, no. 4, pp. 668–675, 1999.
18. B. Teulat, C. Aldam, R. Trehin et al., "An analysis of genetic diversity in coconut (Cocos nucifera) populations from across the geographic range using sequence-tagged microsatellites (SSRs) and AFLPs," Theoretical and Applied Genetics, vol. 100, no. 5, pp. 764–771, 2000.
19. L. Perera, J. R. Russell, J. Provan, and W. Powell, "Use of microsatellite DNA markers to investigate the level of genetic diversity and population genetic structure of coconut (Cocos nucifera L.)," Genome, vol. 43, no. 1, pp. 15–21, 2000.
20. M. K. Rajesh, P. Nagarajan, B. A. Jerard, V. Arunachalam, and R. Dhanapal, "Microsatellite variability of coconut accessions (Cocos nucifera L.) from Andaman and Nicobar Islands," Current Science, vol. 94, no. 12, pp. 1627–1631, 2008.
21. C. Remany, Cataloging and categorization of unexploited ecotypes of coconut grown in Kerala [Ph.D. thesis], Mahatma Gandhi University, Kerala, India, 2003.
22. M. K. Rajesh, B. A. Jerard, P. Preethi et al., "Development of a RAPD-derived SCAR marker associated with tall-type palm trait in coconut," Scientia Horticulturae, vol. 150, no. 4, pp. 312–316, 2013.
23. J. Sambrook, E. F. Fritsch, and T. Maniatis, Cloning: A Laboratory Manual, Cold Spring Harbor laboratory, 1989.
24. L. Dice, "Measures of the amount of ecologic association between species," Ecology, vol. 26, pp. 297–302, 1945.
25. F. J. Rohlf, "NTSYS-pc numerical taxonomy and multivariate analysis system version 2.0. owner manual," 1997.
26. P. H. A. Sneath and R. R. Sokal, Numerical Taxonomy: The Principles and Practice of Numerical Classification, W. H. Freeman, San Francisco, Calif, USA, 1973.
27. R. Peakall and P. E. Smouse, "GenAlEx 6.5: genetic analysis in excel. Population genetic software for teaching and research-an update," Bioinformatics, vol. 28, pp. 2537–2539, 2012.
28. P. K. Gupta, H. S. Balyan, P. C. Sharma, and B. Ramesh, "Microsatellites in plants: a new class of molecular markers," Current Science, vol. 70, no. 1, pp. 45–54, 1996.
29. Y. B. Linhart and M. C. Grant, "Evolutionary significance of local genetic differentiation in plants," Annual Review of Ecology and Systematics, vol. 27, pp. 237–277, 1996.
30. M. J. Lawrence and N. Rajanaidu, "The genetic structure of natural populations and sampling strategy," in Proceedings of the International Workshop on Oil Palm Germplasm and Utilization, pp. 15–26, Selangor, Malaysia, 1985.

CHAPTER 11

Importance and Characterization of Coffee Berry Disease (Colletotrichum kahawae) in Borena and Guji Zones, Southern Ethiopia

Abdi Mohammed* and Abu Jambo

Bule Hora University, Bule Hora, 144, Ethiopia

ABSTRACT

COFFEE (Coffea arabica L.) is one of the most important cash crops in Ethiopia. Coffee Berry Disease (CBD) caused by Colletotrichum kahawae is the severe disease threatening coffee production in most coffee-growing regions of the country. Field survey was conducted in three major coffee growing districts (Abaya, Bule Hora and Kercha) of Borena and Guji zones during 2012 cropping season to determine the incidence, severity and prevalence of CBD. CBD was prevalent in all the surveyed districts with the overall mean incidence and severity of 49.3 and 14.7%, respectively. Laboratory experiment was conducted at Haramaya University to investigate the characteristics of C. kahawae and other fungal **PATHOGENS** associated with coffee berries. The proportion frequencies of infected and non-infected coffee berries were ranged from 24-42 and 3-21%, respectively. C. kahawae, F. lateritium and Phoma spp. of fungal pathogens were isolated from infected coffee berries with the proportions of 89.2, 15.2 and 3%, respectively. In general, the study revealed high occurrence, distribution and **CONTAMINATION** of CBD in the study areas. Therefore, more extensive empirical assessment of the amount of yield loss due to CBD on most grown varieties in the areas and developing or introducing resistant varieties need to have high priority for developing management options which may enhance sustainable coffee production in the country.

Keywords: *Coffea arabica; Colletotrichum kahawae*; Disease intensity; Fungal identifications; Ethiopia

INTRODUCTION

COFFEE belongs to the family *Rubiaceae* and the genus *Coffea*. The genus *Coffea* consists approximately 100 coffee species. The two main species of coffee cultivated on a world scale are Arabica coffee (*Coffea arabica* L.) and Robusta coffee (*Coffea canephora* L.), which account for about 98-99% of the world coffee production. In fact, Ethiopia is the only center of origin and diversity of Arabica coffee [1]. Arabica coffee is cultivated in most parts of the tropics, accounting for 80% of the world coffee market, and about 70% of the production [2]. Coffee plays an important role in the world economy. It is the second most valuable exported **COMMODITY** on earth after oil [3]. More than 80 countries in the developing world depend on coffee as a major source of their foreign currency earnings. For instance, coffee generated about US$13 billion in 1983 [4], and until 2000, it contributed to 80% of Burundi's, 55% of Uganda's and 30% of Nicaragua's hard currency earnings from exported yields [5].

Coffee is the major source of foreign currency for **ETHIOPIA** and contributes more than 35% of the total export earnings. Thus, it is a cornerstone in the export economy of the country [6]. Its cultivation, processing, trading, transportation and marketing provide employment for millions of people [7]. It is also the defining feature of the national culture and identity, with 44% of the production consumed domestically [8]. Ethiopian people drink coffee at least three times per day followed to their meals during fasting days.

In Ethiopia, coffee grows over a wide range of agro-ecological zones and geographical regions. Mainly it is grown in the Southern, Southwestern and Eastern parts of the country under different types of production systems. These systems are mostly forest-based traditional coffee production systems which is followed by **PLANTATION** coffee productions [9,10]. The estimated area of land covered by coffee is about 600,000 hectares, and the estimated annual national production of clean coffee is about 350,000 tones [11].

Despite the high production potential and economic importance of coffee; production and productivity of the crop is still limited due to several factors, such as shortage of adaptive cultivars and soil fertility status, diseases, insect pests and weeds. Among these factors, Coffee Berry Disease (CBD) caused by *Colletotrichum kahawae*, Coffee Wilt Disease (CWD) caused by *Gebberella xylarioides* and Coffee Leaf Rust (CLR) caused by *Hemileia vastatrix* are the three most important and destructive diseases threatening coffee production in Ethiopia. However, CBD is the most serious problem of coffee production commonly prevailing in many coffee growing regions including Southern parts of the country. The disease cause a national yield loss estimated to 25-30%, but on susceptible cultivar under favorable environment, 100% loss could occur [12,13].

Survey data are useful to gain insight in to the occurrence, distribution and relative importance of diseases [14]. In Southern Ethiopia, especially in the Borena and Guji areas, such data, under varying small-scale production situations and systems are lacking. Understanding the association of disease intensity and different management practices will help to identify the most important variables and focus efforts in developing integrated and sustainable management package [14].

Therefore, this study was proposed with the following objectives:

- To determine incidence, severity and prevalence of CBD, and
- To investigate the characteristics of *C. kahawae* and other fungal pathogens associated with coffee berries.

MATERIALS AND METHODS

The study consisted of two components viz. disease survey to assess occurrence, distribution and relative importance of Coffee Berry Disease, and laboratory test to characterize *C. kahawae* and other fungal species associated with coffee berries.

DISEASE SURVEY

DESCRIPTION OF THE SURVEY AREAS: The survey of Coffee Berry Disease (CBD) was conducted in three major coffee growing districts (Abaya, Bule Hora and Kercha) of the Borena and Guji Zones during the 2012 summer cropping season. The Abaya and Bule Hora districts are in the

Borena Zone of Oromia Regional State while Kercha is in the Guji. Relatively the areas differ in their land features and cropping systems (**TABLE 1**). A total of 27 fields (9 from each district) were inspected for disease intensity from 20-25 September 2012 (**TABLE 2**). All sample fields belonged to small, private farmers. Each field was visited once.

Table 1: Characteristic features of the surveyed areas.

Zone	District	Altitude (m.a.s.l)	Rainfall (mm)	Temperature (°C)	Cropping system
Borena	Abaya	1200-2060	900-1400	16-24	Varied
	BuleHora	1300-2600	500-800	15-35	Sole
Guji	Kercha	1450-2300	500-1800	15-32	Varied

Table 2: Mean incidence, severity and prevalence of CBD and other diseases of coffee plant in surveyed areas during 2012 cropping season.

District	Altitude (m.a.s.l)	CBD		Disease prevalence (%)		
		Incidence (%)	Severity (%)	CBD	CW	CLR
Abaya	1630	53.3	21.4	100.0	33.3	11.1
BuleHora	1950	42.2	8.0	100.0	55.6	0.0
Kercha	1875	52.2	14.9	100.0	33.3	11.1
Mean		49.3	14.8	100.0	40.7	7.4

CBD: Coffee Berry Disease, CW: Coffee Wilt, and CLR: Coffee Leaf Rust.

DISEASE AND CROP ASSESSMENT: From each of the three selected district, three coffee growing peasant associations were selected. From each of the three peasant associations, three coffee fields were randomly selected at intervals of 1-2 km along the main and accessible rural roads and assessed for the disease survey. The assessment was conducted on a total of 10 randomly taken plants within each field. Random selection of the plants for assessment was done by moving in 'X' fashion within the field by taking plants at a distance of about 6-8 m intervals based on the size of the field. Sample size consisting of all the berries in four simple branches of 4-6 plants was reported to have least coefficient of variation in a plot of 100 plants [15]. Based on these, stratified sampling i.e. three pairs of branches from each upper, middle and lower canopy layers of the coffee plant were selected to assess CBD severity. The occurrence and severity of CBD was recorded from each sample plants.

The incidence of the disease within each coffee field was then determined as the percentage of diseased plants out of the 10 assessed plants. Likewise, CBD severity was computed as the average of 10 plants by estimating the percentage of diseased berries (damaged barriers over on all barriers of sampled branch times 100) from each of the plants assessed. It was rated using standard disease scales of 0-6; where, 0: 0%, 1: ≤2%, 2: 2-5%, 3: 6-10%, 4: 11-50%, 5: 51-99%, and 6: ≥99% of diseased berries. The scores were changed into Percentage Severity Index (PSI) for the analysis using the formula of Wheeler.

$$PSI = \frac{Sum\ of\ numerical\ rating}{Number\ of\ plant\ rated\ x\ \max imum\ score\ of\ the\ scale} \times 100$$

Disease prevalence was determined as the proportion (%) of total field assessed for each district. For each field assessed, other diseases of coffee, variety type, and plant age and growth habit were noted (**TABLE 2** and Appendix **TABLE 1**). The experiment was structured as nested design (tree was nested under location).

STATISTICAL ANALYSIS: The mean incidence and severity of CBD were calculated for each district, peasant association and field. Data were then subjected to nested random effects analysis of variance. The Statistical Analysis System (SAS) Version 9.2 software [16] was used. Significance of associations between incidence and severity on the occurrences of CBD intensity in the plantation coffee populations were examined using simple correlation analysis. Determined Pearson correlation coefficients were used as indices for strength of the association.

LABORATORY EXPERIMENT

This experiment was carried at Plant sciences laboratory (Haramaya University) to characterize *C. kahawae* and other fungal pathogens associated with coffee berries. In this study a total of 27 samples of coffee beans, each containing about 50-100 berries were collected from coffee fields of the surveyed areas (i.e. one sample of coffee beans per field, 100 gm per sample).

FUNGAL ISOLATION FROM COFFEE BERRY: The collected coffee berry samples were premature green berries and partially ripe beans with slightly sunken and dark brown lesions. The diseased berry samples were surface disinfested with Clorox solution of 70% for three minutes three times consequently and then rinsed with sterilized distilled water twice for 1 min. Totally 15 coffee berries were tested from each sample, five coffee beans by three replications per sample was plated on the prepared media of Potato Dextrose Agar (PDA) and made under 25°C for 5-7 days. Frequency of infected and non-infected berry was taken, the contaminated fungal isolates were counted and the proportions of sprouted fungal species were examined as:

$$\%\ \text{fungal species} = \frac{\text{Number of isolates fungi}}{\text{Total number of sprouted isolates fungi}} \times 100$$

TRANSFERRING AND ISOLATION OF *C. kahawae* FUNGI: This was done in sterile Biological Safety Cabinets (laboratory hood) in order to avoid contaminations of other air borne pathogens. All coffee berries which showed contaminations of fungal growth was counted and recorded as the frequency per location. The fungal colonies of *C. kahawae* and other fungal pathogens species were sub-culturing or transferred to new freshly prepared media of MEA for further identifications by sterilized toothpicks and flamed loops through slightly touching one of the colony and stabbing the center of petriplate according to their species characterizations separately. A new sterile stick was used only for each berry. The name of the isolate and the date were recorded.

IDENTIFICATION OF FUNGAL PATHOGENS: Isolates were grown on the prepared medium and incubated at 25°C for 5-7 days for proper identifications at species level. Isolates were examined for conidial morphology, conidial lengths, colony size, shape, colors and by their textures (smooth and/or rough). Colony and conidial characters were recorded after 6 to 7 days of growth at 25°C in darkness to induce sporulation. Samples of conidia were taken and mounted in water. The conidial shape was observed under a light microscope (Leica Microsystems) and photographs were taken with a Digital camera. For each isolate, length and width of 100 randomly chosen conidia were measured through calibrating the ocular and stage micrometer of microscope by using the image analysis program Leica Qwin Standard (Leica Microsystems) of 40 × objectives. Then identified isolates cutting out plugs from the plates using sterilize plunger in to vials containing 2.5 ml of sterile distilled water. The vials were tightly closed, correctly labeled and stored at temperature of between 4-8°C pending for any further mycological analysis.

DATA ANALYSIS: Data on the frequencies of coffee berry infection by *C. kahawae* and other associated fungal species for samples collected from different districts of peasant associations across the farmer's fields were analyzed by SPSS 16.0. Ink models [17]. The proportions of infected berry by each fungal colony were also determined. All data on identifications and characterizations of *C. kahawae* and other fungal pathogens were analyzed using descriptive statistics.

RESULTS AND DISCUSSION

IMPORTANCE OF CBD IN BORENA AND GUJI ZONES

INCIDENCE AND SEVERITY OF CBD: A total of 27 fields were surveyed in three districts of Borena and Guji Zones. Disease incidence of CBD in the fields was computed for peasant associations, districts and the entire study areas. The indicated analysis shows that CBD incidence among peasant associations and districts appeared similar (Appendix **TABLE 2**). Mean incidence ranged from 42.2% in Bule Hora fields to 53.3% in Abaya. The overall mean incidence at district level was 49.3% within the surveyed areas (**TABLE 2**).

Like disease incidence, severity of CBD in the fields was also computed for peasant associations, districts and the entire study areas. There were no observed significant variation in disease severity among the surveyed peasant associations and districts (Appendix **TABLE 3**). Mean severity ranged from 8.0% in Bule Hora fields to 21.4% in Abaya. The overall mean severity at district level was 14.8% within the surveyed areas (**TABLE 2**). Generally, this level of disease intensity might be due to the use of homogeneous coffee cultivars and crop management practices in the surveyed areas (Appendix **TABLE 4**). The spread and scale of epidemics was suggested to be due to the use of a limited number of varieties and the predominant self-fertility of *C. arabica* and uniform cultural practices over a long period. The importance of CBD might also be judged from the high cost incurred to control the disease using chemicals [13]. Therefore, the use of locally adapted and high yielding coffee cultivars with high level of resistance to CBD is of paramount importance in Arabica coffee production in different regions.

Table 3: Frequency of fungal species from coffee berry samples across the districts.

Districts	*C.kahawae*	*F.lateritium*	*Phoma* **spp.**
Abaya	35.33 ± 2.082	5 ± 4.6	1 ± 1.73
BuleHora	36 ± 3	2 ± 3.46	0 ± 0
Kercha	27.67 ± 6.81	5.33 ± 1.16	1 ± 1.73
Total	33 ± 5.57	4.11 ± 3.33	0.67 ± 1.32
Means ± Standard deviations.			

Table 4: Percentages of infected berry by fungal isolates from those three districts.

Zones	**Districts**	**Fungal species**	**% of berry contaminated**
Borana	Abaya	*C. kahawae*	85.7
		F. lateritium	11.9
		*Phoma*spp.	2.3
	BuleHora	*C. kahawae*	89.2
		F. lateritium	10.8
		Phoma spp.	0
Guji	Karcha	*C. kahawae*	81.8
		F. lateritium	15.2
		*Phoma*spp.	3

PREVALENCE OF CBD AND OTHER DISEASES OF COFFEE PLANT: The computed prevalence of CBD was 100% in each of the surveyed district with an overall mean of 100%. In addition to CBD, prevalence of other coffee diseases, such coffee wilt (caused by *Gibberlia xylorioids*) and Coffee Leaf Rust (caused *Hemileia vastatrix*) were recorded for each district. The overall mean prevalence of coffee wilt and leaf rust were 40.7 and 7.4%, respectively, within the entire study areas (**TABLE 2**). Previously it was also reported that CBD, coffee wilt and Coffee Leaf Rust are the three major economically important diseases that limit coffee production in Ethiopia [12,13].

Simple linear correlations analysis between CBD incidence and severity using mean values revealed the existence of significant ($P<0.001$) interrelationships. Significant positive correlation was exhibited between CBD incidence and severity ($r=0.49$; $P=0.001$). Considering incidence as predictor variable and severity as response and fitting linear model explained 39.0% of the variations.

FREQUENCY OF COFFEE BERRY CONTAMINATED AND FUNGAL OCCURRENCE

A total of 27 samples were collected from 27 farmers' coffee fields in three districts of Borena and Guji Zones. The frequency of coffee berry contamination with the fungal pathogens across the districts within the peasant associations shows that the minimum proportions of infected berry ranges from 24% in samples from Kercha to the maximum of 42% from Abaya and Kercha districts, whereas the proportions of noninfected berry ranges from 3% to samples from Abaya, Bule Hora and Kercha districts up to 21% from Kercha district (**FIGURE 1** and Appendix **TABLE 5**).

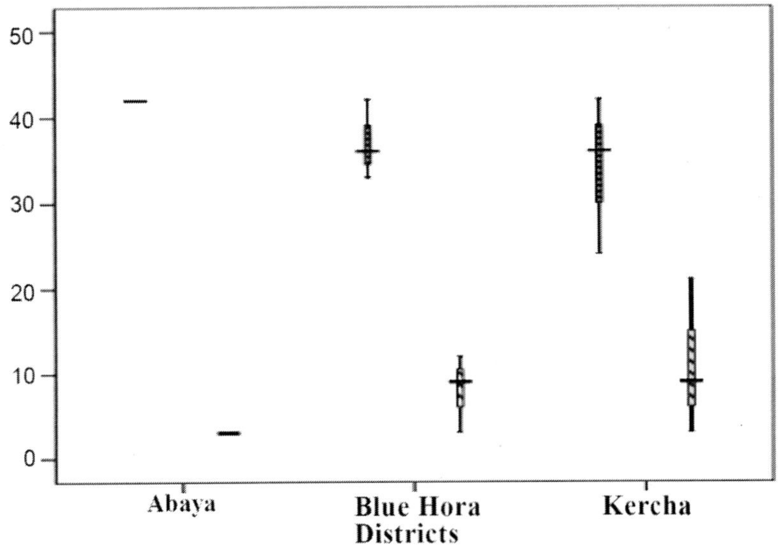

Figure 1: Frequency distributions of infected and non-infected coffee berry contaminations with fungal pathogens (N=45) in the study areas.

This study confirm that the coffee berry samples collected from Abaya was highly contaminated compared to Bule Hora and Kercha districts. It was infected more than 15% and 24% from Bule Hora and Kercha districts, respectively while sample from Bule Hora was contaminated 9% more than Kercha district. This may be due to several factors like; common coffee variety which are grown in the surveyed areas, altitude of the district, adopted cultural practice, weather conditions of the area, presence of the fungal races and other factors, the assortments of such scenario may create a favorable condition for the contamination of coffee berry with the diseases.

Coffee tree as a monoculture is harbor for several fungal disease where ever it was grown such as, Coffee Leaf Rust (*Hemileia vastatrix*), Coffee Berry Diseases (*C. kahawae*), black rot (*Corticium koleroga*), brown eye spot (*Cercospora coffeicola*), anthracnose (*C. gloeosporioides*), American leaf spot (*Mycena citricolor*), Coffee Wilt Diseases (*Gibberella xylarioides*), dieback (*Phoma species*) and pink disease (*Contricium salmonicolor*) [18]. However, in Ethiopia coffee (*Coffea arabica* L.) plants are highly infected by numerous diseases that reduce its production and productivity significantly, those are; CBD, rust, vascular wilt, bean darkening, root rot, damping off and others, among these diseases CBD is the major top diseases by causing a yield loss of 24-30%, but sometimes it reduce yield loss of 50-100% where there is no any management options [19].

In the current study, based on the morphological characteristics (macroscopic and microscopic cultures induces) three different genera of fungal species were crop up from the tested samples of coffee berries. There are *C. kahawae, F. lateritium* and *Phoma* species (**TABLE 3** and **FIGURE 2**). The basic keys for identification of these fungi were based on the growth rate, colony culture and shape and types of conidia (lengths of conidiophores) according to [12,20]. (The distinct features of each fungal pathogen isolated from coffee berries were mentioned under section 3.3). When all collected coffee berry samples were analyzed at district levels the frequency distributions of different fungal species isolated from the diagnosed beans were significantly varied at ($p<0.05$) (ANOVA table, Appendix **TABLE 6**).

The proportions of fungal species from collected coffee berry samples in across the districts of two zones from the infected beans were as follows (**TABLE 4**).

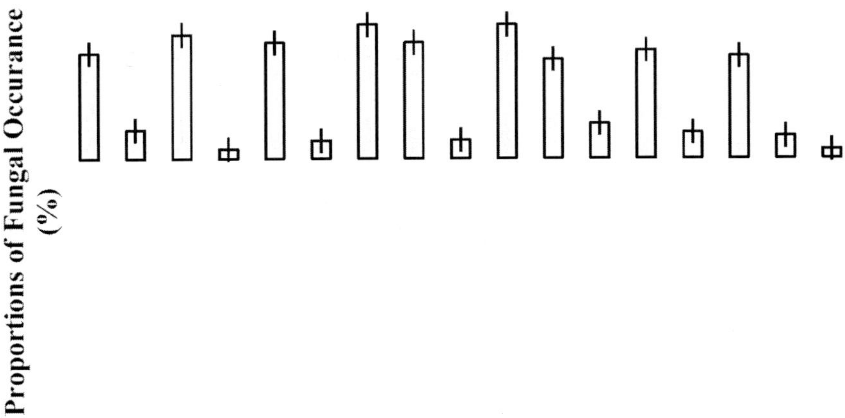

Figure 2: Proportions of fungi isolated from coffee berry across the districts with in the PAs of the two zones of surveyed districts. Abaya; Gangua, Foge and Hafursa Waro. Bule Hora; Saraji Elabada, Kilenso Makanisa and Burka Ebala. Kercha; Guracha Jeldu, Banko Guduba, and kercha Inshe.

In this result, isolates of *C. kahawae* species was highly contaminated to coffee berry samples obtained from all over the surveyed areas. The proportions of isolates ranged from 81.8% of samples from Kercha to 89.2% of Bule Hora districts, while other *Colletotrichum* species were not isolated from the diagnosed berries. The samples from Bule Hora and Kercha districts were highly contaminated by CBD (*C. kahawae*) followed by *Fusarium* species of *F. lateritium* isolates, which was ranged, from 10.8% of samples from Bule Hora to 15.2% of Kercha district. Both *C. kahawae* and *F. lateritium* species were isolated from samples obtained throughout the surveyed regions. In contrary to this, other pathogen also known as *Phoma* spp. was less contaminated to coffee berry samples collected from Abaya with the proportions of 2.3% and Kercha districts 3% and not isolated to samples from Bule Hora district. Perhaps the environmental conditions and agronomic practices of Bule Hora district does not permit for the pathogens, may be the ecology of Abaya and Kercha districts are encouraging factors for *Phoma* species.

However, the fungal species contaminations to berries from each PA were ranged from 6.7% by *Phoma* species to 100% by *C. kahawae* from diagnosed samples at kebele levels (N=3).

The previous reports on associations of *F. lateritium* species on coffee berry were limited in Ethiopia, whereas species of *F. xylarioides* the wilting pathogens on *Coffea arabica* was first observed in Ethiopia (in Keffa province in 1957) [21]. Such studies were showed that, previously most isolates of Fusarium species were associated with coffee wilt; however, in the current study *F. lateritium* species was isolated from green coffee berry samples. It was supported by some research which stated that, isolates of Clades IIA and IIB of the *F. lateritium* isolated from citrus and green coffee berry in New Caledonia, New Zealand and Philippines [22].

Phoma species was reported as an important disease of coffee die-back which was occurring worldwide [23]. It was also a disease attacking leaves, branches and fruits of fruit crops or (green berries) of coffee plant and mainly occurs in nurseries and field plantations. The pathogen inhabit areas such as; higher altitudes with low temperature, high humidity and cold winds. It needs water for infection [18,23]. In this investigation, isolates of this species was isolated from samples of Abaya and Kercha districts which has higher altitudes with low temperatures accompanied with high humidity and cold winds during the survey and sample collection times, this scenario perhaps create a conducive environmental conditions for the out breaks of *Phoma* species associated to coffee green berry.

There was limited report on contamination of coffee berry by *Phoma* species in our country even in the world; however, in Kenya, there is a report on coffee endophytes after surface-sterilizing healthy leaves, pedicels, stems, and green berries with a mercuric chloride saponin solution, and He diagnosed Coffee Berry Disease, *Colletotrichum coffeanum*, (currently known as *C. kahawae* Waller and Bridge), and species of *Phoma* and *Phomopsis* [24]. This result was agreed with our study because samples collected during the survey were green berries. Another report from Puerto Rico mentioned *Botryosphaeria, Colletotrichum, Guignardia,* and *Xylaria* species as fungal endophytes in coffee plants [25,26] and identified 13 *Penicillium* species as endophytes in *C. arabica, Coffea congensis, Coffea dewevrei* and *Coffea liberica*, including *Penicillium coffeae*, a new species described by [27]. Another report in Ethiopia declared that, *C. kahawae* and *C. gloeosporioides* were the only two species of fungus isolated from coffee tissue of samples collected from Habro and Kuni districts in Hararghe region [28].

IDENTIFICATION OF FUNGAL SPECIES ASSOCIATED WITH COFFEE BERRIESIDENTIFICATION OF *C. kahawae***:** In this study, the first species isolated from the collected coffee berry sample was *C. kahawae*. The major distinction currently separates *C. kahawae* from the other species of *Colletotrichum* was the old colony of *C. kahawae* produce pale yellowish to pinkish with dense whitish-grey aerial mycelium and a few bright orange conidial masses on the tips of the active growing hyphae on MEA media (**FIGURE 3**). The young colony produce grey, becoming grey to dark, olivaceous grey, dark greenish in reverse side of plates. The previous report reveals, a colony of *C. kahawae* isolated from Hararghe coffee shows the same color and ability to form saltation but some colonies demonstrate dark mycelium while other produces grayish mycelium [28].

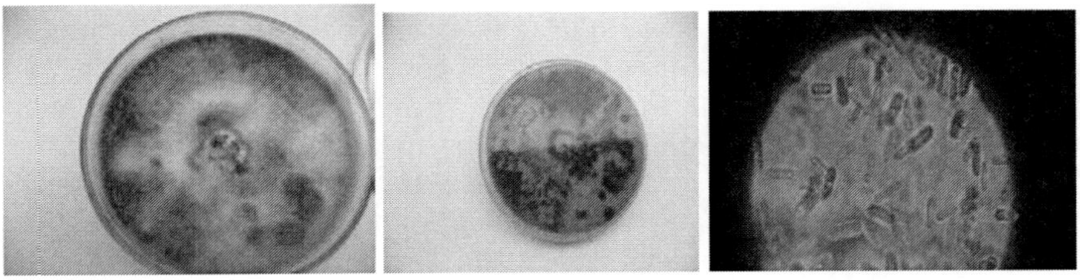

A, Front B, Reverse C, Conidial Features

Figure 3: Species of *C. kahawae* isolated from coffee berries. A and B, colony features and C, conidial features.

The colony had also slow growth rate compared to other species. The cultures produce perithecia ascocarp. The conidial shape was hyaline, straight and generally cylindrical with rounded ends and biguttulates (**FIGURE 3C**). For further confirmations, about 100 conidial of *C. kahawae* were taken from different colonies of 8 days old cultures and measured for the length under light microscopes (*Leicia*), by calibrating the ocular micrometer and stage micrometer (Appendix **TABLE 7**). The length and width of conidia was varying from 7.5-18 × 3-5 µm respectively, this result was exactly agreed with the work of [29,30], but *Colletotrichum fructicola* produces smaller conidia as compared to *C. kahawae* which ranges from 9.7-14 × 3-4.3 µm in length and width respectively.

IDENTIFICATION OF *F. lateritium***:** Based on growth and morphology of isolates, *F. lateritium* was produce colonies of white and light mycelia by both sides of plates on the PDA and MEA media, and later it was become colorless. The report from Uganda demonstrates, *F. lateritium* colony isolates were examined morphologically on PDA and MEA. Colonies on PDA and MEA were white, pale to colorless after 3 days and pale orange 4-14 days later depending on the isolate [22,31]. The orange pigmentation was more intense in the center of the colony and faded to pale orange to white at the growing margin then becomes colorless. The same morphological features of the colony were examined in the current study.

For more description of the species 100 spore containing conidiophores were measured the length by light microscopes (*Leicia*), by calibrating the ocular micrometer and stage micrometer (Appendix **TABLE 6**), and record the lengths, and it was vary from 18-34 µm. *F. lateritium* conidia measuring to 19-42 µm in length according to the number of septa and 3-5 µm in width, sometimes the maximum conidial lengths was up to 82 µm [32].

However, the maximum length very far apart, it likely due to some factors likes media nutrient contents, incubation temperatures or other factors which induces for the growth of the conidiophores.

IDENTIFICATIONS OF *Phoma* SPECIES: *Phoma* species is yet another ubiquitous, cosmopolitan fungus which is commonly found in soils. As a known plant pathogen, it may also be recovered from infected plant materials of different vegetables, fruit crops and green coffee berry. It may be found in the laboratory as an environmental contaminant [33]. However, based on conventional identifications which conducted in the current study, it was a rapidly growing fungus which usually reaches maturity within five days. It was fungal isolates produce dark brown to black mycelia with tiny conidiophores on MEA and PDA media and also not feasible to measure the length and width of conidiophores under light microscopes like another results of *C. kahawae* and *F. lateritium*. In Kenya, report that, green coffee berries were infected by *Phoma* species which demonstrate mycelium colony of brown to dark colors morphologically with fine conidiophores [24]. Another study confirm that, colony colors of *Phoma* species ranged from olivaceous green to dark brown or black, the reverse was brown to dark brown to black with some species producing a diffusible reddish-brown pigment and colony growth rate was similar for all *Phoma* isolates in culture media of Oat Meal Agar (OMA) and MEA with the conidial length ranged from 1.6-3 μm which is very fine and smaller [33].

SUMMARY AND CONCLUSIONS

Coffee plant is a major cash crop, which earns a hard currency for all growing countries on the entire world especially for African developing countries. In Ethiopia, coffee plays a vital role to maintain the livelihoods of small scale farmers and creates a major job opportunity for millions of peoples. It is successfully produced under rain fed conditions in different climatic regions of the country by small scale farmers, private investors, and state enterprises. In spite of the high production potential and economic importance; its production and productivity is constrained by several abiotic and biotic factors. Among biotic factors, CBD is one of the most severe coffee diseases in most coffee-growing regions including Borena and Guji areas of the county, where further information on the occurrence, distribution and relative importance of the disease and its characterization are lacking. Therefore, this study was undertaken to determine the incidence, severity and prevalence of CBD, and investigate the characteristics of *C. kahawae* and other fungal pathogens associated with coffee berries.

In the present work, field surveys were conducted in three major coffee growing districts (Abaya, Bule Hora and Kercha) of Borena and Guji Zones of Oromiya Region by examining and recording of several coffee diseases. The survey revealed that CBD was prevalent in all the surveyed districts. The overall mean incidence and severity at district level were 49.3 and 14.8%, respectively. Perhaps, this spread and scale of epidemics could be due to the use of a limited number of varieties and uniform cultural practices in the entire study areas.

Three fungal pathogens (*C. kahawae, F. lateritium* and *Phoma* spp.) were isolated from coffee berries by diagnosing the berries based on conventional methods (cultural and morphological features). A molecular technique was not conducted for this work, since it needs higher costs. Among the three fungal pathogens, the minimum and maximum proportions of occurrences were recorded from *C. kahawae* isolate (81.8-89.2%) followed by *F. lateritium* (10.8-15.2%) and *Phoma* spp. (2.3-3%) from all the districts, whereas the proportions of infected were ranged from 6.7-100% in the field levels. In general, the present study revealed high occurrence and distribution of CBD in the study areas. Thus, empirical assessment of the amount of yield loss due to CBD on most grown variety in the areas and developing or introducing resistant varieties need to have high priority.

ACKNOWLEDGEMENT

We would like to acknowledge the assistances of Borena and Guji Zones Agricultural Office for their kind collaboration in providing necessary information during survey work. The authors also thank Haymanot Bezuneh (Haramaya University, Plant Pathology Lab.) for her assistance in laboratory works. Our special appreciation and deepest thanks go to Bule Hora University for funding the study.

REFERENCES

1. Anthony F, Bertrand B, Quiros O, Lashermes P, Berthaud J,et al. (2001) Genetic diversity of wild coffee (Coffeaarabica L.) using molecular markers. Euphytica 118: 53-65.
2. Woldemariam T, Denich M, Teketay D,Vlek PLG (2002) Human impacts on Coffeaarabica L. genetic pools in Ethiopia and the need for its in situ conservation. In: Engels JMM, Rao VR, Brown AHD, Jackson MT (eds.) Managing Plant Genetic Diversity. AB International/IPGRI, Rome. pp. 237-247.
3. Pendergrast M (2010) Uncommon grounds: the history of coffee and how it transformed our world. Basic Books, New York, USA

4. Cannell MGR (1983) Coffee. Biologist 30:257-263.
5. Oxfam (2002) Bitter coffee: how the poor are paying for the slump in coffee prices. Oxfam International, London.
6. FAO/WFP (2010) Special report FAO/WFP crop and food supply assessment mission to Ethiopia.
7. EEA (Ethiopian Economic Association) (2001) Annual Report on the Ethiopian Economy.
8. Mayne R, Tola A, Kebede G (2002) Crisis in the birth place of coffee, Oxfam International research paper, Oxfam International.
9. Workafes WT, Kassu K (2000) Coffee production systems in Ethiopia. Proceedings of the workshop on the control of coffee berry disease (CBD) in Ethiopia. (GhionHotel), 13-15 August 1999, Addis Ababa Ethiopia, pp. 99-106.
10. Taye K, Tesfaye S (2002) Organic coffee production: hope for small-scale farmers in Ethiopia. Proceedings of the 19th International Conference on Coffee Science (ASIC), May 14-18, 2001, Trieste, Italy.
11. Teshome A, Kebede E, Kebede K (2008) Coffee Development and Marketing Improvement Plan. In: Proceedings of A National Work Shop Four Decades of Coffee Research and Development in Ethiopia. 14-17 August 2007, EIAR, Addis Ababa, Ethiopia, pp. 375-381.
12. BiratuT (1995) Studies on Colletotrichum population of Coffeaarabica L. 74 in Ethiopia and evaluation of the reactions of coffee germplasms. Doctoral Dissertation. HohenLandwirtschaftlichenFakultat der Rheinschen Friedrich-Wilelms-Universit_tZu Bonn, p. 231.
13. Eshetu D, Teame G, Girma A (2000) Significance of minor coffee diseases of Coffeaarabica L. in Ethiopia; a review. Proceedings of the workshop on control of coffee berry disease in Ethiopia, 13-15 August 1999, Addis Ababa, Ethiopia, pp. 58-64.
14. Rusuka G, Buruchara RA, Gatabazi M, Pastor-Corrales MA (1997) Occurrence and distribution in Rwanda of soil borne fungi pathogenic to the common bean. Plant Disease 8:445-449.
15. Kushalappa AC (1989) Biology and epidemiology. In: KushalappaAC, Eskes AB (eds.) Coffee rust: Epidemiology, Resistance and Management; CRC Press Inc, Florida, pp. 13-80.
16. SAS (Statistical Analysis System) software (2009) Version 9.2.Inc. Carry, North California, USA.
17. Peck R (2005) SPSS Manual for Introductory Statistics.
18. Anna-Elisabeth (2005) Fungalendophytes diversity in coffee plants from Colombia, Hawai'i, Mexico and Puerto Rico.
19. EARO (Ethiopian Agricultural Research Organization) (2003) Facilitating the Implementation and Adoption of Integrated Pest Management (IPM) in Ethiopia. Melkassa Agricultural Research Center. Planning Workshop 13-15 October 2003. Economic botany 12: 111-139.
20. Gassert WL (1979) Research on coffee berry disease in Ethiopia: Epidemiology and control. GTZ, Eschborn, p. 65.
21. Kranz J, Mogk M (1973) Gibberellaxylarioides Heim et Saccas on arabica coffee in Ethiopia. Phytopath Z 78: 365-366.
22. Gerlack W, Nirenberg HI (1982) The Genus Fusarium-a Pictorial Atlas. Berlin-Dahlem: Paul Parey,p.406.
23. Lutzeyer HJ, Pülschen L, Compart W, Scholaen S (1994) Recent developments in plant protection for plantation crops as exemplified by coffee" bmz, p.144.
24. Rayner RW (1948) Latent infection in Coffeaarabica L. Nature 161: 245-246.
25. Santamaría J, Bayman P (2005) Fungal epiphytes and endophytes of coffee leaves (Coffeaarabica). MicrobEcol 50: 1-8.
26. Vega FE, Posada F, Peterson SW, Gianfagna TJ, Chaves F (2006) Penicillium species endophytic in coffee plants and ochratoxin A production. Mycologia 98: 31-42.

27. Peterson SW, Vega FE, Posada F, Nagai C (2005) Penicilliumcoffeae, a new endophytic species isolated from a coffee plant and its phylogenetic relationship to P. fellutanum, P. thiersii and P. brocae based on parsimony analysis of multilocus DNA sequences. Mycologia 97: 659-666
28. Biratu T, Hulluka M (1989) Colletotrichum species associated with coffee berry disease in Hararge. EthJ AgriSci 11: 1-6.
29. Waller JM, Bridge PD, Black R, Hakiza H (1993) Characterization of the coffee berry disease pathogen, Colletotrichumkahawae sp. nova. Mycological Research 97: 989-994.
30. Prihastuti H (2009) Characterization of Colletotrichum species associated with coffee berries in northern Thailand. Fungal Divers 39: 89-109.
31. Geiser DM, Lewis Ivey ML, Hakiza G, Juba JH, Miller SA (2005) Gibberellaxylarioides (anamorph: Fusariumxylarioides), a causative agent of coffee wilt in Africa, is a previously unrecognized member of the G. fujikuroi species complex. Mycologia 97:191-201.
32. Booth C (1971) The Genus Fusarium. Surrey, England: Commonwealth Mycological Institute,p. 237.
33. Boerema GH, JdeGruyter ME, Noordeloos, Hamers MEC (2004) Phoma Identification Manual: Differentiation of Specific and Infra-specific Taxa in Culture. CABI Publishing. Wallingford, Oxford shire, UK.

CHAPTER 12

Fast Growing Plantations for Wood Production – Integration of Ecological Effects and Economic Perspectives

Michael Bredemeier [1*], Gerald Busch[2], Linda Hartmann[1], Martin Jansen[1], Falk Richter[1] and Norbert P. Lamersdorf[1]

[1]*Forest Ecosystems Research Section, Center for Biodiversity and Sustainable Land Use (CBL), University of Göttingen, Göttingen, Germany*

[2]*BALSA Service for Applied Landscape Ecology and Scenario Analysis, Göttingen, Germany*

ABSTRACT

Biomass crops are perceived as a feasible means to substitute sizeable amounts of fossil fuel in the future. A prospect of CO_2 reduction (resp. CO_2 neutrality) is credited to biomass fuels, and thus a potential contribution to mitigate climate change. Short rotation coppices (SRCs) with fast growing poplar and willow trees are an option for producing high yields of woody biomass, which is suitable for both energetic and material use. One negative effect that comes along with the establishment of SRC may be a decrease in groundwater recharge, because high rates of transpiration and interception are anticipated. Therefore, it is important to measure, analyze, and model the effects of SRC-planting on landscape water budgets. To analyze the effects on the water budget, a poplar SRC plot was studied by measuring hydrological parameters to be used in the hydrological model WaSim. Results reveal very low or even missing ground water recharge for SRC compared to agricultural land use or grassland, especially succeeding dry years. However, this strong effect on plot level is moderated on the larger spatial scale of catchment level, for which the modeling was also performed. In addition to water, nutrient fluxes and budgets were studied. Nitrogen is still a crucial issue in today's agriculture. Intensive fertilization or increased applications of manure from concentrated livestock breeding are often leading to high loads of nitrate leaching, or enhanced N_2O emissions to the atmosphere on arable crop fields. SRC or agroforestry systems on former crop land may offer an option to decrease such N losses, while simultaneously producing woody biomass. This is mainly due to the generally smaller N requirements of woody vegetation, which usually entail no need for any fertilization. The trees supply deep and permanent rooting systems, which can be regarded as a "safety net" to prevent nutrient leaching. Thus, SRC altogether can help to diminish N eutrophication. It is important to offer viable and attractive economic perspectives to farmers and other land managers besides of the potential ecological benefits of SRCs. For this reason, an integrated tool for scenario analysis was developed within the BEST project ("BEAST – Bio-Energy Allocation and Scenario Tool"). It combines ecological assessments with calculations of economic revenue as a basis for a participative regional dialog on sustainable land use and climate protection goals. Results show a substantial capacity for providing renewable energy from economically competitive arable SRC sites while generating ecological synergies.

Keywords: short rotation coppice, energy wood plantations, water budget, nitrogen leaching, integrated landscape assessment

INTRODUCTION

Since fossil fuels are finite and may become exploited to a critical extent in the near future, there are strong efforts to substitute them as much as possible with renewable energies, and as fast as this could be achieved. Among the diverse renewables, energy from biomass plays an important role. Between the various biomass energy carriers, which are available, wood plays an important part, particularly in thermal conversion. Energy wood can be gained from different sources. Classical forestry has a long tradition, but additional options and strategies to harvest enough energy wood for meeting an increasing demand are needed. The most important option are new energy wood plantations with fast growing woody species and very short rotation periods compared to "classical" forestry. These "short rotation coppices (SRC)" are the focus of our present study, as they may have significant effects on habitat provision, biodiversity, hydrology, and biogeochemical cycling in cultural landscapes. Interdisciplinary investigations of SRC effects were facilitated by the joint integrated project BEST ("Bioenergie-Regionen stärken" – Boosting Bioenergy Regions), which ran from 2010 to 2014 and was funded by the German Ministry of Education and Research (BMBF).

Short rotation coppice does not only provide woody biomass, but frequently additional ecosystem services. If SRC should come to cover larger parts of our landscapes in the future, it is important to know how they will limit or augment ecosystem services in comparison to other land use systems. One known shortcoming of SRC are possible negative effects on groundwater recharge (GWR), because higher rates of transpiration and interception evaporation may be expected compared to annual crops (Lasch et al., 2010; Schmidt-Walter and Lamersdorf, 2012). Due to the change in land use and introduction of trees, new factors occur, which influence water fluxes and their interactions at a site. One factor is the all year vegetation cover in combination with a longer growing season in comparison to annual crops (Petzold et al., 2010). As a result, trees are able to develop a deeper root system and therefore exploit additional water and nutrient resources in deeper soil layers. The more intensive root penetration as well as the more extensive management practice are also affecting soil structure and thus hydrological fluxes, resulting in a higher soil infiltration (Wahren et al., 2012) and reduced runoff components. Compared to typical annual crops, tree dominated land use types like SRC have a higher leaf area index (LAI). LAI has a control on transpiration and additionally influences soil evaporation and interception evaporation. Furthermore, the latter is increased by the interception on stem and branch surfaces, especially outside of the vegetation period. The magnitude of these hydrological effects of SRC highly depends on local climate conditions and soil water availability (Dimitriou et al., 2009), determining actual evaporative demand and water supply. Water availability is the main growth limiting factor for poplar and willow plantations (Linderson et al., 2007). Due to these effects, SRC influences the water fluxes at the plantation site, as well as those of the surrounding landscape.

An assessment using hydrological models can help to identify hydrological effects at regional and local scales, e.g., to regional climate and/or to adjacent catchments. The objectives of the hydrological part of this study were to investigate the effects of an enhanced cultivation of SRC on local and regional scale using measurements and the raster-based hydrological model system WaSim. We used our own measurements of plant physiological parameters like LAI, stomata resistance (Rsc), and the length of growing season to calibrate sensitive parameters for hydrological modeling a the site scale. For SRC, such parameters are quite rare in literature, due to the fact that this land use form scarcely came into focus of investigation before onset of the renewable energy discussion during the last years (Surendran Nair et al., 2012).

In a second step of the hydrological study, we transfer our model parameterization to a whole catchment area, in order to investigate the effects of land use change on a regional scale by analyzing model simulations with different fractions of agricultural area converted to SRC.

Hydrological studies of SRC were complemented by biogeochemical investigations. The focus here was on nitrogen, since nitrogen leaching from the soil to aqueous ecosystems and groundwater is a well-known problem in intensive agricultural land use. N fertilization is often causing serious problems with respect to nitrate leaching and N_2O emissions in today's conventional agriculture (Pappa et al., 2011; Snyder et al., 2014). The main reasons for such "N-leaking" are enhanced N doses applied to reach economically attractive production targets of crops, and the inability to exactly synchronize N needs for plant growth with the time frames of N applications (Di and Cameron, 2002). Furthermore, calculations of N demands are related to the needs of sufficient plant growth rather than to the capacities of soils to intermediately store excessive loads of N from fertilization or crop residue mineralization. Thus, most N leaching can be expected on rather poor and light soils, where high production targets should be accomplished by high inputs of N fertilizers. In addition, the persistent use of heavy agricultural machinery is often causing serious soil compaction on intensively used crop fields which is, together with a surplus of nitrate, favoring the genesis and emission of N_2O (Ruser et al., 2006).

With respect to the leaching of N, SRC is known to generally reduce the output of nitrate on former cropland due to the smaller net N demand of woody perennials (Dimitriou et al., 2009; Dimitriou and Aronsson, 2011; Schmidt-Walter and Lamersdorf, 2012). Thus the cultivation of SRC, in the form of size restricted monoculture blocks in-between the existing N intensive crop applications, or as strips in the form of windbreaks (alley cropping systems), riparian buffers, or other agroforestry measures may help to restrict the above mentioned negative impacts of large and N intensive monocultures (Tsonkova et al., 2012). However, the specific implementation of SRC as buffer strips for nitrate leaching would be a fairly new application in central European agricultural land-use. In this context, our main goal was to investigate key hydrological conditions and terms of the N cycling for a newly installed SRC plantation with poplar and willow on former cropland. Finally, we were focusing on setting up first N-budgets for those young SRC plantations to better understand and evaluate their performance under initial growth conditions.

Perennial crops like SRC can effectively reduce the risk of water erosion, since they provide a soil surface cover of living vegetation and plant detritus. Surface runoff is considerably reduced and infiltration is supported due to an extensive rooting system, humus incorporation into the soil, and lower management frequency (e.g., harvest operations; Kort et al., 1998; Scholz et al., 2001; Deumlich et al., 2006). Since soils in the district of Göttingen are very exposed to water erosion (Mosimann et al., 2009), owing to their soil texture characteristics and the hilly landscape, SRC could play a substantial role in erosion prevention.

Ecological effects are an important concern with respect to establishment of new SRC. However, economic questions and considerations are also essential. Farmers will never adopt SRC cultivation if they do not perceive this land use form as an economically sound and promising solution. It is therefore highly important to offer viable and attractive economic perspectives to farmers and other land managers besides of the potential ecological benefits of SRCs. To address this issue, an exemplary scenario study was carried out focusing on two major objectives: (a) to derive the maximum amount of economically competitive primary energy production from SRC that could be generated from arable land in the municipality of Friedland (78 km^2, southern district of Göttingen), and (b) to analyze whether ecological synergies in terms of erosion protection could be achieved with an economically guided site selection.

"Economically competitive" means that SRC has to generate a positive annual economic return (annuity) compared to the reference crop rotation of wheat–oilseed rape–barley within a 20-year time period 2011–2030 (Forstliche Versuchs- und Forschungsanstalt Baden Württemberg (FVA), 2012).

MATERIALS AND METHODS

Description of Study Site, Catchment, and Measurements

Short Rotation Coppice research plantations was established in the Göttingen district, Central Germany. The site considered here is located next to the village Reiffenhausen (51.67°N, 10.65°E, 325 m a.s.l.), and is named "Reiffenhausen" accordingly.

The soil substrate at the Reiffenhausen site is characterized by sedimentary deposits of Triassic Sandstone material, partly mixed with clay stone material, and covered by loess sediments. The soil types are stagnic cambisols (pseudovergleyte Braunerde) and haplic stagnosols (Pseudogley). The soil texture is dominated by a loamy sand or silty clay fraction [for further details see also Hartmann et al. (2014)]. The SRC plantation was established in March 2011 and consists of two pure SRC plots, one planted with the poplar clone "Max 1" (*Populus nigra* × Populus maximowiczii), hereafter called P-SRC, and another one planted with the willow clone "Tordis" [(Salix viminalis × Salix schwerinii) × Salix viminalis)] hereafter called W-SRC.

The P-SRC plot was planted with 0.2 m long poplar cuttings on 0.4 ha area in double row spacing system, comprising alternating inter-row distances of 0.75 and 1.50 m, and a distance of 1.0 m within the rows. The overall planting density was 8890 cuttings ha^{-1}. The W-SRC plot has a dimension of 0.6 ha and a planting density of 11,850 cuttings ha^{-1}. Willow clone "Tordis" cuttings of 0.2 m length were planted also in double rows with the same inter-row distance as poplar, but the distance within the rows was only 0.75 m.

In 2013, the most intensive measurements and investigations took place. Measured parameters were LAI, stomatal resistance (Rsc), plant height, and diameter at breast height (DBH), both as annual development, rooting depth, as well as start and length of the growing season. Besides these plant characterizing parameters, also local climate and soil physical parameters were recorded. A detailed description of the investigation of these values is given by Richter et al. (2015); information about soil chemistry and biomass information is given by Hartmann et al. (2014). According to the meteorological data provided by the German Weather Service (Deutscher Wetterdienst, DWD), for the station Göttingen (DWD Station-ID: 01691), nearest to the study site, the climate is characterized by an average temperature of 9.1°C (±0.7°C), and a mean annual precipitation sum of 635 mm (±122 mm) for the period 1971–2010.

For regional scale modeling, the catchment of the Dramme River is used, which is a small stream southeast of Göttingen, flowing into the Leine River. The catchment has an area of 45 km^2, and elevation is ranging from 198 to 476 m a.s.l. The catchment used here is defined down to the discharge gage Mariengarten (Niedersächsischer Landesbetrieb für Wasserwirtschaft, Küsten- und Naturschutz (NLWKN), 2013a), where runoff data is available. Half of the catchment area is used for agriculture, 36% is covered by forest, where 23% of the total area is deciduous forest. Grassland covers 9% and the remaining area are settlements or water bodies. This landscape characteristic is typical for central Germany (Figure 1). A detailed description is given by Richter et al. (in preparation).

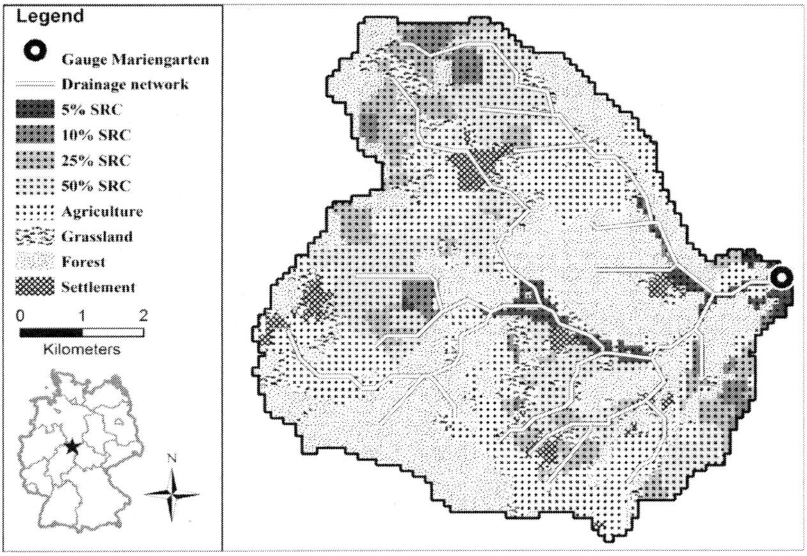

Figure 1. Raster map of the land use composition in the Dramme catchment. Percentages of agricultural area changed to SRC are additive, i.e., 10% SRC includes the 5% SRC areas, etc.

A software tool named "BEAST" was developed within the BEST project ("Bioenergy assessment and scenario tool"; Busch and Thiele, 2015). The BEAST scenario application referred to the municipality of Friedland (Figure 2), which is located in the southern parts of the Göttingen district. The municipality is dominated by arable land use, covering 54% of the total area (76 km^2). Natural growth conditions are suitable for SRC, given an average annual precipitation of around 700 mm, a mean annual temperature of 8.5°C (1981–2010 - derived from DWD 1 km grid information; DWD German Weather Service, 2013), and a majority of medium to high productive soils.

Hydrological Modeling Approach

To investigate the effects of an enhanced cultivation of SRC on hydrology, we applied a three-step approach. The first step was the measurement and derivation of model relevant parameters, especially describing the land use form "poplar SRC" on the plot level. The measurements of plant physiological parameters and a sensitivity analysis of how different vegetation parameterizations can influence hydrological model results are presented in Richter et al. (2015).

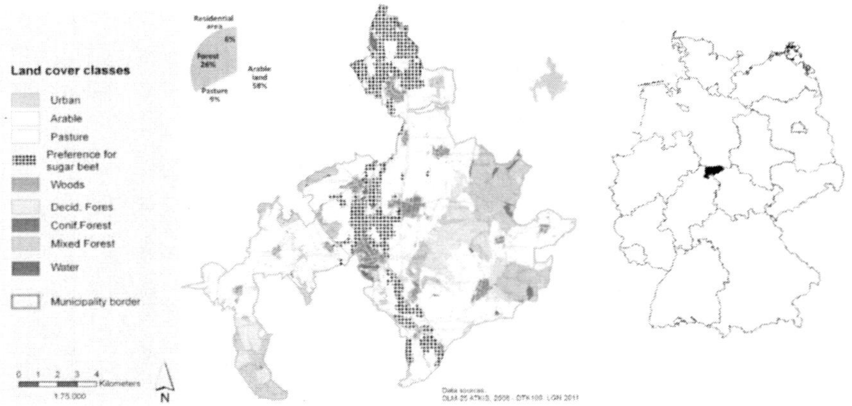

Figure 2. Landscape mosaic in the municipality of Friedland including preference sites for sugar beet–wheat rotations that are excluded from the scenario assessment.

In the second step, WaSim was applied at the plot scale to Reiffenhausen P-SRC, where measurements were done to calibrate and evaluate the model using independent measurements, i.e., soil moisture quantities in our case. This plot model approach of WaSim was also used to perform long-term simulation to study the effects of climate variability as well as different land use and soil types on model results (Richter et al., 2015). The third step was the investigation of hydrological effects of SRC cultivation on the regional scale, by example of the small catchment area of the river Dramme. Here, the parameterization for poplar SRC, derived on the plot level, is used together with other land use types to investigate local and regional changes of hydrological properties for different scenarios of SRC cultivation. Scenarios were selected for different area fractions converted from agriculture to SRC, i.e., 5, 10, 25, and 50% (Figure 1). The rationale behind taking the conversion steps of 5–25% was to reflect biomass supply goals of existing policy scenarios (Landkreis Göttingen, 2013). For hydrological reasons, i.e., to create a strong hydrological signal, a simulation comprising 50% of converted arable land was carried out. The selection process of the most suitable arable sites to be converted to SRC built on a site-specific comparison (performed with the help of BEAST, see below) between yield levels of annual crops (reference crop rotation of oilseed-rape, wheat, and barley) and a poplar SRC (Max1 clone, 5-year rotation). Yield levels were derived from a statistical modeling approach applied to the Göttingen district, which in turn built on field experiments in Lower Saxony and Thuringia (Busch and Thiele, 2015). By running a MIN–MAX optimization routine, sites were selected that ideally show both, comparatively low average agricultural crop yields and at the same time sufficiently high average annual SRC yield levels, respectively.

N Budget Measurements and Calculations

To estimate the initial N-budget as well as nitrogen use efficiencies of the growing SRC plantations, the net-N-mineralization rates (NNM), plus the atmospheric deposition (AD) were taken as input terms. The net N uptake by trees (NTU), i.e., the amount of N which would leave the system through harvesting measures (including bark), as well as the current seepage water output of nitrate (NSWO) were taken as proxies for the output term. The rates of N_2O emissions, determined by Walter et al. (2012) through a parallel running project, were only recorded on the P-SRC plot but were also taken for the budget calculations of the W-SRC plot. For NNM, the buried bag method according to Eno (1960) and Gasche et al. (2002) was applied. Given rates are means of eight sampling dates between March and October 2012. During each 6-week incubation period, air-permeable and watertight PE-bags were buried in 15 cm soil depth.

NNM in [mg kg^{-1} d^{-1}] was calculated as follows:

$$NNM = \frac{(Ae+Ne) - (Ai+Ni)}{t} \quad (1)$$

where Ae, respectively Ne, is the amount of K_2SO_4 (0.5 mol L^{-1}) extractable ammonium (Ae) and nitrate (Ne) at the end of the incubation period, Ai and Ni are respective amounts at the beginning of the incubation period, and *t* denotes the applied incubation period in days [d]. All samples were transported and stored in a cold chain, and were immediately extracted in the lab after the field sampling. NH+4NH4+ and NO−3NO3− were determined by using continuous flow injection calorimetry (Cenco/Skalar Instruments, Breda, the Netherlands).

Atmospheric deposition was measured bi-weekly with four replicated bulk samplers per plot between November 2011 and October 2013. NTU was estimated by determining total N concentrations in 15 shredded trunks per plot. After drying at 60°C and grinding the plant material to a fine powder, total N was analyzed with an automated CHN-O-analyzer (VarioEL, Elementar, Germany). Subsamples were also dried at 105°C to determine the remaining water content, and reported results are related to the absolute dry weight. For the estimation of the above-ground biomass production, the DBH (1.3 m) was determined for each plot at randomly selected trees (n = 240). Those measurements were applied after the second growth period, i.e., in winter 2012/13. Fifteen trees per plot were harvested at 10 cm above soil level and the fresh-weight was determined *in situ*. After dry-weighing to constant mass (105°C), allometric equations given by Röhle and Skibbe (2012) were used to estimate the dry mass of trees. Finally, these estimates were scaled up with the number of survived trees to calculate dry matter (DM) yields per hectare [for further details on biomass estimates, see also Hartmann et al. (2014)]. The NSWO was determined by (i) measuring monthly nitrate concentrations in the soil solution from four replicated suction lysimeter cups per plot in 60 cm soil depth between December 2011 and November 2012, and (ii) multiplying monthly mean nitrate concentrations with respective soil water fluxes, gained from the modeling activities described in the section above. N_2O emissions were recorded on the P-SRC plot between October 2011 and October 2013 (Walter et al., 2012). Furthermore, the N-uptake efficiencies (NUP-E) as well as N-utilization efficiencies (NUT-E) were calculated according to Moll et al. (1982). NUP-E is the rate of tree N uptake per sum of N inputs [NTU/(NNM + AD)], whereas NUT-E is the rate of biomass produced per unit N uptake by trees (DM/NTU). The respective units for those efficiencies are [kg kg^{-1}] but may be also given as [%] for NUP-E.

Model Scenario Setting for a 20-Year Period (2011–2030) Case Study Simulation

Site Selection for the Scenario Application

Arable sites that are preferable for a wheat–wheat-sugar beet rotation were identified and excluded from the assessment procedure since sugar beet production will still be subsidized within the next years and the economic return out-competes other crops, including SRC. Thus, by taking soil quality and slope as parameters for site selection, around 883 ha (21% of 4265 ha of arable land) were identified as preference sites for sugar beet rotations in the study area (Figure 2). For the remaining 3381 ha of the arable land, the scenario setting (Table 1) was applied in order to determine both, economic feasibility and options for erosion protection. Spatial modeling and economic calculation took place on the arable site level (i.e., 1109 arable fields).

Further, for reasons of government payments ("NAWARO-Bonus"), only arable sites with an area >0.3 ha and <10 ha were considered. To avoid potentially negative effects from SRC on humid biotopes, protection buffer zones of 200 m were applied to these biotopes. For reasons of nature conservation, nature conservation areas and FFH (Flora-Fauna-Habitat) zones were excluded from SRC allocation. Information on humid biotopes and target areas for nature conservation were taken from datasets provided by Niedersächsischer Landesbetrieb für Wasserwirtschaft, Küsten- und Naturschutz (NLWKN) (2013b).

Table 1. Scenario setting for the BEAST assessment in the municipality of Friedland.

Item	Description/rule	Expected value	RMSE
Production criteria			
Yield level crops/yield variation	Average yield level (2011) for reference crops (model results) in decitons (dt) in the study area	84 (wheat) 80 (barley) 41 (oilseed rape)	4.1 dt (W) 4, 4 (B) 2. 1(OR)
Yield increase crops	Trend analysis (1981–2011) of annual yield increase for reference crops in the Göttingen district	0.7% (wheat) 0.5% (barley) 0.5% (oilseed rape)	–
Yield level SRC	Mean annual increment (MAI) over a 20-year period (5-year rotation) for MAX-1 poplar SRC with 7, 000 cuttings in the study area	13.5 oven dry tons (odt)	–
Economic criteria			
Commodity prices/price changes	Input price level 2011 (net) in € per dt	20, 5 (wheat) 18, 8 (barley) 44, 2 (oilseed rape)	3, 9(W) 5,1 (B) 5, 3 (OR)
	Input price level 2011 (net) in € per odt (oven dry ton) of wood chips	91	4
Crop production costs (€)	Yield-specific functions for labor costs, variable costs, fix machinery costs (2011 level)	5, 1133× + 752, 72 (W) 3, 5015× + 774,93 (B) 3, 2467× + 1014,4 (OR) (x = annual yield)	–
SRC production costs (€)	Establishment	2650	–
	Harvest operation and site management	12× +600 (x = yield per rotation period)	–
	Re-conversion	1950	–
Ecological state criterion			
Disposition to water erosion	Sites with a potential soil erosion >30t ha^{-1} a^{-1} according to EU Cross Compliance regulation		
Spatial restrictions			
Restricted areas	FFH, nature sanctuaries		
Buffer zones	Distance to humid biotopes >200 m		
Objectives			
Economic return	Annuity difference for SRC >0 € compared to annual reference crops		
Ecological synergies	Potential synergies of erosion protection, renewable energy production, and economic return		

Cost and Price Calculation

The information used to calculate yield-related costs for the annual reference crops and for a poplar SRC in a 5-year rotation originated from the Niedersachsen Chamber of Agriculture (LWK, 2002–2013), the German Farmers' Association (Wagner et al., 2012), and various other sources [e.g., Nahm et al. (2012), Strohm et al. (2012), Forstliche Versuchs- und Forschungsanstalt Baden Württemberg (FVA) (2012), Kröber et al. (2010), and Landesanstalt für Entwicklung der Landwirtschaft und der ländlichen Räume (LEL) (2010)].

For the cost calculation, a distinction was made between variable costs, labor costs, and fixed costs for machinery. Fixed costs for arable crop production were considered since the cost calculation available for SRC (planting, harvesting, re-conversion) was based on subcontracting, and, thus, included fixed costs for machinery. Overhead costs, rents, and EU single farm payments were not considered as these parameters do not differ between SRC and annual crops. Commodity prices were derived from regional and national statistics (C.A.R.M.E.N. e.V, 2014; EUWID, 2014; Federal Statistical Office – Statistisches Bundesamt, 1980–2001; LWK, 2002–2013). These prices were calculated as net prices without VAT (Table 1).

As drying, storage, and transportation to the farm were considered when calculating SRC-related costs, the prices were adjusted by subtracting 15 € per odt of wood chips (Wagner et al., 2012) from C.A.R.M.E.N price statistics. The net price for woodchips from SRC was further adjusted to include drying losses (−20%), in accordance to Bärwolff and Hering (2012) and Kuratorium für Technik und Bauwesen in der Landwirtschaft (KTBL) (2012), respectively. Net commodity prices for wheat, barley, and oilseed rape were directly taken from Federal State statistics of Lower Saxony (LWK, 2002–2013). For both price and cost calculation, 2011 numbers were taken as reference (Table 1).

Yield Modeling and Potential Primary Energy Supply from SRC

The yield data for the annual reference crops (i.e., wheat, oilseed rape, and barley) reflect modeling results of average decadal yields, whereas the SRC yield data (poplar SRC, 5-year rotation, 7000 saplings) refer to the simulated mean annual increment of woody biomass per rotation period. Modeling results are provided for four rotation periods (Busch and Thiele, 2015). Crop yield data for the annual reference crops (i.e., wheat, oilseed rape, and barley) were generated by a statistical yield model based on climate and soil parameters as well as yield information from 52 sites all over Lower Saxony, Germany (LWK, 2002–2013). Model results were calibrated and validated for the Göttingen district (Busch and Thiele, 2015).

Mean annual increment of the MAX-I poplar SRC was simulated according to a modified approach published by Petzold et al. (2014). The approach is based on the relation between mean annual increment of biomass and available soil water capacity, which provides a function of potential annual biomass production for a specific plantation age. Busch and Thiele (2015) adapted the approach to three long-term data sets of yield levels (Max-I poplar) from Thuringian sites (Thüringer Landesanstalt für Landwirtschaft (TLL) (2010, 2012)). They introduced an age-specific modifier to the yield function that allows to model yields for various ages and rotation periods of SRC.

Short Rotation Coppice yields were transformed to numbers of primary energy supply by using a conversion factor of 4.95 MWh per oven dry ton of biomass yield according to Fachagentur Nachwachsende Rohstoffe (FNR) (2012).

Dynamic Investment Calculation over a 20-Year Period (2011–2030)

An economic comparison between perennial and annual cropping systems needs a dynamic investment cost calculation because timing of cash flows in these systems is different. As a result of this approach, annuities for both systems (reference crop rotation and poplar SRC) were calculated and related. Annuities in turn describe the annual constant cash flows over the investment period.

To determine economic return of both, the reference crop rotation and the SRC system, some boundary conditions were set: (a) reference prices and costs are the numbers of 2011, (b) to allow for a better comparison, no price and cost increase over the simulation period was assumed, (c) yield increase of annual crops was considered according to trend analysis (Busch and Thiele, 2015 – see Table 1), (d) price and yield fluctuations were considered according to recent trends (2002–2011) via Monte Carlo simulation (MC).

Taking the 2011 yield- and price-levels as input values for the MC, the RMSE (root mean square errors) of the underlying yield and price models for the period 2002–2011 were used to determine plausible variations of change (see economic and production criteria in Table 1). The RMSE were interpreted as equivalents to the SDs of Gaussian distributions. That way MC simulation randomly changed the expected input values (i.e., yields, prices) within the range of RMSE and iterated 10,000 times to generate statistically valid output. The results from the MC simulation provided a frequency distribution and, thus, allowed for deriving the probability of a positive economic return from SRC compared to the reference crop rotation.

Commodity prices for the annual reference crops as well as the wood chip prices were highly correlated within the 2002–2011 decade. Thus, for each iteration, one Gaussian random number of price change within the MC simulation was applied for all commodities, but scaled according to the crop-specific RMSE (Table 1).

Identification of Arable Sites which are Prone to Water Erosion

For the ecological synergy analysis, only arable sites were considered that show a potential water erosion rate higher than 30 t ha^{-1} a^{-1} since this is the threshold where Farmers have to take protective measures according to EU Cross Compliance (EU, 2009) regulation.

Disposition to water erosion was derived by building on input information from LBEG (Müller and Waldeck, 2011). As a federal state agency, LBEG provides reference methodologies for soil assessments. The assessment on disposition to water-induced soil erosion is based on the German adaptation of the Universal soil loss equation (USLE), and refers to soil information of the German Soil Survey (1:5.000) and a digital elevation model with 12.5 m grid. Details can be found in Schäfer et al. (2010).

RESULTS

Hydrological Field Measurements and Plot Level Modeling

Figure 3A shows precipitation volumes in Reiffenhausen from April to November 2013. The annual precipitation in 2013 amounted to 640 mm (DWD station Göttingen), which is close to the long-term mean of 635 mm (period 1971–2010). The measured plant available water down to 1 m soil depth (PAW1.0) reflects the sequence of precipitation events as illustrated in Figure 3D. After intense rain events in May 2013, only minor or short events occurred which were not able to prevent the soil from drying. Starting from August, drought stress occurred, which is also the reason for the increase in Rsc (stomatal resistance), illustrated in Figure 3B. Besides drought stress, also air temperature is affecting Rsc, which is illustrated by comparing the Rsc measurements with the daily maximum temperature (Figure 3A). At the beginning of September 2013, the wilting point was already reached, affecting biomass growth as illustrated in Figure 3C, as well as changes in plant height and diameter at breast height. Figure 3D compares PAW1.0 as measured and modeled using a parameterization of poplar SRC based on measurements, such as the annual course of LAI (Figure 3B). As an integrated soil hydrological value, PAW1.0 indicates a good model agreement, resulting in a Nash–Sutcliffe model efficiency coefficient of 0.9.

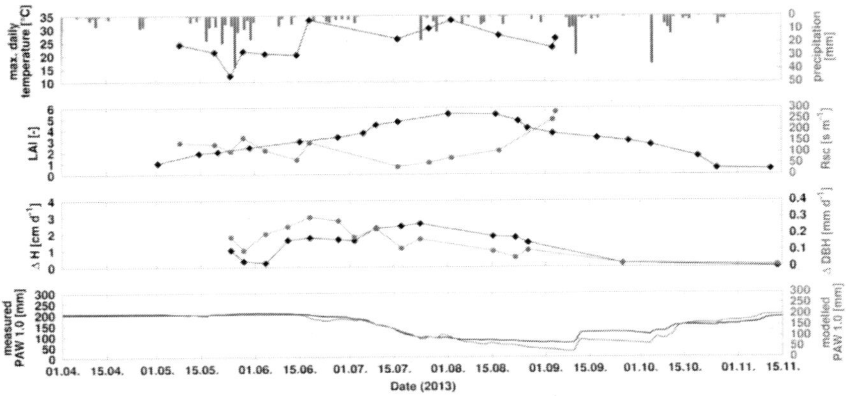

Figure 3. Time series of daily max. temperature, precipitation, leaf area index, stomatal resistance, growth parameters, and soil water changes of a poplar SCR in Reiffenhausen (Lower Saxony) in 2013. **(A)** Daily max. temperature and precipitation; **(B)** measured course of leaf area index and stomatal resistance; **(C)** measured growth rate of tree height (ΔH) and diameter at breast height (ΔDBH); **(D)** measured and modeled course of plant available water, calculated until 1 m soil depth (PAW1.0). Black drawings with diamonds belong to the right axis, gray ones with circles to left axis.

Figure 4 shows results of long term simulations performed on the plot level for Reiffenhausen using this calibrated and parameterized model setup by forcing the model with climate data from the DWD station Göttingen. Figure 4A shows the annual precipitation. Lines in Figure 4B compare the modeled GWR of agriculture and poplar SRC, with no rotation considered, i.e., permanently parameterized as a 3-year-old poplar SRC. The GWR for poplar SRC is lower compared to agriculture. There is an increase in years with reduced or even missing GWR for poplar SRC, especially in succeeding dry years. The gray shading in Figure 4B represents the spread of five simulations reflecting the growth of poplar under consideration of a 5 years rotation period. Each simulation starts with one of the 5 rotation years to cover all possible combinations of rotation years and climate forcing present in this period.

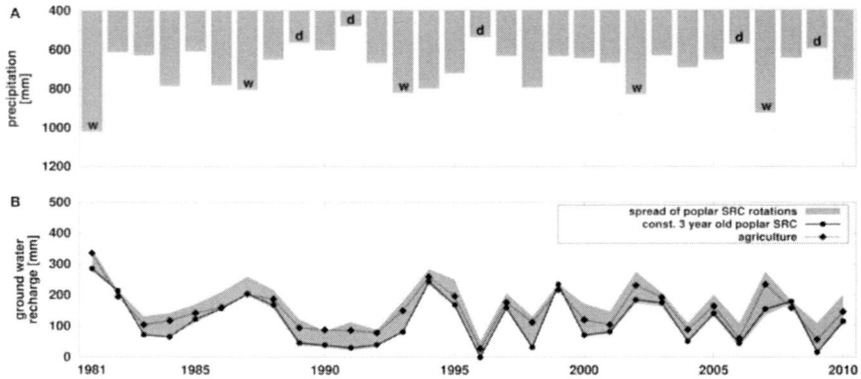

Figure 4. Comparison of plot model results for the two land use types agriculture and poplar SRC – with and without rotation. (A) Sums of precipitation for hydrological years (Oct–Nov), "d" marks the 5 driest years and "w" the 5 wettest years in the period 1981–2010; **(B)** sums of ground water recharge for hydrological years for agriculture and poplar SRC, whereas poplar SRC is kept constant in time (no rotation, 3-year-old) and the spread over simulations using a 5-year rotation period, starting in each of these 5 years.

Figure 4B shows a wide spread of GWR for these five simulations, where especially the first and second year of rotation are characterized by lower evapotranspiration (ETR) and therefore resulting in GWR rates comparable to or only slightly lower than for agriculture (Richter et al. in preparation). Due to the consideration of the 5-year rotation period, GWR rates can be higher or lower, compared to the SRC simulation with no rotation depending on the overlap of dry or wet years with the distinct rotations years.

On average, over all five rotation simulations ETR is lower, and GWR is higher than in the simulation with no rotation. A detailed analysis of the five rotation simulations shows that the highest GWR occurs not in the first rotation year, when the poplar SRC is small in height and LAI and therefore ETR is less, but in the second rotation year. The first year of rotation is needed to refill the soil water storage, which is exploited during the 3rd, 4th, and 5th year of rotation due to the high ETR. The soil water storage of the silty soil in Reiffenhausen is high; therefore, the refilling takes some time and only after that GWR can occur, i.e., mostly in the second rotation year. Only when precipitation is high, GWR is largest in the first year of rotation, for instance in 1981.

Regional Scale Hydrological Modeling

Figure 5 shows the changes of averaged GWR and catchment runoff for the simulations with different fractions of agricultural area changed to SRC relative to the reference run, i.e., no SRC. The changes of spatially and temporally averaged GWR and catchment runoff are smaller for the simulations considering the SRC rotation. The effects for the 5 and 10% land use change simulations are moderate, <10% reduction of GWR, and <3% for the runoff, regardless of the rotation consideration. Starting from a 25% land use change, the effects of considering SRC rotations become larger. Replacing half of the agricultural area with poplar SRC reduces the GWR by 57% (± 16%) without rotation and by 14% (± 9%) with rotation. Catchment runoff is reduced by 19% (± 6%) without rotation and by 2% (± 2%) with rotation. Besides these averaged and aggregated signals, large local changes of GWR are present at specific locations in the catchment. These local reductions of GWR are comparable to the plot model results shown in the former section; however, they are of different magnitude due to different soil physical properties in the catchment, compared to the plot Reiffenhausen (Richter et al. in preparation).

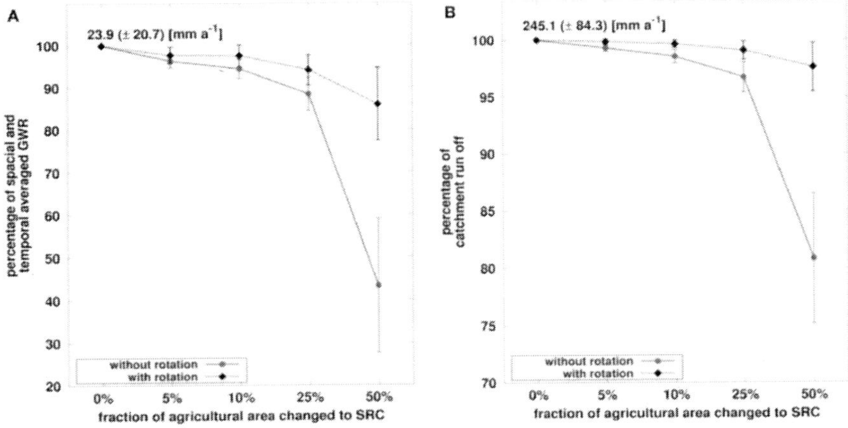

Figure 5. Change of ground water recharge (GWR, spatial mean over catchment area) and catchment run off relative to reference simulation (0% SRC). Mean values and SDs of relative changes are calculated over all simulated years, e.g., 30 for the simulations without rotation and 5 × 30 for the simulations with rotation. **(A)** Percentage of spatial and temporal averages ground water recharge (GWR). **(B)** Percentage of catchment runoff.

Nitrogen Budget

Total input of N, defined here as the mean rate of NNM plus the input rates of AD for the observation period, summed up to almost 30 kg ha^{-1} a^{-1} N for P-SRC and 40 kg N ha^{-1} a^{-1} for W-SRC (Table 2). The higher N input for W-SRC is obviously caused by enhanced rates of NNM, but the differences are not significant due to high spatial and temporal variability. NTU is obviously lower for W-SRC compared to P-SRC, even if biomass production (DM) was almost the same on both plots (P-SRC 1.0 t, W-SRC 1.1 DM ha^{-1} a^{-1}), measured after the second growth period, i.e., in winter 2012/13 [for more details, see Hartmann et al. (2014)]. On the other hand, the output of N, primarily driven by nitrate leaching at both plots and obviously less important for N_2O, was significantly higher under willow, compared to poplar. Considering the time pattern of the nitrate leaching under W-SRC over the entire observation period (December 2011–November 2012; not shown), it becomes obvious that those enhanced mean leaching rates under W-SRC were driven by high initial NO_3-N-concentration in soil solution (spring 2012 with max. of 33 ± 8.7 SD mg NO_3-N L^{-1}). But already since August 2012, nitrate concentrations under willow strongly decreased to the level of the P-SRC plot, i.e., to a range between 0.1 and 2 mg NO_3-N L^{-1}. Enhanced initial nitrate concentrations under Poplar occurred only in May 2012 (7.3 ± SD 8.7 mg NO3-N L^{-1}) but remained almost constantly under the detection limit toward the end of the observation period.

Aspects of SRC Allocation According to the Case Study Scenario

Economic Competitiveness

Monte Carlo simulation results (10,000 runs) revealed that the majority of arable sites in the case study area are not capable to provide a positive economic return for SRC under all circumstances (Figure 6A). Numbers presented in Figure 6A reflect the 100% probability situation for all scenario runs, i.e., the minimum annuity difference of SRC that could be generated on each particular arable site. Thus, in 81% of all cases simulated in this scenario, the minimum annuity difference of SRC is lower than the reference crop rotation of wheat, oilseed rape, and barley. In 19% of the scenario cases, however, MC shows that SRC out-competes the reference crop rotation with a 100% probability. Here, the minimum annuity difference which a farmer could count on ranges between 0 and 180 € (Figure 6A). Note, however, that even sites with a negative minimum annuity difference could reveal positive annuity differences – but with lower probabilities (Figure 6B). Since MC was carried out for each arable site in the study area, it is possible to depict "probability profiles" of annuity differences (Figure 6B) and to support farmer's decision making.

Table 2. Mean (±SD) initial N-budget and N-use efficiency estimates for SRC in Reiffenhausen comprising poplar (P-SRC) and willow (W-SRC) plantations.

Plot	**P-SRC**	**W-SRC**
Input	**[kg ha^{-1} yr^{-1}]**	
Net-N-Mineralization (NNM)	18.6 (13.3)	28.1 (25.5)
Atmospheric deposition (AD)	11.0 (0.4)	11.0 (0.4)
Σ Input	**29.6**	**39.1**
Output		
Net uptake by trees (NTU)	7.4 (0.9)	5.5 (0.9)
NO_3 seepage water output (NSWO)	2.5	22.3
N_2O emission	≤2	≤2
Σ Output	**11.9**	**29.8**
N-Budget (Input–Output)	**17.7**	**9.3**
	[kg kg^{-1}]	
N-uptake efficiency [NTU/NNM + AD]	0.25	0.14
N-utilization efficiency [DM/NTU]	135	198

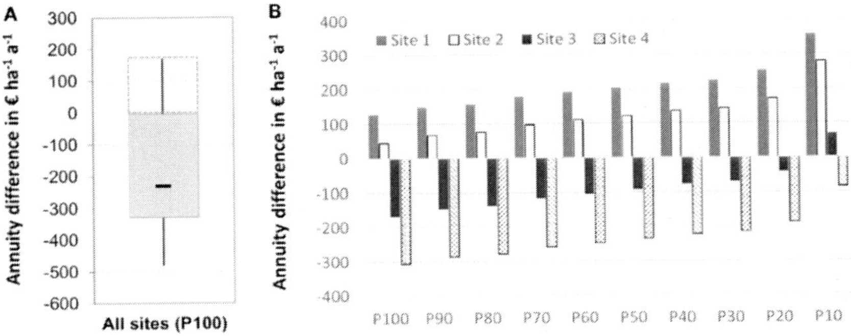

Figure 6. Minimum annuity difference (i.e., the annuity difference that was simulated with a 100% probability – P100) between poplar SRC and a reference crop rotation (wheat-oilseed rape-barley) for (A) all arable sites being subject to the reference crop rotation – dashed green box indicates the sites with a positive annuity difference as shown in more detail in Figure 7A, and (B) probability related annuity differences for exemplary sites to illustrate "probability profiles" of economic return (P10–P100 illustrate the percentiles of probability).

If farmers want to avoid the risk of a negative annuity difference under all circumstances, they have to stick to these 19% of arable sites shown in Figure 6A. Figure 7A depicts the value distribution of these particular sites with a median value of 47 €, and 75% of the numbers ranging from 30 to 75 €. In terms of area, this selection accounts for 541 ha (16%) of the arable land in the case study region.

Again, farmers could use the probabilities of annuity differences delivered by the MC simulation (Figure 6B) to get a more detailed impression of site-specific economic SRC potential, and e.g., to weigh minimum annuity differences (P100) against numbers of median probability (P50) or the complete "probability profile."

Erosion Protection

Taking the 541 ha with a minimum annuity difference >0 € as the spatial reference, intersection with erosion mapping revealed that about 20% (110 ha) of these areas are prone to a very high risk of soil erosion (CC-2 level) and need protective measures according to Cross Compliance regulation. Figure 7B shows that SRC implementation and erosion protection on these sites could be combined without hampering economic return. The median value only slightly decreases to 42 € and the box-range of minimum annuity difference values widens a bit compared to the spatial reference situation (Figure 7A). Since erosion protection is an environmental protection goal that is subsidized by government payments, minimum economic return would increase, which in turn could be an additional incentive for a SRC implementation on these particular sites.

Primary Energy Supply

Taking into account the 541 ha with a 100% probability of positive economic return, the average annual biomass production within a 20 year time frame is 9635 odt of wood which, in turn, is equivalent to 48.2 GWh of annual primary energy production. These sites show an average yield level of 17.4 odt a^{-1}, which corresponds to an energy density of around 85 MWh ha^{-1} a^{-1}, and is, thus, equivalent to maize production on a 500 dt annual yield level (Kuratorium für Technik und Bauwesen in der Landwirtschaft (KTBL), 2012). If the spatial reference is restricted to sites which are subject to high risk of soil erosion, the primary energy supply diminishes to 9.5 GWh a^{-1}, while the average yield level only slightly declines to 17.2 odt a^{-1}.

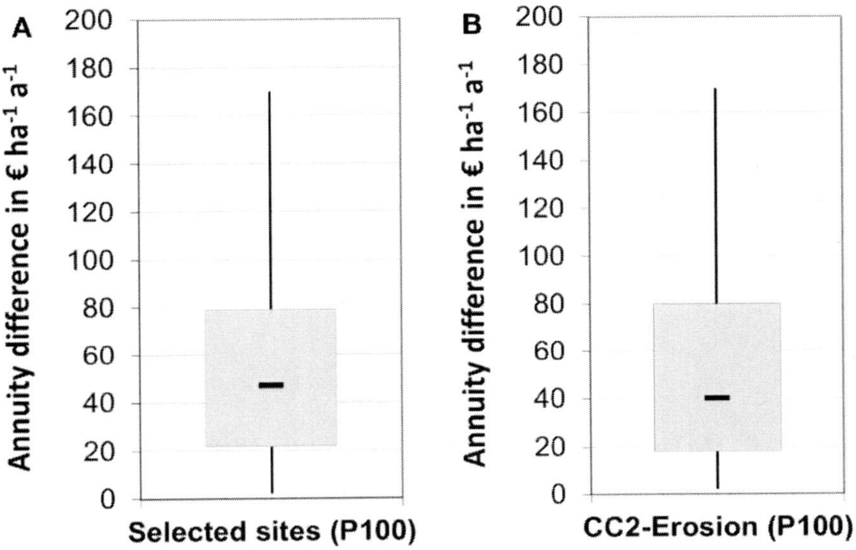

Figure 7. Annuity difference between poplar SRC and a reference crop rotation (wheat-oilseed rape-barley) for (A) selected economically competitive sites with positive economic return in all scenario cases (P100), and (B) ecological synergy of erosion protection on these economically competitive sites.

Potential Impact on Humid-Sensitive Habitats

Locally, SRC can have a negative impact on habitats that rely on high groundwater level or soil interflow from neighboring areas. Around 58 ha of the potential 541 ha of SRC sites were identified within a 200 m buffer radius of riparian FFH areas (Figure 8A), and another 38 ha could have an impact on neighboring humid-sensitive habitats. Thus, a potentially negative effect of SRC on surrounding habitats has to be carefully considered. However, 25 ha of these particular sites show a very high disposition to soil erosion which needs to be taken into consideration as well (Figure 8B).

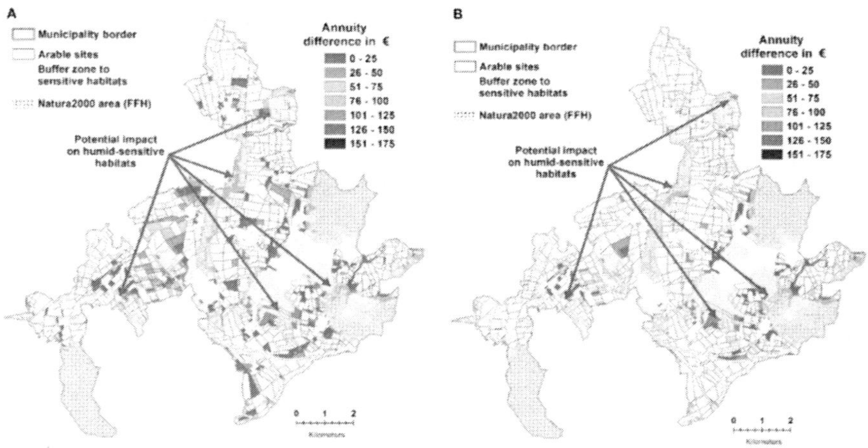

Figure 8. **(A)** Economically competitive SRC sites, **(B)** economically competitive SRC sites that provide cross compliance-relevant erosion protection. **(A)** additionally depicts layers addressing potential restrictions for SRC allocation.

DISCUSSION AND CONCLUSION

More cultivation of SRC changes water fluxes on both the plot and the landscape scale. On the plot scale, it was shown that the water consumption of SRC is all together higher in comparison to annual crops. Locally, this can result in a significant reduction up to the ceasing of GWR. So, the high water demand of poplar SRC could be confirmed by the measurements and simulations on the plot scale, as well as the link between sufficient water availability and optimal biomass growth.

The higher water demand results from the higher evapotranspiration of SRC compared to annual crops. This is especially due to a longer vegetation period, a deeper and more developed rooting system, as well as higher interception evaporation, because of larger LAI values. However, these effects differ in a wide range depending on the site conditions and the SRC management (Stone et al., 2010). The latter has the potential to shift site-specific changes in water fluxes into the desired direction. Reduced GWR leads to a lower mass transfer from soil to ground water and can reduce nitrate leaching. But the higher water demand of SRC also involves the risk of water shortage, which could be critical in areas where a certain water supply is needed (Le et al., 2011), e.g., drinking water reservoirs or ecosystems dependent on a minimum ecological runoff. However, to observe a significant reduction signal on the catchment scale, (averaged GWR and run off) more than 25% of the catchment would have to be converted from agriculture into SRC. Nevertheless, effects of SRC cultivation on site water fluxes can be locally significant, especially for sensitive ecosystems and in dry years.

With respect to biogeochemical effects, the conversion of conventional crop fields into SRC plantations may also cause significant changes. One of the most prominent and initial effects is the end of N input by fertilization, because for SRC a sufficient N supply can usually be generated from internal sources (mineralization of soil organic N, N_{org}) or, at least in most parts of central Europe, even from the input of atmospheric N (Lamersdorf et al., 2010). On the other hand, site preparation for an SRC plantation (i.e., plowing, harrowing etc.) will stimulate mineralization and nitrification pulses of formerly accumulated N_{org}. In consequence, an initial nitrate leaching pulse may occur under newly installed SRC plantations, and was already previously described (Goodlass et al., 2007; Lamersdorf et al., 2010; Schmidt-Walter and Lamersdorf, 2012).

For the present study, it is concluded that soil disturbances by site preparation and cultivation measures may have caused such initial nitrification pulses, especially under willow, because site conditions here are to some extent more favorable for such processes [higher total carbon contents due to enhanced percentage of clay and silt texture; for further details see Hartmann et al. (2014)]. Thus, the enhanced mean NNM under willow could also be taken as an indication for higher biological activities at this plot, and thus the risk for initial nitrification pulses after soil disturbances by cultivation measures. However, both SRC plantations already accumulate N in the range of almost 10–18 kg ha^{-1} a^{-1} in their initial growth phase. In addition, it is noticeable that the investigated poplar plantation is obviously more efficient in taking up nitrogen from the input (25%), compared to willow (14%). Vice versa, the willow plantation seems to be much more efficient in using the nitrogen taken up for wood production, as almost 200 kg DM of woody biomass were produced per 1 kg N, whereas poplar produced only 135 kg of DM with the same amount of consumed N. It can be summarized that the poplar and willow plantations investigated here show already in the initial growth stage obvious differences in their N-cycling and N-use efficiencies.

It has to be pointed out that the given N-budget calculations are only first estimates, i.e., the underlying measurements are just covering the first one or 2 years of the entire life of such SRC plantations. Furthermore, there might be clone specific effects, which were not further considered here. On the other hand, if such identified "traits" of certain SRC plantations with respect to the nitrogen cycling could be confirmed by further investigations, they could also be used for a target-oriented usage of SRC in the landscape.

If for instance an N excess from the previous crop cultivation should be reduced, poplar as considered here (clone Max 1) seems to be more favorable compared to willow, due to its higher N-uptake efficiency. On the contrary, if less N is available on, e.g., naturally poor or long-time abandoned sites but high woody biomass production is still in focus, the here considered willow SRC (clone Tordis) seems to be more advisable.

Scenario results of the case study simulation illustrated that the vast majority of arable sites were not capable of providing economic return from SRC that outcompetes annual reference crops in all scenario cases. At first sight, this may lead to the fallacy that SRC is no option for farmers in the Göttingen district. But, despite of a relatively high price level of annual crops used in the case study scenario, MC revealed that around 19% of the arable sites provide conditions under which economic return grom SRC beats a reference crop rotation of wheat, oilseed rape, and barley under all circumstances. These sites are considered as “first choice” options since they fulfill the criterion of a low-risk alternative to annual crops and could be used to enlarge the regional renewable energy mix.

However, with a range between 0 and 180 €, the median minimum annuity-difference to the reference crop rotation is only +47 € ha^{-1}, which raises the question if a risk-averse farmer will perceive this as an incentive to shift to SRC cropping.

Implementing SRC for a 20-year period results in a loss of flexibility and provides high initial investment costs for farmers. Thus, the farmer needs apart from spatial explicit information that allows for a thorough site selection, incentives to opt for perennial crops. Apart from much better adapted investment, subsidies toward lower thresholds of farmers implementation investments (Bemmann and Butler Manning, 2013) the most important prerequisite is confidence in reliable political boundary conditions, which in turn are the basis for the establishment of value- and supply chains which again support market penetration. This “vicious circle” or “chicken-and egg” problem is not only described for Germany (Bemmann et al., 2010; Thrän et al., 2011; Bemmann and Butler Manning, 2013) but was investigated for the UK as well (Sherrington et al., 2008; Sherrington and Moran, 2010). Overregulation for SRC compared to conventional crops in combination with only marginal consideration of ecological services provided by SRC in environmental payment schemes (Bemmann and Butler Manning, 2013) additionally hampers farmers to implement SRC.

However, as pointed out in the abovementioned studies, local and regional solutions could be an important step-stone to overcome some of the political and perception barriers. At this point, regional politics and actors play a crucial role. The Göttingen district, as many regions in Germany, is focusing on renewable energy strategies for climate protection and needs a mix of renewable energy carriers to fulfill the ambitious policy targets. Promoting the economic feasibility of proper SRC sites could provide an important incentive for farmers and a further renewable energy source helping to reach regional climate protection goals. This could be especially interesting on sites where ecosystem services can be combined with a positive economic return. In terms of erosion protection, case study results depicted that SRC could be a competitive option for around 3% (110 ha) of the arable sites in the study area. To breach the abovementioned "vicious circle," regional policy must initiate more demonstration projects to illustrate if and how regional supply chains could work. For the Göttingen district, public buildings (e.g., schools) should be used as part of the climate protection agenda to establish local heating networks. Participatory approaches – as applied for the development of the integrated climate protection strategy (Landkreis Göttingen, 2013) – turned out to be very helpful to build social networks and to address the concerns of regional actors. When it comes to woody biomass supply for renewable energy generation, the authors deem the integration of ecological and economic considerations in a joint tool for scenario development and analysis as a promising way to develop a participatory decision support on a regional scale. Building on previous work by Busch (2012), BEAST combines ecological assessments with calculations of economic revenue as a basis for a participative regional dialog on sustainable land use and climate protection goals [see Busch and Thiele (2015) for further details]. Introduced on regional stakeholder workshops in 2014, BEAST will be applied in the regional LEADER process and case studies of the district energy agency as a first step toward regional implementation strategies of renewably energy supply from SRC in 2015.

ACKNOWLEDGMENTS

The work presented in this study was funded by the German Federal Ministry of Education and Research (BMBF), grant number 033L033A, and is part of the BEST-Research Framework (http://www.best-forschung.de). We gratefully acknowledge this support. We give our special thanks to D. Fellert for valuable advices and assistance in handling the meteorological equipment; to D. Böttger for his assistance in field measurements. Special thanks to J. Schulla for advising in all WaSim questions which arose during the study, solving problems, and providing new model versions.

REFERENCES

1. Bärwolff, M., and Hering, T. (2012). *Fremdenergiefreie Trocknungsvarianten für Holz aus Kurzumtriebsplantagen*. Dornburg: Thüringer Landesanstalt für Landwirtschaft.
2. Bemmann, A., and Butler Manning, D. (Hrsg.). (2013). *Energieholzplantagen in der Landwirtschaft*. Hannover: Agrimedia.
3. Bemmann, A., Nahm, M., Brodbeck, F., and Sauter, U. (2010). Holz aus Kurzumtriebsplantagen: Hemmnisse und Chancen. Wood from short rotation coppice: obstacles and chances. *Forstarchiv* 81, 246–254. doi: 10.2376/0300-4112-81-246
4. Busch, G. (2012). GIS-based tools for regional assessments and planning processes regarding potential environmental effects of poplar SRC. *Bioenerg. Res.* 5, 584–605. doi:10.1007/s12155-012-9224-0
5. Busch, G., and Thiele, J. C. (2015). "The Bio-Energy Allocation and Scenario Tool (BEAST) to assess options for the siting of short rotation coppice (SRC) in agricultural landscapes – Tool development and case study results from the Göttingen district," in *Bioenergy from Dendromass for the Sustainable Development of Rural Areas*, Chap. 4, ed. B. D. Manning. (Wiley-VCH), p. 23–44 (in press).
6. C.A.R.M.E.N. e.V. (2014). Available at: http://www.carmen-ev.de/infothek/ preisindizes /hackschnitzel/jahresmittelwerte (accessed May 16, 2014)
7. Deumlich, D., Funk, R., Frielinghaus, M., Schmidt, W., and Nitzsche, O. (2006). Basics of effective erosion control in German agriculture. *J. Plant Nutr. Soil Sci.* 169, 370–381. doi:10.1002/jpln.200621983
8. Di, H. J., and Cameron, K. C. (2002). Nitrate leaching in temperate agroecosystems: sources, factors and mitigating strategies. *Nutr. Cycl. Agroecosys.* 64, 237–256. doi:10.1023/A:1021471531188
9. Dimitriou, I., and Aronsson, P. (2011). Wastewater and sewage sludge application to willows and poplars grown in lysimeters – plant response and treatment efficiency. *Biomass Bioenergy* 35, 161–170. doi:10.1016/j.biombioe.2010.08.019
10. Dimitriou, I., Busch, G., Jacobs, S., Schmidt-Walter, P., and Lamersdorf, N. (2009). A review of the impacts of short rotation coppice cultivation on water issues. *Landbauforsch. Voelkenrode* 59, 197–206.
11. DWD German Weather Service. (2013). *Precipitation and Temperature Data*. Available at: http://www.dwd.de/ (accessed November 18, 2013)
12. Eno, C. F. (1960). Nitrate production in the field by incubating the soil in polyethylene bags. *Soil Sci. Soc. Am. J.* 24, 277–279. doi:10.2136/ sssaj1960.03615995002400040019x
13. European Union – EU. (2009). Richtlinie 2009/28/EG des europäischen Parlamentes und Rates. Amtsblatt der Europäischen Union. L140/16. Available at: http://eur-lex.europa.eu/LexUriServ/LexUriServ.do?uri=OJ:L:2009: 140:0016: 0062:DE:PDF
14. EUWID – Europäischer Wirtschaftsdienst (2014). Available at: http://www.euwid-energie.de/ (accessed 20 March 2014).
15. Fachagentur Nachwachsende Rohstoffe (FNR). (2012). *Energieholzproduktion in der Landwirtschaft*. 5. Gülzow: Auflage.
16. Federal Statistical Office – Statistisches Bundesamt. (1980–2001). *Fachserie 3 Reihe 2.1.1: Land- Und Forstwirtschaft, Fischerei*. Wiesbaden: Betriebe mit Waldflächen Landwirtschaftszählung/Agrarstrukturerhebung.
17. Forstliche Versuchs- und Forschungsanstalt Baden Württemberg (FVA). (2012). *KUP-Ernteplaner – Version 1.1.* Freiburg: FVA.
18. Gasche, R., Butterbach-Bahl, K., and Papen, H. (2002). Development and application of a method for determination of net nitrification rates. *Plant Soil* 240, 57–65. doi:10.1023/A:1015860322656

19. Goodlass, G., Green, M., Hilton, B., and McDonough, S. (2007). Nitrate leaching from short-rotation coppice. *Soil Use Manag.* 23, 178–184. doi:10.1111/j.1475-2743.2006.00080.x
20. Hartmann, L., Richter, F., Busch, G., Ehret, M., Jansen, M., and Lamersdorf, N. (2014). Etablierung von Kurzumtriebsplantagen im Rahmen des Verbundprojektes BEST in Süd-Niedersachsen und Mittel-Thüringen – Standorteigenschaften und anfängliche Erträge. *Forstarchiv* 85, 134–150. doi:10.4432/0300-4112-85-134
21. Kort, J., Collins, M., and Ditsch, D. (1998). A review of soil erosion potential associated with biomass crops. *Biomass Bioenergy* 14, 351–359. doi:10.1016/S0961-9534(97)10071-X
22. Kröber, M., Heinrich, J., Wagner, P., and Schweinle, J. (2010). "Ökonomische Bewertung und Einordnung von Kurzumtriebsplantagen in die gesamtbetriebliche Anbaustruktur," in *Agrowood. Kurzumtriebsplantagen in Deutschland und europäische Perspektiven*, eds A. Bemmann and C. Knust (Berlin: Weißensee Verlag), 217–229.
23. Kuratorium für Technik und Bauwesen in der Landwirtschaft (KTBL). (2012). *Energiepflanzen. Daten für die Planung des Energiepflanzenanbaus.* 2A. Darmstadt: KTBL.
24. Lamersdorf, N., Petzold, R., Schwärzel, K., Feger, K., Köstner, B., Moderow, U., et al. (2010). "Bodenökologische Aspekte von Kurzumtriebsplantagen," in *AGROWOOD: Kurzumtriebsplantagen in Deutschland und Europäische Perspektiven*, eds. A. Bemmann., and C. Knust, (Hrsg.) (Berlin: Weissensee-Verlag), 170–188.
25. Landesanstalt für Entwicklung der Landwirtschaft und der ländlichen Räume (LEL). (2010). *KUP-Rechner – Version 1.1.* Schwäbisch-Gmünd: LEL.
26. Landkreis Göttingen. (2013). *Landkreis Göttingen – Integriertes Klimaschutzkonzept für den Landkreis.* Available at: http://www.landkreis-goettingen.de (accessed November 7, 2014)
27. Landwirtschaftskammer (LWK) Niedersachsen. (2002–2013). *Richtwer tdeckungsbeiträge Niedersachsen 2002-2012.* Hannover: LWK.
28. Lasch, P., Kollas, C., Rock, J., and Suckow, F. (2010). Potentials and impacts of short-rotation coppice plantation with aspen in Eastern Germany under conditions of climate change. *Reg. Environ. Change* 10, 83–94. doi:10.1007/s10113-009-0095-7
29. Le, P. V. V., Kumar, P., and Drewry, D. T. (2011). Implications for the hydrologic cycle under climate change due to the expansion of bioenergy crops in the Midwestern United States. *Proc. Natl. Acad.Sci. U.S.A* 108, 15085–15090. doi:10.1073 /pnas.1107177108
30. Linderson, M.-L., Iritz, Z., and Lindroth, A. (2007). The effect of water availability on stand-level productivity, transpiration, water use efficiency and radiation use efficiency of field-grown willow clones. *Biomass Bioenergy* 31, 460–468. doi:10.1016/j.biombioe.2007.01.014
31. Moll, R. H., Kamprath, E. J., and Jackson, W. A. (1982). Analysis and interpretation of factors which contribute to efficiency of nitrogen utilization. *Agron. J.* 74, 562–564. doi:10.2134/agronj1982.00021962007400030037x
32. Mosimann, T., Bug, J., Sanders, S., and Beisiegel, F. (2009). *Bodenerosion sdauerbeobachtung in Niedersachsen 2000–2008. Methodik, Erosionsgeschehen, Bodenabträge und Anwendung der Ergebnisse, Geosynthesis*, 14, Hannover, 101 S.
33. Müller, U., and Waldeck, A. (2011). "Auswertungsmethoden im Bodenschutz. Dokumentation zur Methodenbank des Niedersächsischen Bodeninfor mationssystems (NIBIS®)," in *Geo-Berichte, Landesamt für Bergbau, Energie und Geologie*, Vol. 19 (Hannover).
34. Nahm, M., Brodbeck, F., and Sauter, H. U. (2012). *Verschiedene Erntemethoden für Kurzumtriebsplantagen. Ergebnisse aus der Praxis.* Freiburg: Forstliche Versuchs- und Forschungsanstalt Baden Württemberg (FVA).
35. Niedersächsischer Landesbetrieb für Wasserwirtschaft, Küsten- und Naturschutz (Lower Saxony Water Management, Coastal Defence and Nature Conservation Agency) (NLWKN). (2013a). *Landesweite Datenbank Niedersachsen (LDB) im*

Fachinformationssystem Wasser (FIS-W). Available at: http://www.wasserdaten.niedersachsen.de/cadenza/
36. Niedersächsischer Landesbetrieb für Wasserwirtschaft, Küsten- und Naturschutz (NLWKN). (2013b). *Biotopkartierung Niedersachsen.* Available at: http://tinyurl.com/k62y5pb (accessed August 21, 2014)
37. Pappa, V. A., Rees, R. M., Walker, R. L., Baddeley, J. A., and Watson, C. A. (2011). Nitrous oxide emissions and nitrate leaching in an arable rotation resulting from the presence of an intercrop. *Agric. Ecosyst. Environ.* 141, 153–161. doi:10.1016/j.agee.2011.02.025
38. Petzold, R., Butler-Manning, D., Feldwisch, N., Glaser, T., Schmidt, P. A., Denner, M., et al. (2014). Linking biomass production in short rotation coppice with soil protection and nature conservation. *iForest Biogeosci. For.* 7, 353–362. doi:10.3832/ifor1168-007
39. Petzold, R., Wahren, A., and Feger, K.-H. (2010). "Steuerungsoptionen des Wasser- und Stoffhaushalts auf Landschaftsebene durch den Anbau von Kurzumtriebsplantagen – ein Forschungsansatz," in *Bodenschutz in Europa – Ziele und Umsetzung* (Marktredwitzer), 91–96. Available at: http://boku.forst.tu-dresden.de/pdf/petzold_wahren_feger_Steuerungsoptionen.pdf
40. Richter, F., Döring, C., Jansen, M., Panferov, O., Spank, U., and Bernhofer, C. (2015). How to predict hydrological effects of local land use change: how the vegetation parameterisation for short rotation coppices influences model results. *Hydrol. Earth Syst. Sci. Discuss.* 12, 405–448. doi:10.5194/hessd-12-405-2015
41. Röhle, H., and Skibbe, K. (2012). "Ertragsschätzung in Kurzumtriebsplantagen aus Pappel und Weide," in *Züchtung und Ertragsleistung schnellwachsender Baumarten im Kurzumtrieb ProLoc und Weidenzüchtung: Erkenntnisse aus drei Jahren FastWOOD*, ed. Nordwestdeutsche Forstliche Versuchsanstalt (Göttingen: Universitätsverlag Göttingen), 105–116.
42. Ruser, R., Flessa, H., Russow, R., Schmidt, G., Buegger, F., and Munch, J. C. (2006). Emission of N_2O, N_2 and CO_2 from soil fertilized with nitrate: effect of compaction, soil moisture and rewetting. *Soil Biol. Biochem.* 38, 263–274. doi:10.1016/j.soilbio.2005.05.005
43. Schäfer, W., Sbresny, J., and Thiermann, A. (2010). Methodik zur *Einteilung von landwirtschaftlichen Flächen nach dem Grad ihrer Erosionsgefährdung durch Wasser gemäß §2 Abs. 1 der Direktzahlungen-Verpflichtungenverordnung in Niedersachsen.* Hannover: Niedersächsisches Landesamt für Bergbau, Energie und Geologie (LBEG), Geozentrum Hannover.
44. Schmidt-Walter, P., and Lamersdorf, N. (2012). Biomass production with willow and poplar short rotation coppices on sensitive areas – the impact on nitrate leaching and groundwater recharge in a drinking water catchment near Hanover, Germany. *Bioenergy Res.* 5, 546–562. doi:10.1007/s12155-012-9237-8
45. Scholz, V., Krüger, K., and Höhn, A. (2001). Vergleichende Untersuchungen zum umweltverträglichen und energieeffizienten Anbau von Energiepflanzen. *Arch. Agron. Soil Sci.* 47, 333–361. doi:10.1080/03650340109366218
46. Sherrington, C., Bartley, J., and Moran, D. (2008). Farm-level constraints on the domestic supply of perennial energy crops in the UK. *Energy Policy* 36, 2504–2512. doi:10.1016/j.enpol.2008.03.004
47. Sherrington, C., and Moran, D. (2010). Modelling farmer uptake of perennial energy crops in the UK. *Energy Policy* 38, 3567–3578. doi:10.1016/j.enpol.2010.02.034
48. Snyder, C. S., Davidson, E. A., Smith, P., and Venterea, R. T. (2014). Agriculture: sustainable crop and animal production to help mitigate nitrous oxide emissions. *Curr. Opin. Environ. Sustain.* 9-10, 46–54. doi:10.1016/j.cosust.2014.07.005
49. Stone, K. C., Hunt, P. G., Cantrell, K. B., and Ro, K. S. (2010). The potential impacts of biomass feedstock production on water resource availability. *Bioresour. Technol.* 101, 2014–2025. doi:10.1016/j.biortech.2009.10.037

50. Strohm, K., Schweinle, J., Liesebach, M., Osterburg, B., Rödl, A., Baum, S., et al. (2012). *Kurzumtriebsplantagen aus ökologischer und ökonomischer Sicht*. Arbeitsberichte aus der vTI-Agrarökonomie, 6. Available at: http://tinyurl.com/k425d7j (accessed August 21, 2014)

51. Surendran Nair, S., Kang, S., Zhang, X., Miguez, F. E., Izaurralde, R. C., Post, W. M., et al. (2012). Bioenergy crop models: descriptions, data requirements, and future challenges. *Glob. Change Biol. Bioenergy* 4, 620–633. doi:10.1111/j.1757-1707.2012.01166.x

52. Thrän, D., Edel, M., Seidenberger, T., Gesemann, S., and Rode, M. (2011). *Identifizierung strategischer Hemmnisse und Entwicklung von Lösungsansätzen zur Reduzierung der Nutzungskonkurrenzen beim weiteren Ausbau der Biomassenutzung*. Leipzig: DBFZ.

53. Thüringer Landesanstalt für Landwirtschaft (TLL). (2010). *Feldversuchsbericht 2008 und 2009. Ölfrüchte und Nachwachsende Rohstoffe*. Erfurt: TLL.

54. Thüringer Landesanstalt für Landwirtschaft (TLL). (2012). *Feldversuchsbericht 2010 und 2011. Ölfrüchte und Nachwachsende Rohstoffe*. Erfurt: TLL.

55. Tsonkova, P., Böhm, C., Quinkenstein, A., and Freese, D. (2012). Ecological benefits provided by alley cropping systems for production of woody biomass in the temperate region: a review. *Agroforestry Syst.* 85, 133–152. doi:10.1007/s10457-012-9494-8

56. Wagner, P., Schweinle, J., Setzer, F., Kröber, M., and Dawid, M. (2012). *DLG-Standard zur Kalkulation einer Kurzumtriebsplantage. DLG-Merkblatt 372*. Frankfurt am Main: Deutsche Landwirtschaft Gesellschaft.

57. Wahren, A., Schwärzel, K., and Feger, K.-H. (2012). Potentials and limitations of natural flood retention by forested land in headwater catchments: evidence from experimental and model studies. *J. Flood Risk Manag.* 5, 321–335. doi:10.1111/j.1753-318X.2012.01152.x

58. Walter, K., Don, A., and Flessa, H. (2012). *Greenhouse Gas Fluxes During Growth of Different Bioenergy Crops*. Vienna: EGU General Assembly, 5883.

CHAPTER 13

A Multiscale Assessment of the Risks Imposed by Plantation Forestry on Plant Biodiversity in the Hotspot Central Chile

Andreas Ch. Braun[1], Joachim Vogt[2]

[1]*Chair of Remote Sensing and Landscape Information Systems (FeLis), University of Freiburg, Freiburg im Breisgau, Germany*

[2]*Institute of Regional Science (IfR), Karlsruhe Institute of Technology (KIT), Karlsruhe, Germany*

ABSTRACT

We assessed the effects of plantations of exotic trees (*Pinus radiata*, *Eucalyptus globulus*, and *Populus nigra*) on plant biodiversity in the temperate zone of the biodiversity hotspot of Central Chile. This region has suffered from intense deforestation in favor of plantation establishment in the major part of the coastal area since the neoliberal turn in 1973. The approach presented aimed to analyze plant biodiversity on the α-, β- and γ-scale. Furthermore, a plantation evaluation index was applied in order to provide quantitative figures on management practices. Species richness is reduced by 50% below plantations. Diversity and evenness index values are also significantly reduced. Analyses on β-similarity indicate that plantations do not host species absent in adjacent native forests, and no additional habitat heterogeneity is gained. On the γ-scale, plantations lower the number of total species observed; especially of endemic species. The abundance of species considered as invasive is significantly higher and frequently, invasive plants dominate the understory. The evaluation index attests rather poor plantation management in Central Chile since plantations are grown as monocultures, natural elements and native species are lacking at specific sites and plantations are insufficiently connected to native plant formations at the landscape scale. Results give much concern since deforestation processes as observed in our study area is about to begin in Patagonia as well. If management practices from the temperate zone are adopted in Patagonia, a considerable decline in plant biodiversity has to be expected there.

Keywords: Biodiversity Decline, Plantations, Biodiversity Hotspot, Pinus radiate, Eucalyptus globules, Chile

1. INTRODUCTION

Deforestation is reported to cause rapid and significant losses in biodiversity and extinction of endemic species worldwide [1] -[5]. Biodiversity decline can be caused by direct removal of habitat and local extinction or by habitat fragmentation. However, secondary effects that arise e.g. from the construction of roads can have much stronger impacts on biodiversity [6] - [8]. Central Chile is considered a biodiversity hotspot, not only because of its total species richness, but due to its high level of endemism [9]. The coastal area of the temperate zone of Central Chile has been altered by a rapid, intense and widespread process of transformation. The transformation from Mediterranean temperate shrublands and forests to exotic tree plantations causes rapid deforestation and forest fragmentation [10] - [16]. After the neoliberal turn in 1973, the Chilean government forcefully promoted forestry with exotic tree species—mainly *Pinus radiata* and *Eucalyptus globulus* [17] - [20]. Between 1973 and 2000, 67% of the original extent of temperate forests was lost [10]. Although some further minor drivers have been identified, deforestation has largely been attributed to replacement with exotic plantations [11] [18] [21] - [24]. Until today, plantations with exotic tree species continue to spread in the temperate zone of Chile leaving less than 4% of the area covered with native forests (own data, unpublished results). Of these remaining forest patches, over 60% are smaller than 1 ha [25]. Since native forest sites available for plantation establishment are almost depleted, forest companies have recently started to establish plantations in Patagonia [26] [27]. Due to the insufficient conservation strategies in Chile, this process gives much concern [15] [20] [28]. Chilean natural parks and reserves (NPR) cover a relatively large portion of the country (19% of the country's area). Despite this large spatial extent, the Chilean system of NPRs is considered rather inefficient, although improvements by appropriate funding could theoretically be realized by the Chilean government [29]. 90% of the NPRs are found in zones with low plant biodiversity and in areas of rather low economic interest for forest companies [20] [21]. Hence, it seems unlikely that the biodiversity hotspots of Central Chile or Patagonia can be conserved with existing instruments.

The motivation of this report is as follows: Central Chile—like all other biodiversity hotspots—is considered a priority region for environmental conservation [9]. When discussing conservation strategies, one has to decide what to conserve—landscapes, ecosystems, communities, populations or species [30]. In Central Chile, native forest ecosystems are rapidly replaced by exotic tree plantations. Thus, a rapid deforestation and forest fragmentation have been observed, which can *principally*, although not *exclusively* be attributed to plantation establishment [10] [14]. Although small remnants are left in between plantations, alternative habitats are hardly being created. Hence, obviously larger units (like entire landscapes or ecosystems) are currently not being conserved, which raises the importance of least conserving the *species* typical for these units. Such a conservation strategy requires alternative habitats for species in the modified landscapes. This implies that Chilean species need to be able to prevail below the canopy of plantations. If this would be the case, e.g. endemism rates in plantations should be similar to those found in native forests. Furthermore, the β-diversity between native forests and plantations should be high. There is another opportunity for survival of the Chilean endemics in plantations. Despite local extinction of individual populations, endemics could survive via their metapopulations[1]. For this opportunity, the connectivity of the landscape presence of suitable stepping stone habitats is crucial [31] [32]. Furthermore, species following this survival strategy face a number of constraints (e.g. an effective reproduction mechanism over relatively large distances). These theoretical mechanisms of survival are to be clarified empirically within this report. Therefore, a mainly descriptive approach is followed at first.

Since many studies on deforestation and biodiversity have been published in literature over recent years, the question is, what makes the Chilean case so relevant? Why is Chile not merely a special case of some general concept? First of all, deforestation in Chile is mainly caused by plantation establishment, and not by other land use change processes like e.g. agricultural expansion. Thus, closed canopy tree ecosystems (forests) are replaced by other closed canopy tree ecosystems (plantations). Especially for this situation, our understanding of the influence deforestation on native (plant) biodiversity is limited [33] [34]. While some authors tend to attribute positive effects of plantation establishment on biodiversity—e.g.

by diversification of the landscape— [35] - [39] , others focus on negative impacts especially for Chile [11] [20] [27] [40] , but also worldwide [41] - [45]. Since plantations grow fast and cover a vast area in Central Chile, this region can be considered a large-scale experiment to study their relationship with biodiversity. Despite the vast distribution of exotic tree plantations, few studies have investigated their effects on biodiversity so far [11] [27].

The main goal of this study is to close existing research gaps by analyzing plant biodiversity in plantations and comparing them to other ecosystems. Some studies by other authors complement our approach. [11] [28] analyze species richness in temperate plant formations, predicting partial declines in species richness for the future. However, plantations are not incorporated into the analysis. [27] compares biodiversity between native forests and plantations on three taxa. The authors find biodiversity in plantations significantly reduced. However, their study analyses the situation in Patagonia and not in central Chile. [39] [46] point out that diversity of the avifauna may be increased within plantations. Yet if plantations are not managed sustainably, the diversity of the avifauna may be reduced as well [47]. [48] points out that the diversity of small mammals is reduced in plantations in contrast to near-natural forests. [49] shows that a reduction of patch size due to forest fragmentation may reduce beetles diversity. [50] sees the habitat of Chilean amphibians endangered by plantation growth, albeit not providing any biodiversity data. Many of these studies focus on the effects of forest fragmentation on biodiversity instead of focusing on biodiversity conditions within plantations themselves. Secondly, with up to 8.15% of *local* annual loss, the deforestation rates reported for Central Chile are among the highest worldwide [14]. If deforestation in favor of plantations continues at these rates, biodiversity in plantations should be managed in a sustainable manner [51]. Otherwise, Central Chile will not qualify as a biodiversity hotspot in the future anymore. Thirdly, the same process of deforestation and plantation establishment is about to begin in Chilean Patagonia [26] [27] —a landscape appreciated for its originality. Lastly, plantation establishment is frequently described in an economically biased and ecologically rather optimist perspective [52]. Given the spatial extent of plantation establishment in Chile, though, it is worthwhile to question this position by having a close look on its ecological impacts. Within this report, results on α-, β- and γ-diversity will be presented to revise this perspective.

The analysis of the impacts of current management practices (huge monotypic tree cohorts, harvesting by clear felling followed by pesticide applications and fire clearance of the soils) on plant biodiversity in Central Chile will finally be used to develop a set of management implications for the recently beginning plantation establishment in Patagonia. If followed, these implications will be useful to lower plant biodiversity loss rates in this region.

2. MATERIALS AND METHODS

2.1. Introduction to the Study Area and Plant Formations Observed

Chile's VII Region (Región Del Maule, capital Talca) and the VIII Region (Región Del Biobío, capital Concepción) together cover around 67338 km^2. Around 2.7 million people inhabit this zone [53]. The area has a warm-temperate, summer dry climate, the mean annual temperature is around 12.2°C and precipitation is around 1330mm (Köppen-Geiger: Csb, [54]). Parallel to the Pacific lies the coastal range, followed by the central depression in the inner side of the area. Towards the east, the depression is followed by the Andes. The coastal range is made of metamorphic rock (Paleozoic age) and some intrusive volcanic rock. This basement is covered with quaternary and tertiary material transported from the Andes [55]. The coastal range is mainly covered with deciduous and sclerophyllous forests, many of which are secondary [56]. The central depression is mainly covered with shrublands of *Acacia caven* [57]. The zone is an area of intensive landscape transformation since the arrival of the Spanish conquerors in the 16th century. The most intensive change in modern times, however, has taken place under the economic transformation of the Pinochet era. The area has been opened for large-scale forestry by the establishment of exotic tree plantations for economic purposes. Within this area, we assessed plant biodiversity of the following plant formations.

2.1.1. Near-Natural Forests

According to [11] , the vegetation in the coastal range of the biodiversity hotspot of Central Chile should be separated into four large units: the deciduous Maulino forest (VII Region), the Nahuelbuta forest (VIII Region), the Valdivian evergreen forest (IX. Region) and the Chiloé evergreen forest (X region). In our study, we assessed the Maulino and Nahuelbuta forests, for an example, see Figure 1(a). Within these large units, we mainly

(a)

(b)

(c)

(d)

(e)

(f)

Figure 1. Examples of the studied ecosystems: (a) near-natural forest, (b) degraded forest, (c) A. cavenshrubland, (d) P. radiata plantation, (e) E. globulus plantation, (f) P. nigra plantation.

analyzed deciduous forests, dominated by *Nothofagus spp., Persea lingue, Drimys winteri* and *Aristotelia chilensis* which are a typical formation of the rather humid part of the Coastal range [58] -[60] and sclerophyllous forests dominated by *Peumus boldus, Maytenus boaria, Lithreacaustica, Escallonia pulverulenta* and *Gevuina avellana* which are typical for the rather dry part of the Coastal range [61] [62]. We furthermore assess some riparian habitats—dominated by *Azara microphylla, Luma apiculata,* and *Rhaphithamnus spinosus*—due to their special importance for plant biodiversity conservation in plantations [39]. The Maulino forest, as well as the Nahuelbuta forest are both strongly fragmented today. Outside of NPRs, like the Reserva Nacional Los Ruiles, they remain as small remnants within plantations. For our analysis, both NPRs and remnants were sampled.

2.1.2. Degraded Forests

Some of the forests we investigated represent degraded forests on formerly deforested areas. These forests are frequent in Central Chile. Degraded forests are characterized by a much larger human influence than near-natural forests. Frequently, these forests represent artificially opened or fragmented patches by anthropogenic activities. Within such patches, native or non-native shrubs like *Chusqueaquila* or *Teline monspessulana* which benefit from human disturbance frequently form dominate strata, covering large amounts of the habitat and casting shadow on the herbaceous stratum, for an example, see Figure 1(b). These effects reduce biodiversity of small shrubs and herbs. Thus, such degraded forests are expected to yield lower biodiversity values than the near-natural forests where the aforementioned shrubs only seldom form dominant strata. We investigated degraded forests of typical *sclerophyllous* species (*Cryptocarya alba, P. boldus*) and sites where plantations of *E. globulus* and *P. radiata* have been abandoned and therefore converted into degraded forests. The first type is interesting since such forests are sometimes discussed as an option to recover plant biodiversity in sustainable forms of forestry or after plantation abandonment [63] , the second type is interesting since it shows plant biodiversity conditions under plantation trees without the influence of management.

2.1.3. Acacia Caven Shrublands

Shrublands of *A. caven* represent a savanna-like succession of the sclerophyllous shrublands, for an example, see Figure 1(c). The native shrublands are replaced by the *A. caven* shrublands after anthropogenic disturbance like intensive grazing or human induces fires [57]. They are composed of a single tree (*A. caven)* and an herbaceous stratum frequently dominated by annual species with European origin [70]. Nowadays, the A. caven shrublands are mainly used for extensive agroforestry [57]. The *A. caven* shrublands are considered as the most persistent formation in the central depression today. A succession from *A. caven* shrublands back to sclerophyllous shrublands is not to be expected [71]. The relevance of the *A. caven* shrublands lies in the fact that if their usage for agroforestry is given up, areas may be acquired by forestry companies and plantations will be established [72].

2.1.4. Exotic Tree Plantations

During the 18th century, large areas of Central Chile suffered from initial deforestation due to Spanish and German settlers that burned areas to create space for farming. Leaving these areas without vegetation cover resulted in heavy erosion which can still be observed today. In order to prevent further erosion, plantations were established from 1900 onwards [18] [64]. For this reason, it is sometimes emphasized that plantations have mainly positive effects since they serve as erosion control [52]. However, it is ignored that since the Decree No. 701 on forest promotion of 1974, plantations are mainly established on areas formerly covered by native forests and thus not affected by erosion [11] [18] [21] - [24]. Nowadays, plantations cover the vast parts of the Coastal Range of the VII and VIII Regions. 64% of the country's plantations are found within these regions. Plantations cover around 20% of the total area of the VII region and 33% of the VIII region. The main tree species used are P. radiata (84.4% of the planted surface in the VII Region, 58.7% in the VIII Region) and *E. globules* (12.1% of the planted surface in the VII Region, 38.6% in the VIII Region) and over 75% of products are for the international market [65]. Besides, some other tree species like *Populus nigra* or *Acacia spp*. are planted (1.4% of the planted surface in the VII Region, 2.7% in the VIII Region) [66] [67] , for examples, see Figures 1(d)-(f). Plantations in the VII and VIII region are mainly monotypic and cover large areas, frequently larger than 100 square kilometers. Trees are grown to reach around 15 years (25 at most) after which entire plantations are cut down by clear felling in a matter of days. Herbicides are applied directly after clear felling and shortly before planting the next generation of plantation trees, usually the next year [18] [64]. In the past, areas have been burned after clear felling, a practice which is abandoned today. In literature, plantations are sometimes referred to as "plantation forests" [68]. However, as [69] point out, plantations should be linguistically distinguished from forests in studies on deforestation and forest degradation; this suggestion will be followed hereafter.

2.2. Braun-Blanquet Vegetation Assessments

In July of 2011, a field campaign in two regions of Central Chile was carried trough (VII Región del Maule, VIII Región del Biobío). The goal of the field campaign was to assess plant biodiversity in natural and managed ecosystems of Chile's temperate zone. We assessed the species number, abundance and minimum areal according to the method of Braun-Blanquet [73]. A total of 62 vegetation relevés was assessed (13 native forest patches, seven degraded forest patches, 38 plantations and four shrublands of *A. caven*). Assessments were assisted by local vegetation experts from the University of Concepción (UdeC). No standard values from literature for the minimum areal were used since transferability to the biodiversity hotspot of Central Chile was not assumed. In order to be able to analyze β-diversity, ten coupled samples (e.g. a plantations relevé in direct vicinity to a native forests relevé) were assessed. Vegetation strata were identified and sampling was carried. According to Braun- Blanquet's recommendations, we started with an area of 5 × 5 m, collecting and identifying all plant species present. Then, the number of species in the 5 × 5 m plot was determined. The coverage of each species according to Braun-Blanquet's scale was estimated, usually by two scientist, who then agreed on a final value. Afterwards, the sampling area was sequentially increased (10 × 10 m, 20 × 20 m etc.). The species-area curve was determined and sampling stopped when this curve reached an asymptote. The respective size of the plots for which the asymptote was reached is considered the minimum areal. The abundance-dominance method of Braun-Blanquet has been subject to some review recently [74]. For instance, it is argued that its non-metric scale makes it difficult to analyze especially with multivariate statistics [75]. However, having performed some adaptations to the abundance-dominance data, data can indeed be analyzed statistically and valid results can be achieved [76]. In order to analyze data statistically, we transformed the abundance scores into relative scores according to [77] : A species with a score of e.g. 2 has a coverage between 25% and 50%. It is given a coverage value r_i of (0.25 + 0.5)/2 = 0.375. Other scores were transformed accordingly. Species with a score of + were given r_i = 0.025 and species with r were given r_i = 0.00625. Then, the relative coverage of a species was calculated as $p_i = r_i / \sum r_j$ for each of the i, j = 1, · · · , S species present in one relevé.

2.3. The Plantation Biodiversity Benefit Score (PBBS)

As [28] [78] point out it is crucial to assess biodiversity issues via standardized quantitative figures in order to enable political stakeholders to develop adequate conservation strategies. Since finding commonly acceptable standards for biodiversity is a hard task, it is proposed to use a variety of indicators to evaluate biodiversity conditions [28] [79]. One of the measures we used in order to meet these requirements is the plantation biodiversity benefit score (PBBS) proposed by [80]. The PBBS is a method specifically designed to assess the value of plantations for biodiversity conservation which consists of two components. The site scale components measures predominantly habitat diversity (e.g. presence of paddock trees, dead wood, age structure, species composition, presence of native trees), at specific sites up to 75 points can be yielded. The landscape scale component measures landscape heterogeneity and the integration of a plantation into a landscape (e.g. surrounding with native vegetation, shape of the plantation), up to 25 points can be yielded. The PBBS is the sum of both which therefore can reach a maximum value of 100.

2.4. Biodiversity Indices

Several biodiversity indices are computed herein which are derived from relative abundance values. [81] proposes a new and consistent concept of defining biodiversity indices. α-diversity ${}^{q}D_{\alpha}$ indices are derived from the basic sum of relative abundances $\sum p_i^q$ using some exponent q, that is by computing $^qD_\alpha = \left(\sum p_i^q\right)^{1/(1-q)}$. From this equation, most biodiversity indices can be derived. The first index used, $^0D_\alpha$, i.e. using q = 0 equals S, the species richness. The second index employed herein $^2D_\alpha$ uses an exponent of q = 2. Note that with this ex-

ponent, the basic sum $\sum p_i^2$ equals the original Simpson index and $^2D_\alpha = \left(\sum p_i^2\right)^{1/(1-2)} = \left(\sum p_i^2\right)^{-1}$ equals the

inverse Simpson index. Besides richness $^0D_\alpha$ and inverse Simpson diversity $^2D_\alpha$ an evenness index should be employed. In order to avoid biases and redundancy in the analysis, evenness should be independent of richness. [82] shows that this is only possible if evenness = diversity/richness is assumed. Thus, evenness is computed as $^2E_\alpha = {}^2D_\alpha / {}^0D_\alpha$.

For β-diversity, following [81] the Sørensen index $^qD_\beta$ of species turnover between two plots A and B is computed. Again, $^qD_\beta$ can be computed on the basis of different values for q. Herein, $^0D_\beta$ with q = 0 is used. Thus, differences of species relative abundances p_i are ignored at the β-scale species presences and absences are considered for β-diversity. $^0D_\beta$ is computed according to: $^0D_\beta = 2 \cdot {}^0D_\beta^{\underline{A1\,B}} / \left({}^0D_\beta^{\underline{A}} + {}^0D_\beta^{\underline{B}}\right)$. In order to facilitate inter pretation of the figures, we computed the maximum Sø rensen index max$^0D_{\beta\ \text{possible}}$ for each plantation/forest couple. We did so by assuming that all species of the site with the lowers pecies number are found on the other site as well, i.e. $^0D_\beta^{\underline{A1\,B}} = \min\left({}^0D_\beta^{\underline{A}}, {}^0D_\beta^{\underline{B}}\right)$. We further more gave the percentage of $^0D_\beta$, i.e. $\%{}^0D_\beta = \left(100/\max {}^0D_\beta\right) \cdot {}^0D_\beta$.

2.5. Statistical Analysis

We analyzed our data with standard statistical test procedures. Some indices of α-diversity, like e.g. richness $^0D_\alpha$, cannot be assumed to be Gaussian distributed. Instead of having to test for Gaussianness and homoscedasticity for each pair of variables, and employing a different test for each case, a consistent test procedure is employed. The nonparametric Mann-Whitney U test, which does not assume Gaussianness is employed for each pair of variables. In order to clarify the amount of reduction in $^0D_\alpha$, we performed Mann-Whitney U test on the hypothesis that the values $^0D_\alpha^A$ of richness $^0D_\alpha$ in plant formation A are lower than some percentage of the values ${}^0D_\alpha{}^B$ in another plant formation B (i.e. $H1 = {}^0D_\alpha{}^A < f \cdot {}^0D_\alpha{}^B$). We

evaluate the values by testing the assumption several times and by determining the critical value f_c, where H0 cannot be rejected anymore. $p_c = f_c \cdot 100$ is assumed to be the statistically significant percentage of reduction. Results were considered as significant on the basis of the p £ 0.05 confidence level. All tests were computed using Matlab R2014a.

2.6. Regression Analysis

The richness data ${}^0D_\alpha$ of plantations is assumed to partly depend on tree height. Young plantations with open canopies impose less harsh competition on understory species than plantations with higher trees and closed canopies. Thus, in younger plantations, higher ${}^0D_\alpha$ are expected than in plantations with higher trees. In very old plantations with the highest trees, more time for ecosystem development has passed, leaving chances for more understorey species to establish. Thus, in plantations with the highest trees, richness ${}^0D_\alpha$ is also expected to be higher. In total, these assumption theoretically motivate using 2nd degree polynomial functions, which were used exclusively herein. The quality of the regression curved is assessed using the R^2 measure. Furthermore, a randomized procedure is employed to assess the quality of regressions. Regressions are computed N = 1000 times on random subsets including 75% of the field data of the respective plantations. For each subset, a regression curve is produced. Then, the standard deviation σ of the N = 1000 regressions is computed and plotted in the graphs. All regressions were computed using the statistical package Matlab R2014a.

3. RESULTS

3.1. Alpha-Scale—Richness, Diversity, Evenness

At first, the characteristics of the four major plant formations (near-natural forests, degraded forests, plantations and shrublands) investigated were outlined. Near-natural forests are found in NPRs and as small remnants within plantations. All three types of near-natural forests which were assessed showed a heterogeneous and diverse aspect. They were composed of three to five strata (trees, shrubs and herbaceous strata, the latter one is missing in riparian forests). For most near-natural forests we sampled the minimum areal size required to reach an asymptote in the species-area curve was 160 ´ 160 m, however, in some forests, up to 640 ´ 640 m had to be assessed. Near-natural forests are composed of numerous herbaceous and woody plant species which grow so densely that accessibility is not assured in some places. Degraded forests also show a heterogeneous and diverse aspect. They consist of two to four vegetation strata and have a guideline minimum areal of 160 ´ 160 m. Under tree species which dissociate allelopathic compounds—e.g. *C. alba* and *P. boldus* [83] —the richness of the herbaceous stratum can be reduced. All sites were more open than the near-natural forests and affected by human presence. The invasive bamboo species *C. quila* was found at sites were light availability and soil moisture were sufficiently high. We investigated one degraded forest of *E. globulus* and three of *P. radiata*. The sites had been established as plantations in the first place but had to remain unmanaged afterwards due to local governmental decisions. Richness within these sites was comparable to the one found under C. *alba* and *P. boldus*. However, sites were partly invaded by *T. monspessulana* or *Rubus ulmifolius* which both are far less abundant or absent in the near-natural forests. The *A. caven* shrublands are composed of a single tree (*A. caven*). Coverage of *A. caven* differs between them. They are composed of two to three vegetation strata and have a minimum areal of 20 ´ 20 m. The sites were used as pastures with differing intensity. Hence, the herbaceous stratum was affected by grazing. Exotic tree plantations show a very homogeneous pattern, which is not surprising given the fact they are managed as monocultures in Central Chile. An area of 20 ´ 20 m should be assumed as the minimum areal, though many plantations had minimum areas of only 5 ´ 5 m. Below the tree stratum, one to four shrub or herbaceous strata are found. In most cases, their coverage is very low. In the cases where coverage is higher, it is frequently composed of invasive shrubs (*T. monspessulana, R. ulmifolius, Rosa rubiginosa* and *Acacia dealbata*).

Boxplots of the richness values ${}^0D_\alpha$ are given in Figure 2(a), Figure 2(d) and Figure 2(g). The average richness ${}^0D_\alpha$ of native forests is 33.3 species. Degraded forests show the second highest richness ${}^0D_\alpha$ with 20.8 species, followed by shrublands with 13.5 species and plantations with 13.1. In the next step, the differences between richness ${}^0D_\alpha$ of different landscape units were tested for significance according to the scheme described in Sec. 2.5. The richness ${}^0D_\alpha$ of near-natural forests was significantly higher than the richness ${}^0D_\alpha$ of plantations (Mann-Whitney U Test $p \approx 8.402^{-7}$). Furthermore, the near-natural forests hosted significantly more species than degraded forests and shrublands (Mann-Whitney U Test $p \approx 0.032/0.0033$). Other differences were found to be insignificant. Hence, a significant reduction of richness ${}^0D_\alpha$ below exotic tree plantations could be observed.

Next, we determined the amount of reduction in ${}^{0}D_{\alpha}$. We evaluated different values of f and found $f_c = 0.50$ to be the critical value. From there, we concluded that the richness ${}^{0}D_{\alpha}$ below plantations was reduced by $p_c = 45\%$ in comparison to near-natural forest.

3.1.1. Rarefaction Analysis

The usage of richness ${}^{0}D_{\alpha}$ in biodiversity considerations is controversially debated. On the one hand, richness ${}^{0}D_{\alpha}$ is an easy interpretable and meaningful figure which has a much more natural interpretation than e.g.

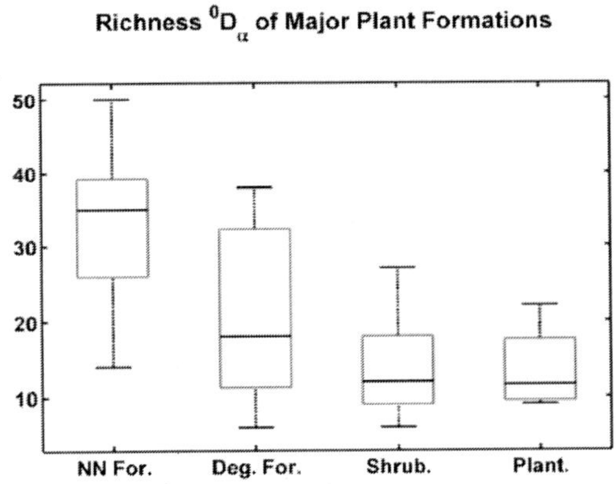

(a)

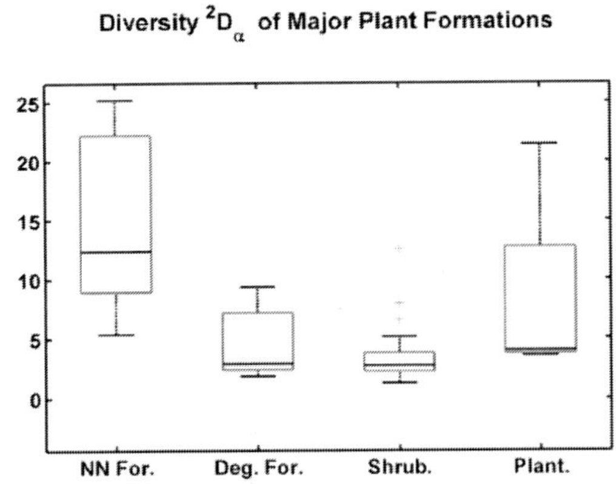

(b)

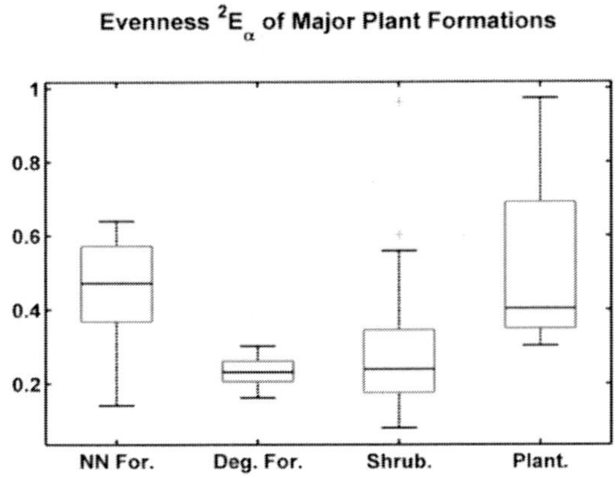
Evenness $^{2}E_{\alpha}$ of Major Plant Formations
1
0.8
0.6
0.4
0.2
NN For.
Deg. For.
Shrub.
Plant.

(c)

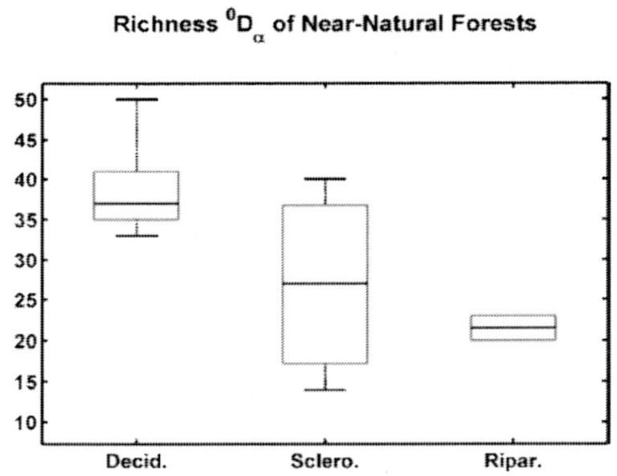
Richness $^{0}D_{\alpha}$ of Near-Natural Forests
50
45
40
35
30
25
20
15
10
Decid.
Sclero.
Ripar.

(d)

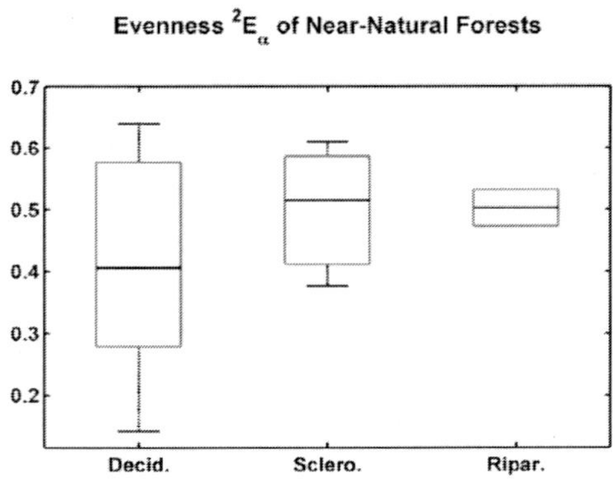
Evenness $^{2}E_{\alpha}$ of Near-Natural Forests
0.7
0.6
0.5
0.4
0.3
0.2
Decid.
Sclero.
Ripar.

(e)

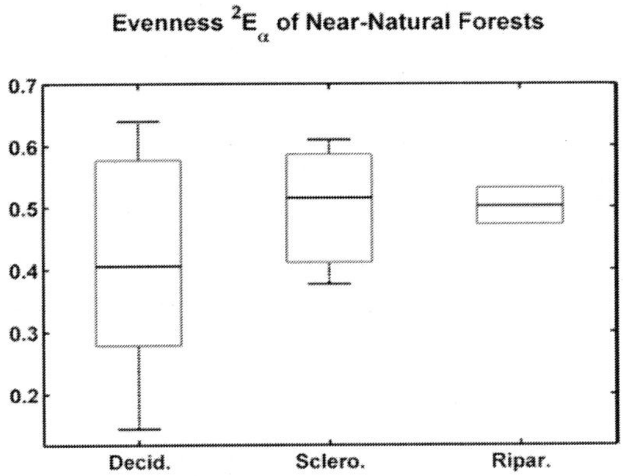
Evenness ²Eα of Near-Natural Forests
0.7
0.6
0.5
0.4
0.3
0.2
Decid.
Sclero.
Ripar.

(f)

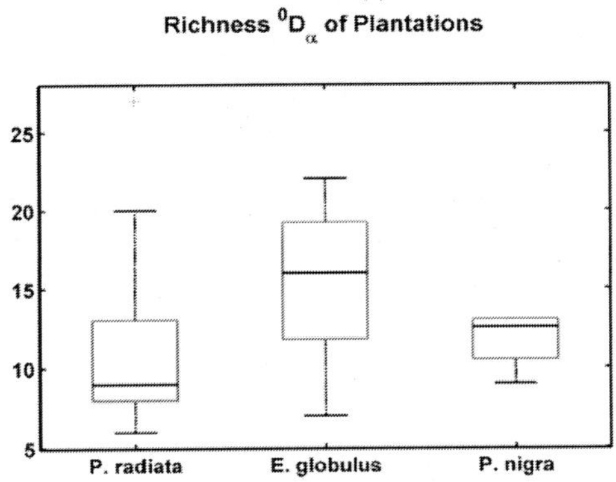
Richness ⁰Dα of Plantations
25
20
15
10
5
P. radiata
E. globulus
P. nigra

(g)

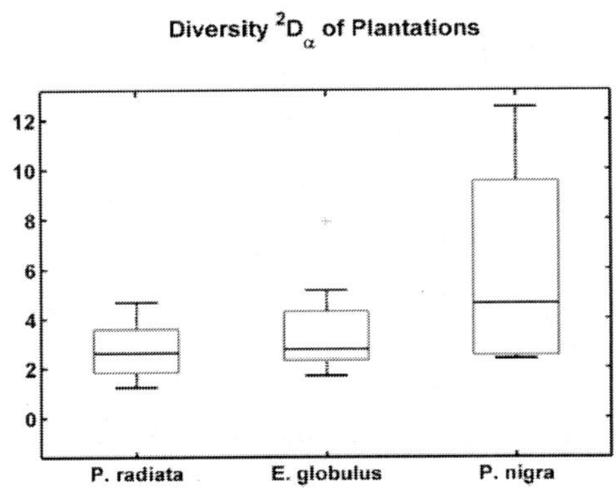
Diversity ²Dα of Plantations
12
10
8
6
4
2
0
P. radiata
E. globulus
P. nigra

(h)

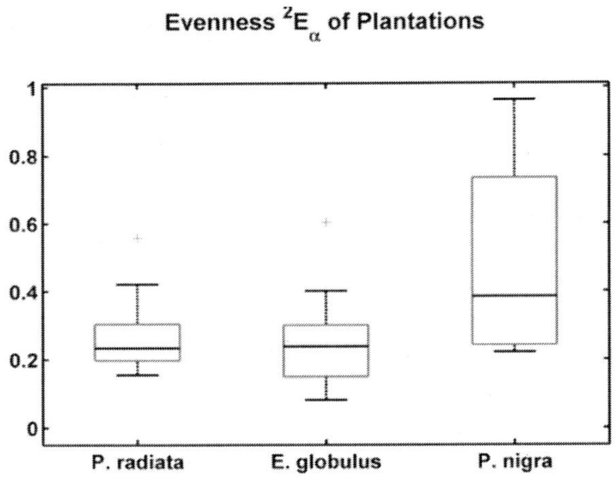

(i)

Figure 2. Boxplots of richness, diversity and evenness indices for the study systems observed. Top: richness $^0D_\alpha$, center: diversity $^2D_\alpha$, bottom: evenness $^2E_\alpha$. Left: major plant formations, center: near-natural forests, right: plantations. Abbreviations are NN For.: near-natural forest, Deg. For.: degraded forest, Shrub.: *A. caven* shrubland, Plant.: commercial plantation, Decid.: deciduous forest, Sclero.: sclerophylluous forest, Ripa.: riparian forest.

higher order $^qD_\alpha$ values [28] [83] [84]. On the other hand, it is dependent on the sample size [30] [84] [85]. Therefore, when arguing with richness$^0D_\alpha$, one should be sure to have sampled a sufficiently large area. Rarefaction analysis can be used to prove this assumption [27] [85]. We computed sample based rarefaction curves according to [85] , which can be seen in Figure 3(a). In case of sufficient sampling, the curves asymptotically converge the expected richness $^0D_\alpha$. However, rarefaction is sensitive to both high numbers of rare and very common species. 38 of the species (~23%) were found in only one relevé, 57 of the species (~35%) were found in up to two relevés. The plantation trees were present in all plantation relevés. Therefore, both rare and common species were present, affecting the convergence of rarefaction curves, which approximately converged for near-natural forests and plantations but not for degraded forests and shrublands.

3.1.2. Diversity

Near-natural forests had an average diversity index $^{2}D_{\alpha}$ of 14.78 indicating that a high number of effective species is present and many of them have higher abundances. Shrublands had the second highest average index (8.25). Degraded forests had an average index value of 4.78. Plantations yield the lowest diversity index $^{2}D_{\alpha}$ values with a value of 3.38, Figure 2(b), Figure 2(e), Figure 2(h). Near-natural forests had a significantly higher diversity index $^{2}D_{\alpha}$ index than plantations and degraded forests did (Mann-Whitney U Test $p \approx 3.580^{-7}/ 0.001$). Other differences were found to be insignificant.

3.1.3. Evenness

Near-natural forest yielded the an average evenness $^{2}E_{\alpha}$ value of 0.44, shrublands yielded a value of 0.52, followed by plantations (0.28) and degraded forests (0.23), Figure 2(c), Figure 2(f), Figure 2(i). Near-natural forests had a significantly higher evenness $^{2}E_{\alpha}$ than plantations and degraded forests (Mann-Whitney U Test $p \approx 0.001/0.007$). Furthermore, the differences between degraded forests and shrublands (Mann-Whitney U Test $p \approx 0.006$) and between shrublands and plantations (Mann-Whitney U Test $p \approx 0.019$) were also significant. Other differences were not significant on the basis of the p £ 0.05 confidence level.

3.2. Alpha-Scale—Differences between Individual Plant Formations

For the three different types of forests, only differences in the richness $^{0}D_{\alpha}$ between mixed deciduous forests and riparian forests were significant (Mann-Whitney U Test $p \approx 0.004$). For the three different types of plantations, only differences in the richness $^{0}D_{\alpha}$ between P. radiata and E. globulus were significant (Mann-Whitney U Test $p \approx 0.012$). Since P. radiate covered the largest areas [67] , the plantation type predominant in Central Chile obviously is far less favourable in terms of biodiversity than less frequent plantation types.

Our field experience from plantations not sampled revealed a clear ranking of richness's $^{0}D_{\alpha}$, which were

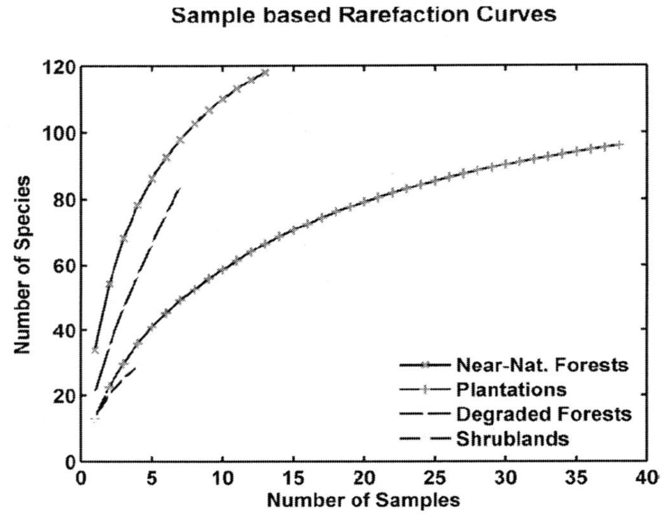

(a)

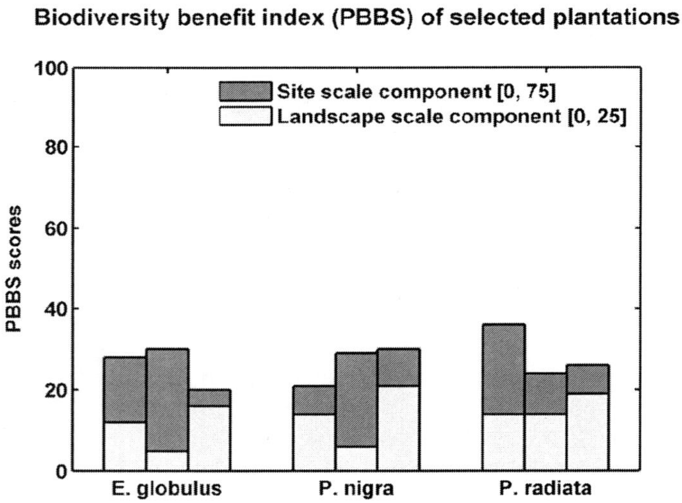

(b)

Figure 3. Left.: Sample based rarefaction curves for the studies ecosystems. Number of samples vs. number of species. Sufficiently sampled ecosystems show asymptotically converging curves. Right: Biodiversity benefit index for sampled plantations (PBBS). Site scale component and landscape scale component.

highest below E. globulus, followed by *P. nigra* and *P. radiata*. For both diversity index ${}^2D_\alpha$ and evenness ${}^2E_\alpha$ values, *P. radiata* plantations yielded lower values than both *E. globulus* and *P. nigra.*

3.3. Alpha-Scale—Relationship between Richness and Tree Height

In the next step, the relationship between richness ${}^0D_\alpha$ and height of the exotic trees within plantations was investigated. Experience from the field survey revealed that the number of species in the shrub and herbaceous strata is higher below young trees of *P. radiata, E. globulus* and *P. nigra*. Once the trees grow higher and close their canopy, richness ${}^0D_\alpha$ in the strata below is reduced. In very old plantations of *P. radiata* and *E. globulus,* a slight rise of richness ${}^0D_\alpha$ can be recorded. However, this observation does not account for *P. nigra*. In order to investigate the relationships statistically, regression curves to the plots species number against tree height were computed in Figure 4. The richness ${}^0D_\alpha$ values of *P. radiata* and *E. globulus* were approximated via a 2nd degree polynomial fit, due to the lower number of relevés, *P. nigra* was linearly fitted. A convincing fit can be recorded for P. radiate (Figure 4(a), $R^2 \approx 0.796$). Richness ${}^0D_\alpha$ is highest[2] in P. radiata tree plantations of 2 m and lowest at 21 m. A similar, but far less strong tendency can be observed for E. globules (Figure 4(b), $R^2 \approx 0.274$). For *E. globulus* plantations richness ${}^0D_\alpha$ is highest at 3 m and lowest at 17 m. The lower determinedness is explained by the large variance in species number at medium tree heights (around 10 m). Unlike *P. radiata* plantations, *E. globulus* plantations do differ much more at intermediate ages and heights. The distances between individual trees differ much more and while some sites are densely covered with pulp *E. globulus,* others are not. These factors seem to impose variable pressure on other species. Therefore the slight rise of the fitting curve at bigger tree heights should be interpreted cautiously. Due to the fact that P. nigra is far less frequently planted in Central Chile, we only took four relevés. Therefore, a linear fit instead of a 2nd degree polynomial fit has been computed (Figure 4(c), $R^2 \approx 0.379$). Richness ${}^0D_\alpha$ is highest in plantations of 8 m and lowest at 25 m. This low value is explained by the fact that herbicides had been applied to the site a couple of months before our relevé. Keeping in mind this fact, a similar tendency as for *P. radiata* and *E. globulus* was expected and confirmed by our field experience.

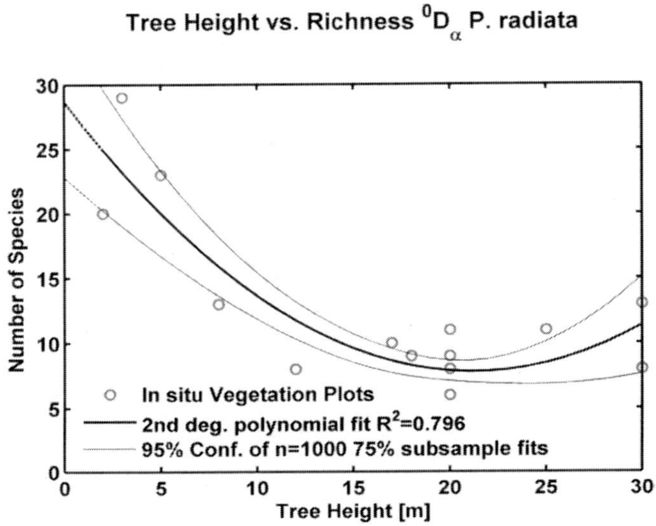
Tree Height vs. Richness $^0D_\alpha$ P. radiata
Number of Species
Tree Height [m]
In situ Vegetation Plots
2nd deg. polynomial fit R^2=0.796
95% Conf. of n=1000 75% subsample fits
0
5
10
15
20
25
30

(a)

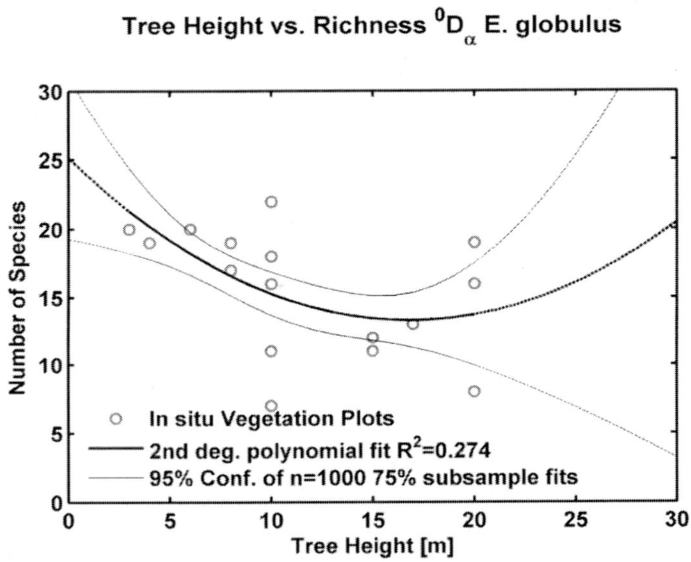
Tree Height vs. Richness $^0D_\alpha$ E. globulus
Number of Species
Tree Height [m]
In situ Vegetation Plots
2nd deg. polynomial fit R^2=0.274
95% Conf. of n=1000 75% subsample fits
0
5
10
15
20
25
30

(b)

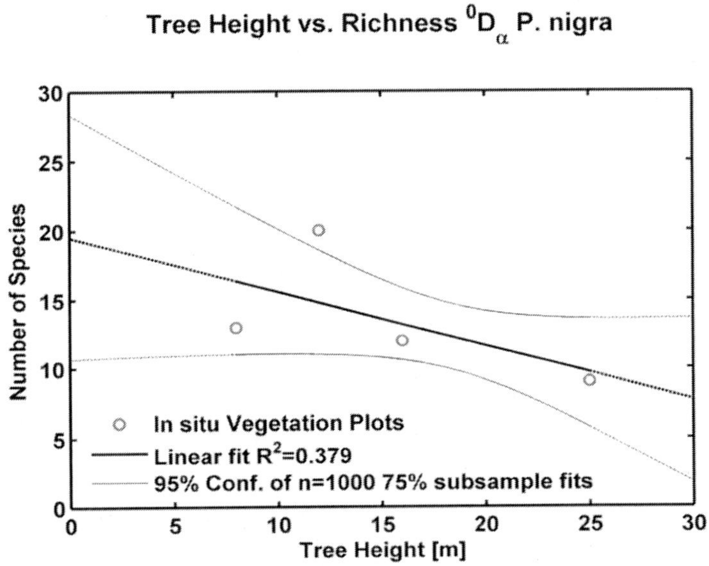

(c)

Figure 4. Regression curves between tree height and richness $^0D_\alpha$ in the plantations and confidence intervals of the regressions. Dashed lines: extrapolated parts. Points: *in situ* assessments. Left: *R. radiata*, center: *E. globulus*, right: *P. nigra*.

3.4. Beta-Scale—Species Turnover between Forest-Plantation Couples

In is frequently argued that plantations can contribute to biodiversity since they raise heterogeneity of landscapes, thus helping to creating diverse habitat conditions raising nichés for species [22] [30] [33] [34] [86]. If this assumption holds true, β-diversity between plantations and native species should be high; or *β*-similarity should be low, respectively. In the VII and VIII region of Central Chile, plantations cover vast areas and native forests are found almost exclusively as remnants within plantations, in riparian habitats along rivers or in national reserves [10] [88]. We chose ten specific sites were plantations actually appear closely linked to native forests. Five sites represented P. radiata plantation and five were *E. globulus* plantations. Individual figures for the index can be found in Table 1 . As can be seen, most plantation sites had a relatively high *β*-similarity with the native forests found nearby. The lowest value of % $^0D_\beta$ yielded was 42.9%, the highest value was 100%, stating that all species found in the plantation also appear in the native forest (it should be noted that the opposite is not the case). The % $^0D_\beta$ of all plantations was 64.5%, indicating that most plantations show a high β-similarity with native forest. Frequently, the only species found in the plantation but not in the native forest were the plantation trees and exotic invaders like *T. monspessulana*, or *A. dealbata*. The average % $^0D_\beta$ of *P. radiata* plantations was 73.0% and 56.0% for *E. globulus* respectively. Thus, it appears that *E. globulus* contributes more to β-diversity than P. radiata does.

3.5. Gamma-Scale—Relative Abundance of Exotic Species and Endemics

As [78] point out, it is important to not only assess biodiversity on the site scale. A scale jump to the landscape scale is crucial in order to draw valid conclusions on biodiversity impacts. By then, we had analyzed and compared the site specific α-diversity of different natural and managed plant formations and analyzed the β-diversity between them. Therefore, in a next step, we aimed to assess the plant biodiversity on the landscape scale via considerations on γ-diversity. Due to the vast extent of deforestation and plantation establishment in Central Chile, it might be argued that the landscape scale of γ-diversity is the most important one. In fact, we found 165 plant species in 63 families in total. Among them, 69 species (~42%) appear exclusively in non-commercial ecosystems (near-natural forests, degraded forests and shrublands).

Table 1. Comparison of near-natural forests and plantations on the β-scale, using the Sorensen $^0D_\beta$ index.

Plot A	Plot B	Plant Formation A	Plant Formation B	Sorensen Index		
				$^0D_\beta$	max$^0D_\beta$	%$^0D_\beta$
No. 13	No. 15	*P. radiata* plantation	Near-natural forest	0.39	0.56	70.0
No. 37	No. 39	*P. radiata* plantation	Near-natural forest	0.37	0.37	100.0
No. 24	No. 35	*P. radiata* plantation	Near-natural forest	0.34	0.66	52.6
No. 38	No. 39	*P. radiata* plantation	Near-natural forest	0.47	0.73	65.0
No. 25	No. 28	*P. radiata* plantation	Near-natural forest	0.27	0.35	77.8
No. 09	No. 10	*E. globulus* plantation	Near-natural forest	0.48	0.85	57.1
No. 27	No. 28	*E. globulus* plantation	Near-natural forest	0.20	0.31	62.5
No. 16	No. 15	*E. globulus* plantation	Near-natural forest	0.13	0.31	42.9
No. 59	No. 56	*E. globulus* plantation	Near-natural forest	0.19	0.41	46.2
No. 12	No. 10	*E. globulus* plantation	Near-natural forest	0.61	0.85	71.4
Mean	of	All plantations		0.35	0.54	64.5
Mean	of	*P. radiata* plantations		0.36	0.53	73.0
Mean	of	*E. globulus* plantations		0.32	0.54	56.0

Only 9 species (~5%) appear exclusively in plantations. Of these, five are introduced in Chile (*Cirsium vulgare, Crysanthemum coronarium, P. nigra, Robinia pseudoacacia,* and *Stellaria media*). Three are native species (*Lobelia tupa, Tropaeolum leptophyllum* and *Vicia magnifolia*). The last one, *Escallonia revoluta,* is endemic to Chile. Among the 165 species, 35 are endemic to Chile (~21%), 56 species have been introduced (~34%). Of the 35 endemics, 34 (~97%) were found in at least one relevé from a natural ecosystems and 14 (~40%) in plantations. In general, many species—endemic or not—are rarely found. On average, species are found in only 7 relevés. Introduced species have been shown to successfully invade natural ecosystems in Chile [89] [90]. Therefore, we aimed to analyze the effects of plantations' invasive species in the next step by applying the manual of invasive plants of Chilean authorities as data basis [90]. From these manual, we selected 18 species which are expected to be found at this time of year. We calculated the relative abundances p_i of each species mentioned therein at each relevé site. We furthermore calculated the sum of relative abundances $\sum p_i$ of invasive species to show to which extent each site is dominated by (potentially) invasive species. Of course, the plantation trees themselves were not included in this analysis since naturally high abundances would strongly bias results. Within plantations, 3.8 of the 18 invasive species were found on average, whereas in native forests, 3.6 invasive plants were found on average. The number of invasive species encountered within plantations was not significantly different from native forest (t-Test $p = 0.7348$). Subsequently, we tested the difference of the sum of relative abundances $\sum p_i$ of the 18 invasive species for significance and revealed that $\sum p_i$ is significantly higher in plantations (Mann-Whitney U Test $p \approx 0.009$). Furthermore, we tested differences in the relative abundances pi of individual species for significance. For most species, no significant differences could be demonstrated, though *T. monspessulana, Plantago lanceolata, A. dealbata* and *Verbascum thapsus* were significantly more abundant in plantations (Mann-Whitney U Test $p \approx$ 0.0038/0.026/0.050/0.040). Furthermore, some invasive species featured significantly higher abundance values in specific types of plantations. *C. vulgare, Eschscholzia californica, Lupinus angustifolius* and *Taraxacum officinale* were significantly more abundant in *P. nigra* plantations, while *P. lanceolata* was clearly more abundant in P. radiata plantations (Mann-Whitney U Test $p \approx$ 0.008/0.032/0.009/0.030/0.001). Since Central Chile is considered an area rich in endemics [9] [91] [92] , an important question is how plantations influence the abundance of endemics. We assessed this influence in the same manner as for invasive species. The information on species origin was derived from references like [93]. The average number of endemic species was 8.30 in native forests and 1.73 in exotic tree plantations. Differences were tested positively for significance. It may be argued that a reduction of relative abundances of endemic trees in plantation is straightforward. Yet, despite the intense discussions on plant biodiversity decline in plantations [10] - [20] , to the knowledge of the authors, the scientific proof for the case of plants in Central Chile is still missing. Moreover, plantations managed in a more sustainable manner may maintain higher relative abundances of endemic trees. Nevertheless, in order to gain information not only on trees, but also on the understory, we analyzed the tree, shrub and herbaceous stratum separately. We computed the relative abundances of each endemic species and tested the sum of relative abundances $\sum p_i$ for significant differences. The sum of relative abundances $\sum p_i$ of endemic species in the tree and shrub strata turned out to be significantly reduced in plantations (Mann-Whitney U Test $p \approx 0.000/0.004$. Differences in the herbaceous stratum were insignificant (Mann-Whitney U Test $p \approx 0.018$). In the tree stratum, the relative abundances pi of C. alba, Myrceugenia obtusa, Nothofagus alessandrii, N. glauca and N. obliqua were significantly reduced (Mann-Whitney U Test $p \approx 0.004/0.017/0.000/0.020$). The relative abundances pi of P. boldus, L. caustica and Quillaja saponaria were lower, yet differences were not significant. These species are frequently found as small shrubs in the understory of plantations. In the shrub stratum, the relative abundance of *Pseudopanax valdiviensis* was significantly reduced (Mann-Whitney U Test $p \approx 0.018$). In the herbaceous stratum, the relative abundances of *Hymenoglossum cruentum* and *Satureja multiflora* were significantly reduced (Mann-Whitney U Test $p \approx 0.016/0.017$). However, note that plantations tend to be composed of less vegetation strata, the same abundance value r_i will result in a higher relative abundance p_i.

3.6. Plantation Biodiversity Benefits Score

Since the PBBS measures biodiversity benefits through plantations, plantations that are claimed to promote—or at least maintain—biodiversity should yield high PBBS values (attesting good management practices to forest companies). In order to evaluate a possible beneficial influence of Chilean plantations on plant biodiversity, the method was applied to three sites of each tree (note that the PBBS focuses on biodiversity in general, not only plant biodiversity). All plantations observed yield rather low values, cf. Figure 3(b), (a P. radiata plantation yielded the highest value of 37). There were no significant differences between the types of plantation according to the Mann-Whitney U test[3]. Since the management of the plantations is very similar, they were not expected either. The landscape scale component yielded rather poor values for all plantations observed. This is explained by the fact that plantations are almost completely surrounded by other plantations. The connection to native vegetation endorsed by the PBBS is very weak.

4. DISCUSSION

This study has evaluated the contribution of exotic tree plantations in Central Chile by comparing their biodiversity to forests and shrublands. It has been shown that plantations have a strong negative impact on plant biodiversity. The results indicate that plant biodiversity would benefit from diversifying plantations w.r.t. species composition. The diversity index ${}^2D_\alpha$ values in plantations were the lowest found in the plant formations observed. The diversity index ${}^2D_\alpha$ of near-natural forests were proven to be significantly higher. The evenness values ${}^2E_\alpha$ of plantations were the lowest among all plant formations observed.

Rarefaction analysis was performed in order to evaluate the sufficiency of the sampling for the analyses on α-diversity. From rarefaction, we conclude that near-natural forests and plantations are sufficiently sampled, for degraded forests and shrublands would benefit from further sampling. Therefore, conclusions based on the α-diversity of these plant formations need to be drawn more cautiously.

Within plantations, a correlation between tree height and richness ${}^0D_\alpha$ was substantiated. While younger plantations represent quite open habitats, light availability decreases rapidly once the canopy closes. At this age, plantations are quite similar, they differ almost exclusively with respect to pruning. However, due to the tight canopy, whether trees are pruned or not has little influence on the light availability. Furthermore, a dense layer of *P. radiata* needles is found within this particular type of plantations imposing further competitive pressure on other species. In plantations with higher trees richness ${}^0D_\alpha$ rises slightly again. Due to the longer time of ecosystem development, habitats are more diversified leaving more niches for other species. These findings reveal that plant biodiversity would benefit from diversifying plantations w.r.t. tree ages.

In order to test the hypothesis that plantations contribute to β-diversity by raising habitat heterogeneity, the Sørensen index of ten plantation/native forest couples was computed. Although the habitat types are structurally different, Sørensen indices were rather high, attesting a high β-similarity between plantations and native forests. However, these results have to be interpreted cautiously and compared with α-diversity. Results do not describe plantations as similar to forests w.r.t. plant biodiversity and richness ${}^0D_\alpha$, but merely demonstrate that almost all of the rather few plant species found in plantations (low α-diversity) also appear in native forests.

Thus, plantations do not seem to contribute to local ecosystem heterogeneity and do not seem to provide habitat to additional species in the case of Central Chile as hypothesized by e.g. [22] [30] [34] [86]. Moreover, it should be kept in mind that the coastal range of Central Chile is not a diverse, small-scale pattern of plantations mixed with other habitat types. Instead, it mainly consists of vast plantations with very few sites of native forests (<4% of the area, own data, unpublished results). Therefore, instead of raising local habitat diversity by incorporating small-scale patches of plantations, forestry in Central Chile vastly replaces a α-diverse habitat (native forests) by a habitat of low α-diversity (plantations).

Concerning γ-diversity in total, most of the species are exclusively found in non-commercial ecosystems, relatively few are exclusive to plantations. Among them, many are foreign species, however, one of them is endemic to Chile. It has been shown that invasive species contribute more to species composition within plantations. Although the individual abundances of many invasive species are not significantly higher, the sum of their abundances is. This is due to the fact that some invasive species (e.g. *T. monspessulana, P. lanceolata, A. dealbata* or *V. thapsus*) cover large areas of the herbaceous or shrub strata below plantations. These local dominances should be monitored carefully since they can be the nucleus of invasions into native ecosystems [94] [95]. The number of endemic tree species found was only slightly lower in plantations. However, since the relative abundance of endemic species was found to be clearly reduced in plantations, it is coherent to see endemic species threatened by plantations. Above, the decline of richness ${}^0D_\alpha$ under older plantations raises the question whether endemic species will be able to survive multiple rotations. While a reduction in relative abundances of endemic trees is straightforward (intentional deforestation), a reduction in their occurrence (number of endemics) gives much concern. If plantations are to be biotopes for the survival of endemic species, these species should be found at least within the shrub stratum of plantations. Yet for many species observed, this assumption does not seem to hold. However, it should be investigated in further research whether these species are present in the seed bank below plantations.

The PBBS is proposed as a measure for biodiversity benefits of plantations. Accepting this method as valid, any plantation which promotes biodiversity should yield high values. However, all nine plantations investigated herein yield quite low values. This situation is quite typical for most plantations in Central Chile. The site scale values yielded were relatively lower. This is explained mostly by the absence of paddock trees, dead wood, and the lack of microhabitat diversity, the low variability of plantations with respect to age and density, the lack of native trees and by the fact that most plantations are monotypic. Note that the PBBS constitutes another measure to perform the scale jump from site scale to landscape scale (sensu [78]) by combining both components. Therefore, it is concluded that management practices in Central Chile do not promote biodiversity, but impair it by contrast. It should be noted that the PBBS focuses on the total biodiversity, not only on plant biodiversity. We though believe that most of the factors that are considered beneficial for the total biodiversity should apply to plant biodiversity as well (e.g. landscape heterogeneity, long rotation times, and planting of local species, site preparation and artificial hollows).

Frequently, a survival of species by the metapopulation is observed despite local extinction [31] [32]. Such a survival mechanism is coupled to landscape connectivity, but also the presence of stepping stone biotopes. There are two possible types of stepping stone biotopes which could allow for a survival of endemics metapopulations. The remnants of native forests and young plantations. Remnants of native forests indeed feature a high α-diver- sity, yet distances between individual patches of these remnants can be quite large. Thus, only species with a reproductive mechanism effective over large distances are able to adapt to this situation. Some of the younger plantations we observed also had a rather high α-diversity. However, herbicide application before planting strongly threatens saplings of native species. We therefore cannot confirm the possibility of survival of endemic species through both stepping stone biotopes. Conservation strategies for Central Chile should not be optimistically based on this option. Both degraded forests and shrublands are frequently discussed as ecosystems maintaining high biodiversity levels after restoration of areas formerly used as plantations [22] [30] [34] [87]. Results on the biodiversity of these ecosystems are ambiguous. More research should be dedicated to clarify the differences in plant biodiversity of degraded forests, shrublands and plantations. Until then, management options should not be based on the assumption that plant biodiversity can be maintained by recreating degraded forests from plantations after a couple of rotations. The data presented delineate the plant biodiversity conditions in a mainly descriptive manner.

Based on our results, the reasons for the plant biodiversity decline can be clearly attributed to the management practices in force. Obviously, plantations of P. radiata, E. globulus and P. nigra are able to suppress the regeneration of native species. Since the degraded forests of P. radiata and E. globulus show a high plant biodiversity, the management seems to be more influential than the species themselves. However, the mechanisms explaining higher competitiveness are to be clarified. Besides herbicide application, light competition seems to be another important factor in the first two decades of plantation growth given the decrease of richness ${}^{0}D_{\alpha}$ with the tree height. In the last phase before clearfelling, richness ${}^{0}D_{\alpha}$ slightly rises. This is explained by the fact that within older stands, the diversity of habitats rises, allowing more species to compete [87] [96] [97]. Besides the Braun-Blanquet analyses, more data will be required to explain the mechanisms of higher competitiveness.

5. CONCLUSION AND OUTLOOK

Results indicate that plantations with exotic trees put the plant biodiversity and endemism in Central Chile under great risk. Plantations of *P. radiata* show the least number of species followed by plantations of *P. nigra* and *E. globulus*. Especially for *P. radiata*, a dependency between richness and height of the plantation trees could be established. The β-similarity between plantations and native forests is high, indicating that no gain in habitat heterogeneity is achieved by plantations. Concerning γ-diversity, it is shown that plantations give habitat to less total species, and are inhabited by more invasive species and less endemic species. Applying the PBBS as a measure for favorable management practices, it is shown that management should be significantly improved w.r.t. to biological conservation. The homogeneity at site scale (age structure, monotypic composition, lack of natural elements and native vegetation) and the lack of connectivity with native plant formations at the landscape scale explain for the low values yielded. The tremendous decline of Chilean native forests affects the land use of entire Central Chile. Management practices are considered as the main reason for the decline of plant biodiversity in the investigated area. Since the forest companies that manage plantations in the VII and VIII region of Chile are currently expanding their business to Patagonia, a similar tendency of plant biodiversity decline has to be expected.

Finally, management implications shall be outlined. Within 35 years, the temperate zone of Central Chile has experienced rapid deforestation [10]. Plantations of exotic trees, embedded in an environmental matrix of native forests have now virtually become small remnants of native forest, embedded in an environmental matrix of plantations sometimes referred to as Green Desert [22] [28] [39] [72]. Only a few percent of the native forest vegetation have remained and almost no more space has been left for creating further plantations in forested areas. Our results attest a considerable reduction of plant biodiversity in exotic tree plantations. Related work by other authors confirm these findings [41] [42] [45]. Apparently, the main driver of plant biodiversity loss due to plantations is the poor management of plantations. Since processes of deforestations that are almost finished in the VII and VIII Region are about to begin in Patagonia—even led by the same forest companies—a further loss of plant biodiversity and endemic taxa has to be expected if similar management methods are applied. The current process of adopting Forest Stewardship

Council (FSC) guidelines [65] [98] is an important first step but hardly sufficient if not substantiated and explicitly formulated into management practices [6] [99] [100]. Therefore, some important management implications shall be stated herein. First of all, harvesting by clear felling should be changed to a less intrusive form of harvesting, e.g. partial logging or gap logging with gaps smaller than 100 m in diameter [101] - [105]. According to literature on biodiversity conserving plantation management, a two-tiered approach would be ideal. Native forests should be conserved both within the landscape and within plantations. Research has shown that landscapes should retain at least 30% of native vegetation to prevent biodiversity decline [106] - [108]. However, in our opinion, deforestation should be strictly limited to a much lesser extent. Here, we agree with [98] that legally binding agreements for all involved parties (public and private) are strictly required. A much larger proportion of native forest remnants should be left untouched within plantations. 10% of native forest remnants within plantations are recommendable [34]. These remnants should be connected in order to enable species to form a functioning and sufficiently large metapopulation [34] [108]. Connectivity could be ensured following water courses which should be aligned by a larger buffer zone of native vegetation to prevent pollution of water supplies. Application of herbicides should be strictly limited to cases of serious pests, preventive application should be prohibited by legally binding agreements. Furthermore, plantations should be diversified w.r.t. age structure and species composition and native trees should be left in between plantations. If these management implications are ignored, Patagonia is bound to convert into an extension of the Green Desert found in the Central zone.

ACKNOWLEDGEMENTS

Acknowledgments are given to scientists of University of Concepción (UdeC), Concepción Chile for assistance during the fieldwork. These persons are Dr. Rafael Garcia, Helmuth Puschmann, Dr. Cristian Echeverria, Dr. Mauricio Aguayo, Dr. Carolina Rojas and Dr. Gunhild Hansen-Rojas.

Furthermore, the authors acknowledge the financial support for the fieldwork, granted by Ph.D. students scholarships granted by Karlsruhe Institute of Technology (KIT), Karlsruhe, which has financed the in situ sampling of this contribution.

Andreas Ch. Braun, Joachim Vogt (DFG) and the Albert Ludwigs University Freiburg in the funding programme Open Access Publishing.

REFERENCES

1. Fearnside, P.M. (2005) Deforestation in Brazilian Amazonia: History, Rates, and Consequences. Conservation Biology, 19, 680-688. http://dx.doi.org/10.1111/j.1523-1739.2005.00697.x.
2. Brook, B.W., Sodhi, N.S. and Ng, P.K.L. (2003) Catastrophic Extinctions Follow Deforestation in Singapore. Nature, 424, 420-426. http://dx.doi.org/ 10.1038/nature01795
3. Pandit, M.K., Sodhi, N.S., Koh, L.P., Bhaskar, A. and Brook, B.W. (2007) Unreported yet Massive Deforestation Driving Loss of Endemic Biodiversity in Indian Himalaya. Biodiversity and Conservation, 16, 153-163. http://dx.doi.org/10.1007/s10531-006-9038-5
4. Cowlishaw, G. (1999) Predicting the Pattern of Decline of African Primate Diversity: An Extinction Debt from Historical Deforestation. Conservation Biology, 13, 1183-1193. http://dx.doi.org/10.1046/j.1523-1739.1999.98433.x
5. Sanchez-Cordero, V., Illoldi-Rangel, P., Linaje, M., Sarkar, S. and Peterson, A.T. (2005) Deforestation and Extant Distributions of Mexican Endemic Mammals. Biological Conservation, 126, 465-473. http://dx.doi.org/10.1016/j.biocon.2005.06.022.
6. Bennett, E.L. (2001) Timber Certification: Where Is the Voice of the Biologist? Conservation Biology, 15, 308-310. http://dx.doi.org/10.1046/j.1523-1739.2001.015002308.x.
7. Putz, F.E., Redford, K.H., Robinson, J.G., Fimbel, R. and Blate, G.M. (2000) Biodiversity Conservation in the Context of Tropical Forest Management. The World Bank, Washington DC.
8. Forman, R.T. and Alexander, L.E. (1998) Roads and Their Major Ecological Effects. Annual Review of Ecology and Systematics, 29, 207-231. http:// dx.doi.org/10.1146/annurev.ecolsys.29.1.207.
9. Myers, N., Mittermeier, R.A., Mittermeier, C.G., da Fonseca, G.A.B. and Kent, J. (2000) Biodiversity Hotspots for Conservation Priorities. Nature, 403, 853-858. http://dx.doi.org/10.1038/35002501.
10. Echeverria, C., Coomes, D., Salas, J., Rey-Benayas, J.M., Lara, A. and Newton, A. (2006) Rapid Deforestation and Forest Fragmentation of Chilean Temperate Forests. Biological Conservation, 130, 481-494. http://dx.doi.org/ 10.1016/j.biocon.2006.01.017.
11. Smith-Ramirez, C. (2004) The Chilean Coastal Range: A Vanishing Center of Biodiversity an Endemism in South American Temperate Rainforests. Biodiversity and Conservation, 13, 373-393. http://dx.doi.org /10.1023/ B:BIOC.0000006505.67560.9f.
12. Braun, A.C., Rojas, C., Echeverria, C., Rottensteiner, F., Bähr, H.P., Niemeyer, J., Aguayo, M., Kosov, S., Hinz, S. and Weidner, U. (2014) Design of a Spectral-Spatial Pattern Recognition Framework for Risk Assessments Using Landsat Data—A Case Study in Chile. IEEE Journal of Selected Topics in Applied Earth Observations and Remote Sensing, 7, 917-928. http://dx.doi.org/10.1109/JSTARS.2013.2293421
13. Grez, A.A., Bustamante, R.O., Simonetti, J.A. and Fahrig, L. (1998) Landscape Ecology, Deforestation and Forest Fragmentation: The Case of the Ruil Forest in Chile. In: Salinas-Chavez, E. and Middleton, J., Eds., Landscape Ecology as a Tool for Sustainable Development in Latin America, Universitario Santiago de Chile, Santiago de Chile, 32-42
14. Bustamante, R.O. and Castor, C. (1998) The Decline of an Endangered Temperate Ecosystem: The Ruil (Nothofagus alessandrii) Forest in Central Chile. Biodiversity and Conservation, 7, 1607-1626.
15. Altamirano, A. and Lara, A. (2010) Deforestation in Temperate Ecosystems of Pre-Andean Range of South-Central Chile. Bosque, 31, 53-64.

16. Cisternas, M., Martinez, P., Oyarzun, C. and Debels, P. (1999) Characterization of the Replacement Processes of Native Vegetation by Forest Plantations in a Lacustrine Watershed in the Nahuelbuta Mountain Range, VIII Region, Chile. Revista Chilena de Historia Natural, 72, 661-676.
17. Altieri, M.A. and Rojas, A. (1999) Ecological Impacts of Chile's Neoliberal Policies, with Special Emphasis on Agroecosystems. Environment, Development and Sustainability, 1, 55-72. http://dx.doi.org/10.1023/A:1010063724280
18. Clapp, R.A. (1995) The Unnatural History of the Monterey Pine. The Geographical Review, 85, 1-19. http://dx.doi.org/10.2307/215551
19. Morales, F.R. (1996) La necesidad de avanzarhaciauna Historia Ecologica para Chile. Ambiente y Desarrollo, 12, 61-68.
20. Armesto, J.J., Villagran, C. and Donoso, C. (1996) Desde la era glacial a la industrial: La historia del bosque templado chileno. Ambiente y Desarrollo, 12, 19-24.
21. Wilson, K., Newton, A., Echeverria, C., Weston, C. and Burgman, M. (2005) A Vulnerability Analysis of the Temperate Forests of South Central Chile. Biological Conservation, 122, 9-21. http://dx.doi.org/10.1016/j.biocon.2004.06.015.
22. Guerrero, P.C. and Bustamante, R.O. (2007) Can Native Tree Species Regenerate in Pinus radiata Plantations in Chile? Evidence from Field and Laboratory Experiments. Forest Ecology and Management, 253, 97-102. http:// dx.doi.org/10.1016/j.foreco.2007.07.006.
23. Guerrero, P.C. and Bustamante, R.O. (2009) Abiotic Alterations Caused by Forest Fragmentation Affect Tree Regeneration: A Shade and Drought Tolerance Gradient in the Remnants of Coastal Maulino Forest. Revista Chilena de Historia Natural, 82, 413-424.
24. Bustamante, R.O. and Simonetti, J.A. (2005) Is Pinus radiata Invading the Native Vegetation in Central Chile? Demographic Responses in a Fragmented Forest. Biological Invasions, 7, 243-249. http://dx.doi.org/10.1007/s10530-004-0740-5.
25. Altamirano, A., Echeverria, C., Lara, A. and Rey-Benayas, J.M. (2007) Effect of Forest Fragmentation on Vegetation Structure of Legrandia concinna (Myrtacea) Threatened Populations in South-Central Chile. Revista Chilena de Historia Natural, 80, 27-42.
26. Langdon, B., Pauchard, A. and Aguayo, M. (2010) Pinus contorta Invasion in the Chilean Patagonia: Local Patterns in a Global Context. Biological Invasions, 12, 3961-3971. http://dx.doi.org/10.1007/s10530-010-9817-5
27. Paritsis, J. and Aizen, M.A. (2008) Effects of Exotic Conifer Plantations on the Biodiversity of Understory Plants, Epigeal Beetles and Birds in Nothofagus dombeyi Forests. Forest Ecology and Management, 255, 1575-1583. http://dx.doi.org/10.1016/j.foreco.2007.11.015.
28. Altamirano, A., Field, R., Cayuela, L., Aplin, P., Lara, A. and Rey-Benayas, J.M. (2010) Woody Species Diversity in Temperate Andean Forests: The Need for New Conservation Strategies. Biological Conservation, 143, 2080-2091. http:// dx.doi.org/10.1016/j.biocon.2010.05.016.
29. Asmüssen, M.V. and Simonetti, J.A. (2007) Can a Developing Country Like Chile Invest in Biodiversity Conservation? Environmental Conservation, 34, 183-185. http://dx.doi.org/10.1017/S0376892907004183.
30. Brockerhoff, E.G., Ecroyd, C.E. and Langer, E.R. (2001) Biodiversity in New Zealand Plantation Forests: Policy Trends, Incentives, and the State of Our Knowledge. New Zealand Journal of Forestry, 46, 31-37.
31. Murphy, H.T. and Lovett-Doust, J. (2004) Context and Connectivity in Plant Metapopulations and Landscape Mosaics: Does the Matrix Matter? Oikos, 105, 3-14. http://dx.doi.org/10.1111/j.0030-1299.2004.12754.x
32. Baum, K.A., Haynes, K.J., Dillemuth, F.P. and Cronin, J.T. (2004) The Matrix Enhances the Effectiveness of Corridors and Stepping Stones. Ecology, 85, 2671-2676. http://dx.doi.org/10.1890/04-0500
33. Barlow, J., Gardner, T.A., Araujo, I.S., Avila-Pires, T.C., Bonaldo, A.B., Costa, J.E., Esposito, M.C., Ferreira, L.V., Hawes, J., Hernandez, M.I.M., Hoogmoed, M.S., Leite,

R.N., Lo-Man-Hung, N.F., Malcom, J.R., Martins, M.B., Mestre, L.A.M., Miranda-Santos, R., Nunes-Gutjahr, A.L., Overal, W.L., Parry, L., Peters, S.L., Ribeiro-Junior, M.A., da Silva, M.N.F., da Silva Motta, C. and Peres, C.A. (2007) Quantifying the Biodiversity Value of Tropical Primary, Secondary, and Plantation Forests. Proceedings of the National Academy of Sciences of the United States of America, 104, 18555-18560. http://dx.doi.org/10.1073/pnas.0703333104.

34. Hartley, M.J. (2002) Rationale and Methods for Conserving Biodiversity in Plantation Forests. Forest Ecology and Management, 155, 81-95. http://dx.doi.org/10.1016/S0378-1127(01)00549-7.
35. Ferns, P.N., Cowie, R.J. and Weir, A.G. (1992) Managing Conifer Plantations for the Benefit of Granivorous Birds and Mammals. Aspects of Applied Biology, 29, 83-88.
36. Allen, R., Platt, K. and Wiser, S. (1995) Biodiversity in New Zealand Plantations. New Zealand Forestry, 39, 26-29.
37. Michelsen, A., Lisanework, N., Friis, I. and Holst, N. (1996) Comparisons of Understory Vegetation and Soil Fertility in Plantations and Adjacent Natural Forests in the Ethiopian Highlands. Journal of Applied Ecology, 33, 627-642. http://dx.doi.org/10.2307/2404991
38. Chey, V.K., Holloway, J.D. and Speight, M.R. (1997) Diversity of Moths in Forest Plantations and Natural Forests in Sabah. Bulletin of Entomological Research, 87, 371-385. http://dx.doi.org/10.1017/S000748530003738X
39. Estades, C.F. and Temple, S.A. (1999) Deciduous-Forest Bird Communities in a Fragmented Landscape Dominated by Exotic Pine Plantations. Ecological Applications, 9, 573-585. http://dx.doi.org/10.1890/1051-0761(1999)009 [0573:DFBCIA]2.0.CO;2
40. Pauchard, A., Aguayo, M., Peña, E. and Urrutia, R. (2006) Multiple Effects of Urbanization on the Biodiversity of Developing Countries: The Case of a Fast-Growing Metropolitan Area (Concepción, Chile). Biological Conservation, 127, 272-281. http://dx.doi.org/10.1016/j.biocon.2005.05.015.
41. Friend, G.R. (1982) Bird Populations in Exotic Pine Plantations and Indigenous Eucalypt Forest in Gippsland, Victoria. Emu, 82, 80-91. http://dx.doi.org/10.1071/MU9820080.
42. Freedman, B., Zelazny, V., Beaudette, D., Fleming, T., Flemming, S., Forbes, G., Gerrow, J.S., Johnson, G. and Woodley, S. (1996) Biodiversity Implications of Changes in the Quantity of Dead Organic Matter in Managed Forests. Environmental Research, 4, 238-265.
43. Potton, C. (1994) A Public Perception of Plantation Forestry. New Zealand Forestry, 39, 2-3.
44. Perley, C. (1994) Biodiversity, Sustainability and a Land Ethic. New Zealand Forestry, 39, 2-3.
45. Fomegas, D.Y., Bayos, N.W. and Atayoc, J.P. (2004) Phytodiversity under Pine, (Pinus insularis) Forest Community in Tadian, Mountain Province. Proceedings: Regional Sectoral/Commodity Review, ATI-CAR, La Trinidad, 15-17 June 2004, 17-26.
46. Vergara, P.M. and Simonetti, J.A. (2004) Avian Responses to Fragmentation of the Maulino Forest in Central Chile. Oryx, 38, 383-388. http://dx.doi.org/10.1017/S0030605304000742.
47. Estades, C.F. (1994) Impacto de la sustitucion del bosque native por plantaciones de Pinus radiata sobre una comunidad de aves en la Octava Region de Chile. Boletin Chileno de la Ornitologia, 1, 8-14.
48. Saavedra, B. and Simonetti, J.A. (2005) Small Mammals in Remnants of Maulino Forest and Adjacent Monterrey Pine plantations, Historia, biodiversidad y ecologia de los bosques de la cordillera de la Costa. Editorial Universitaria, Santiago de Chile.
49. Barbosa, O. and Marquet, P.A. (2002) Effects of Forest Fragmentation on the Beetle Assemblage at the Relict Forest of Fray Jorge, Chile. Oecologia, 132, 296-306. http://dx.doi.org/10.1007/s00442-002-0951-3
50. Peña-Cortes, F., Rebolledo, G., Hermosilla, K., Hauenstein, E., Bertran, C., Schlatter, R. and Tapia, J. (2006) Dinámica del paisaje para el período 1980-2004 en la

cuencacostera del Lago Budi, Chile. Consideraciones para la conservación de sushumedales. Ecología Austral, 16, 183-196.
51. Pawson, S.M., Brockerhoff, E.G., Watt, M.S. and Didham, R.K. (2006) Maximizing Biodiversity in Plantation Forests: Insights from Long-Term Changes in Clearfell-Sensitive Beetles in a Pinus radiata Plantation. Biological Conservation, 144, 2842-2850.
52. Toro, J. and Gessel, S.P. (1999) Radiata Pine Plantations in Chile. New Forests, 18, 33-44. http://dx.doi.org/10.1023/A:1006597823190.
53. Instituto Nacional de Estadisticas (2002) Censo 2002: Sintesis de Resultados. Instituto Nacional de Estatisticas, Santiago de Chile.
54. Peel, M.C., Finlayson, B.L. and McMahon, T.A. (2007) Updated World Map of the Köppen-Geiger Climate Classification. Hydrology and Earth System Sciences, 11, 1633-1644. http://dx.doi.org/10.5194/hess-11-1633-2007.
55. Veit, H. and Garleff, K. (1995) Evolucion del paisaje Cuaternario y los suelos en Chile centro-sur. In: Armesto, J.J., Villagran, C. and Arroyo, M.K., Eds., Ecologia de los bosquesnativos de Chile, Editorial Universitaria, Santiago, 363-387.
56. Donoso, C. and Lara, A. (1995) Utilizacion de los bosquesnativos en Chile, pasado, presente y future. In: Armesto, J.J., Villagran, C. and Arroyo, M.K., Eds., Ecologia de los bosquesnativos de Chile, Editorial Universitaria, Santiago, 363-387.
57. Ovalle, C., Avedaño, J., Aronson, J. and Del Pozo, A. (1996) Land Occupation Patterns and Vegetation Structure in the Anthropogenic Savannas (Espinales) of Central Chile. Forest Ecology and Management, 86, 129-139. http://dx.doi.org/ 10.1016/ S0378-1127(96)03786-3.
58. Luebert, F. and Pliscoff, P. (2005) Sinopsis bioclimatica y vegetacional de Chile. Editorial Universitaria, Santiago de Chile.
59. Amigo, J., San Martin, J. and Garcia Quintanilla, L. (2000) Estudio fitosociologico de los bosques de Nothofagus glauca (Phil.) Krasser del Centro-Sur de Chile. Phytocoenologia, 30, 193-221.
60. Hildebrand-Vogel, R. (2002) Structure and Dynamics of Southern Chilean Natural Forests with Special Reference to Their Relation of Evergreen versus Deciduous Elements. Folia Geobotanica, 37, 107-128. http://dx.doi.org/10.1007/BF02803194.
61. Armesto, J.J. and Pickett, S. (1985) A Mechanistic Approach to the Study of Succession in the Chilean Matorral. Revista Chilena de Historia Natural, 58, 9-17.
62. Armesto, J.J., Gutierrez, J. and Martinez, J. (1979) Las comunidades vegetales de la region mediterranea de Chile: Distribucion de especies y formas de vida en ungradiente de aridez. Medio Ambiente, 4, 62-70.
63. Echeverria, C. and Lara, A. (2004) Growth Patterns of Secondary Nothofagus obliqua—N. alpina Forests in Southern Chile. Forest Ecology and Management, 195, 29-43. http://dx.doi.org/10.1016/j.foreco.2004.02.034.
64. Camus, P. (2003) Federico Albert: Artifice de la Gestion de los Bosques de Chile. Revista de Geografía Norte Grande, 30, 55-63.
65. Sheppard, H.E. (1999) Certificacion forestal (II). Una opcion medioambiental de vanguardia para Chile. Ambiente y Desarrollo, 15, 54-57.
66. CONAF (2011) Catastro de los Recursos Vegetacionales Nativos de Chile: Monitoreo de Cambios y Actualizaciones. Chilean Forest Service CONAF, Santiago de Chile.
67. Instituto Nacional de Estadisticas (2002) Censo 2002—Sintesis de Resultados. Comision Nacinoal del XVII Censo de Poblacion y VI de Vivienda, Santiago de Chile, Chile. http://www.ine.cl/cd2002/sintesiscensal.pdf
68. Salt, D., Lindenmayer, D. and Hobbs, R. (2004) Trees and Biodiversity: A Guide for Australian Farm Forestry. Joint Venture Agroforestry Program 2004, Australian National University, Canberra
69. Sasaki, N. and Putz, F.E. (2009) Critical Need for New Definitions of "Forest" and "Forest Degradation" in Global Climate Change Agreements. Conservation Letters, 2, 226-232. http://dx.doi.org/10.1111/j.1755-263X.2009.00067.x

70. Deil, U., Alvarez, M. and Paulini, I. (2007) Native and Non-Native Species in Annual Grassland Vegetation in Mediterranean Chile. Phytocoenologia, 37, 769-784. http://dx.doi.org/10.1127/0340-269X/2007/0037-0769

71. Van de Vouw, W., Echeverria, C., Rey-Benayas, J.M. and Holmgren, M. (2011) Persistent Acacia Savannas Replace Mediterranean Sclerophyllous Forests in South America. Forest Ecology and Management, 262, 1100-1108. http://dx.doi.org/10.1016/j.foreco.2011.06.009.

72. Aronson, J., Del Pozo, A., Ovalle, C., Avedaño, J., Lavin, A. and Etienne, M. (1998) Land Use Changes and Conflicts in Central Chile. In: Rundel, P.W., Rizzardini, G.M. and Jaksic, F.M., Eds., Landscape Disturbance and Biodiversity in Mediterranean-Type Ecosystems, Springer, Heidelberg, 155-168. http://dx.doi.org/10.1007/978-3-662-03543-6_9.

73. Braun-Blanquet, J. (1964) Pflanzensoziologie: Grundzüge der Vegetationskunde. 3rd Edition, Springer, Berlin. http://dx.doi.org/10.1007/978-3-7091-8110-2.

74. Chytry, M., Schaminée, J.H.J. and Schwabe, A. (2011) Vegetation Survey: A New Focus for Applied Vegetation Science. Applied Vegetation Science, 14, 435-439. http://dx.doi.org/10.1111/j.1654-109X.2011.01154.x.

75. Podani, J. (2006) Braun-Blanquet's Legacy and Data Analysis in Vegetation Science. Journal of Vegetation Science, 17, 113-117. http://dx.doi.org/10.1111/j.1654-1103.2006.tb02429.x.

76. Walker, D.A., Kuss, P., Epstein, H.E., Kade, A.N., Vonlanthen, C.M., Raynolds, M.K. and Danie F.J.A. (2011) Vegetation of Zonal Patterned-Ground Ecosystems along the North America Arctic Bioclimate Gradient. Applied Vegetation Science, 14, 440-463. http://dx.doi.org/10.1111/j.1654-109X.2011.01149.x.

77. Pellisier, V., Gallet, S. and Roze, F. (2004) Comparison of the Vegetation of Seed Bank on Hedge Banks of Different Ages in Brittany, France. Environmental Management, 34, 52-61.

78. Spellerberg, I.F. and Sawyer, J.W.D. (1996) Standards for Biodiversity: A Proposal Based on Biodiversity Standards for Forest Plantations. Biodiversity and Conservation, 5, 447-459. http://dx.doi.org/10.1007/BF00056390.

79. Duelli, P. and Obrist, M.K. (2003) Biodiversity Indicators: The Choice of Values and Measures. Agriculture, Ecosystems & Environment, 98, 87-98. http://dx.doi.org/10.1016/S0167-8809(03)00072-0.

80. Cawsey, E.M. and Freudenberger, D. (2008) Assessing the Biodiversity Benefits of Plantations: The Plantation Biodiversity Benefits Score. Ecological Management and Restoration, 9, 42-52. http://dx.doi.org/10.1111/j.1442-8903.2008.00386.x

81. Tuomisto, H. (2010) A Consistent Terminology for Quantifying Species Diversity? Yes, It Does Exist. Oecologia, 164, 853-860. http://dx.doi.org/10.1007/s00442-010-1812-0.

82. Tuomisto, H. (2012) An Updated Consumer's Guide to Evenness and Related Indices. Oikos, 121, 1203-1218. http://dx.doi.org/10.1111/j.1600-0706.2011.19897.x.

83. Verdeguer, M., Garcia-Rellan, D., Boira, H., Perez, E., Gandolfo, S. and Blazquez, M.A. (2011) Herbicidal Activity of Peumus boldus and Drimys winterii Essential Oils from Chile. Molecules, 16, 403-411. http://dx.doi.org/10.3390/molecules16010403.

84. Magurran, A.E. (2004) Measuring Biological Diversity. Blackwell Publishing, Oxford.

85. Hurlbert, S.H. (1971) The Nonconcept of Species Diversity: A Critique and Alternative Parameters. Ecology, 52, 577- 586. http://dx.doi.org/10.2307/1934145.

86. Gotelli, N.J. and Colwell, R.K. (2001) Quantifying Biodiversity: Procedures and Pitfalls in the Measurement and Comparison of Species Richness. Ecology Letters, 4, 379-391. http://dx.doi.org/10.1046/j.1461-0248.2001.00230.x.

87. Lugo, A.E. (1992) Comparison of Tropical Tree Plantations with Secondary Forests of Similar Age. Ecological Monographs, 62, 1-41. http://dx.doi.org/10.2307/2937169.

88. Rojas, C., Vivanco, M., Opazo, S., Peters, S. and Constanza, V. (2011) Pre and Post Earthquake Land Use and Land Cover Identification in Concepción. In: Krisp, J.M.,

Meng, L., Pail, R. and Stilla, U., Eds., Earth Observation of Global Changes, Springer, Heidelberg, 223-231.
89. Holmgren, M., Aviles, R., Sierralta, L., Segura, A.M. and Fuentes, E.R. (2010) Why Have European Herbs so Successfully Invaded the Chilean Matorral? Effects of Herbivory, Soil Nutrients and Fire. Journal of Arid Environments, 44, 197-211. http://dx.doi.org/10.1006/jare.1999.0589.
90. Figueroa, J.A., Castro, S.A., Marquet, P.A. and Jaksic, F.M. (2004) Exotic Plant Invasions to the Mediterranean Region of Chile: Causes, History and Impacts. Revista Chilena de Historia Natural, 77, 465-483.
91. Quiroz, C.L., Pauchard, A., Marticorena, A. and Cavieres, L.A. (2009) Manual de Plantas Invasoras del Centro-Sur de Chile. Laboratorio de Invasiones Biologicas, Concepción.
92. Muñoz-Moreira, A. (2011) Plant Geography of Chile. Springer, Heidelberg.
93. Omarzabal, C.S. (1993) The Conservation of Biodiversity in Chile. Revista Chilena de Historia Natural, 66, 383-402.
94. Levine, J.M., Vila, M., Antonio, C., Dukes, J.S., Grigulis, K. and Lavorel, S. (2003) Mechanisms Underlying the Impacts of Exotic Plant Invasions. Proceedings of the Royal Society B: Biological Sciences, 270, 775-781. http://dx.doi.org/ 10.1098/rspb.2003.2327.
95. Hierro, J.L., Maron, J.L. and Callaway, R.M. (2005) A Biogeographical Approach to Plant Invasions: The Importance of Studying Exotics in Their Introduced and Native Range. Journal of Ecology, 93, 5-15. http://dx.doi.org/10.1111/j.0022-0477.2004.00953.x
96. Donald, P.F., Haycock, D., Fuller, R.J., Evans, A.D. and Gough, S.J. (1998) Effects of Forest Management and Grazing on Breeding Bird Communities in Plantations with Broad-Leaved and Coniferous Trees in Western England. Biological Conservation, 85, 183-197. http://dx.doi.org/10.1016/S0006-3207(97)00114-6.
97. Keenan, R., Lamb, D., Woldring, O., Irvine, T. and Jensen, R. (1997) Restoration of Plant Biodiversity beneath Tropical Tree Plantations in Northern Australia. Forest Ecology and Management, 99, 117-131. http://dx.doi.org/10.1016/S0378-1127(97)00198-9.
98. Lara, A. (2004) Conservacion de los ecosistemas boscosos: Algunas lecciones de los ultimos 20 años. Ambiente y Desarrollo, 20, 111-115.
99. Bennett, E.L. (2001) The Joint Effort of Timber Certification. Conservation Biology, 15, 318-319. http://dx.doi.org/10.1046/j.1523-1739.2001.015002318.x.
100. Cauley, H.A., Peters, C.M., Donovan, R.Z. and O'Connor, J.M. (2001) Forest Stewardship Council Forest Certification. Conservation Biology, 15, 311-312. http://dx.doi.org/10.1046/j.1523-1739.2001.015002311.x
101. Neyland, M.G., Wilkinson, G.R. and Edwards, L.G. (1999) The Forestier Silvicultural Systems Trial: Alternatives to Clearfelling. Tasforests, 11, 35-48.
102. Siira-Pietikäinen, A., Haimi, J. and Siitonen, J. (2003) Short Term Responses of Soil Macroarthropod Community to Clear Felling and Alternative Forest Regeneration Methods. Forest Ecology and Management, 172, 339-353. http:// dx.doi.org /10.1016/S0378-1127(01)00811-8
103. Baker, S.C., Grove, S.J., Forster, L., Bonham, K.J. and Bashford, D. (2009) Short-Term Responses of Ground-Active Beetles to Alternative Silvicultural Systems in the Warra Silvicultural Systems Trial, Tasmania, Australia. Forest Ecology and Management, 258, 444-459. http://dx.doi.org/10.1016/j.foreco.2009.03.044
104. Taylor, R.J. and Haseler, M.E. (1995) Effects of Partial Logging Systems on Bird Assemblages in Tasmania. Forest Ecology and Management, 72, 131-149. http://dx.doi.org/10.1016/0378-1127(94)03470-H
105. Lefort, P. and Grove, S. (2009) Early Responses of Birds to Clearfelling and Its Alternatives in Lowland Wet Eucalyptus Forest in Tasmania, Australia. Forest Ecology and Management, 258, 460-471. http://dx.doi.org/ 10.1016 /j.foreco.2009.04.022 [Citation Time(s):1]

106. Radford, J.Q., Bennett, A.F. and Cheers, G.J. (2005) Landscape-Level Thresholds of Habitat Cover for Woodland- Dependent Birds. Biological Conservation, 124, 317-337.

107. Watson, J., Freudenberger, D. and Paull, D. (2001) An Assessment of the Focal Species Approach for Conserving Birds in Variegated Landscapes in Southeastern Australia. Conservation Biology, 15, 1364-1373. http://dx.doi.org/10.1046/j.1523-1739.2001.00166.x

108. MacIntyre, S., McIvor, J.G. and MacLeod, N.D. (2000) Principles for Sustainable Grazing in Eucalypt Woodlands: Landscape-Scale Indicators and the Search for Thresholds. In: Hale, P., Petrie, A., Maloney, D. and Sattler, P., Eds., Management for Sustainable Ecosystems, Centre for Conservation Biology, University of Queensland, Brisbane, 92-100.

NOTES

[1]Note that this study does not analyse metapopulations explicitly (e.g. by presenting data on connectivity, dispersal rates, source and sink populations). Nonetheless, for two reasons, it is important to incorporate the notion of metapopulations in the argumentation presented herein. Firstly, it has to be kept in mind that while local extinctions of species may occur which alter α-diversity, species may still survive at larger scales (e.g. at the γ-scale or even higher scales). Thus, the proof of absence of a species in a certain plant formation may not be mistaken as a proof for its total extinction. Secondly, in order to have species survive via such a rescue effect by metapopulations, a certain connectivity between local populations is crucial. Exactly this connectivity is reduced by plantation establishment, c.f. [10]. Both aspects need to be kept in mind when drawing conclusion on Chilean plantation forestry discussed herein.

[2]Decisions taken by determining the minimum and maximum values of ${}^{0}D_{\alpha}$ and the respective tree heights within the interpolated data ranges of Figure 4.

[3]It should be kept in mind though that the results of the tests, performed to accomplish a reviewer's comment, which are based on only three plantations of each type, should not be over interpreted.

CHAPTER 14

Vegetables, Herbs and Spices: The Importance of Family and Tasting

Pamela A. Heinrichs[1,2], Cassandra J. Nikolaus[1], Brenna Ellison[1], Sharon M. Nickols-Richardson[1], Karen Chapman-Novakofski[1]

[1]*University of Illinois at Urbana-Champaign, Champaign, IL, USA*

[2]*Currently Address: The Hersey Company, Hershey, PA, USA*

ABSTRACT

Objective: To explore why and how consumers utilize vegetables, herbs and spices. **Methods**: Focus groups were conducted with adult participants (n = 54) to determine attitudes, self-efficacy, and behaviors related to liking and usage of vegetables, herbs and spices as well as strategies to increase vegetable intake. **Results**: Seasonal availability, and family-centered memories and practices were motivation for vegetable intake. A diverse number of vegetables (n = 39) were listed as favorites based on taste and versatility, but seasonings were not commonly included in responses regarding preparation. Participants felt that vegetable dish names should reflect ingredients explicitly; that seasonings enhanced vegetable flavors; and that salt and pepper were most commonly used seasonings. The preferred strategy to increase vegetable consumption was through tasting or sampling opportunities; seasoning use was not mentioned. However, participants perceived that suggestions or demonstrations on how to prepare vegetables at home would be helpful. Personal gardens were mentioned frequently, but were not connected to strategies for increasing intake. **Conclusions and Implications**: Vegetable intake has a family focus that should be emphasized to increase intake. Efforts to increase intake could emphasize flavor and versatile preparation methods through tasting opportunities or educational demonstrations.

Keywords: Dietary Habits, Focus Groups, Vegetables, Spices, Herbs

1. INTRODUCTION

Vegetable intake recommendations are largely unmet in the United States (US). Only 10% to 30% of US adults meet the recommended intake of two to three and a half cups of vegetables per day [1]. The average intake is one and one half cups per day, according to the National Health and Nutrition Examination Survey, 2009-2010 database (n = 9042 adults) [2]. In a nationally representative sample of 373,580 adults, 9% reported meeting vegetable recommendations [3].

Explorations into factors related to greater vegetable intake have evaluated various behavioral theories and how theoretical constructs relate to behavior. Choice architecture has been applied in a buffet-line intervention, where re-arranging the display successfully increased the quantity of salad vegetables selected by patrons [4]. When applying Stages of Change theory [5] to vegetable intake, self-efficacy best predicted stage movement [6] and moderated intake [7] [8]. Within the Theory of Planned Behavior [9] , affective attitude and perceived behavioral control were significant predictors of intention to consume five vegetables and fruits per day, with intentions significantly linked to behavior [10].

One area that has not received much attention is use of herbs and spices with vegetables as a method of increasing intake. A sample of parents reported adding flavoring to just over half of vegetable preparations; however, spices were less frequently used than salt, butter or oil [11]. Interest in frequency of spice use was previously alluded to in analyses of over 2000 vegetable recipes from 36 countries that found vegetable dishes included spices less frequently than meat dishes. The most frequently included spices in vegetable dishes were black pepper, followed by onion, garlic and chili pepper [12].

While herbs and spices continue to be investigated for health benefits [13] [14] [15] , the practical application of enhancing flavor and increasing vegetable intake has been less well studied. Previous sensory studies found that herb and spice use increased preference for reduced-fat meals and reduced-sodium tomato soup [16] [17]. Only one published intervention protocol that used herbs with vegetables was identified [18] , but the focus was on impacting bone resorption, leaving a gap in knowledge regarding the ability of herbs and spices to facilitate vegetable intake. Therefore, the aim of the current study was to explore why and how consumers utilize vegetables, herbs and spices through qualitative focus groups.

2. METHODS

Adults, aged 18 years or older, were recruited through a university social media website. Recruits had to have eaten carrots, broccoli, cauliflower, peas, squash, zucchini, Brussel sprouts, corn, potatoes, okra, or tomatoes at least once in the previous two months to be considered for the study. There were no other exclusion criteria. Sex, age and residence questions were asked during screening to stratify participants across focus groups as needed. Invitations to participate were sent by email, asking for availability. Four to eight participants were assigned to each conference call focus group time. Participants were called simultaneously by the moderator. Participants selected fictitious names to preserve confidentiality. Focus groups lasted an average of 38 minutes, after which participants were compensated with a $70 gift card.

2.1. Focus Group Script

The research team developed focus group questions, according to guidelines by Krueger and Casey [19] , to learn about participants' vegetable-related perceptions and behaviors. Questions reflected affective attitudes, perceived behavioral control, and self- efficacy, as well as environmental influences (Table 1). Herbs and spices were not operationally defined for participants to encourage open and unrestricted focus group discussions. Seven experts in nutrition education were invited to review the focus group script. Six professors of nutrition or nutrition-related disciplines returned comments, and the script was refined. Flow and clarity were tested with investigator-led practice sessions.

Procedures for telephone focus groups were followed, including limiting the number of questions, keeping the focus group discussion to one hour or less, and calling on par ticipants who were not actively engaged [19]. All procedures were approved by the University of Illinois Institutional Review Board.

Table 1. Focus group questions for why and how consumers utilize vegetables, herbs and spices.

Theory Variable	Focus Group Questions
Attitudes	Would you tell me about the 3 vegetables you like the most? Things like why you like them, how often you eat them, how you like them prepared, or what meal you eat them?
	Next, I'd like your opinion on names: recipe names, vegetable products you may buy (like a frozen package), or vegetable dish names on a menu. There are generally 4 types of names: ones that describe the cooking method, like "roasted", ones that describe the ingredients, like "rosemary", ones that describe the general taste, like "spicey", and ones that create an image or suggest ingredients or taste, such as southwestern or Italian. Are any of these names more compelling for you to choose a vegetable than others, and why or why not? Or in other words, What types of names do you think would be most effective at getting people to select vegetables? Why? Next, I'd like to get your opinions on spices and herbs, particularly in regards to vegetables. For instance, do they make vegetables more, less or the same degree of appealing as a plain steamed or roasted vegetable? If there are combinations of vegetables and seasonings that come to mind or certain spices or herbs, please bring those up as well.
Environment and behavioral control	Now I'd like you to take a moment and think about the meal or meals you usually eat vegetables during. Could you tell me about that meal? Things like, where the meal is, who is there if anyone, who prepared it, and also why a vegetable is part of that meal?
	Now, I want to talk about eating in restaurants or cafeterias. Could you tell me about yourself in terms of eating in restaurants, in general, like how often and what types of restaurants, and then about vegetables ordered or eaten in restaurants. Could you also discuss whether the vegetables you eat in restaurants are the same as those you eat at home.
Self-efficacy	I want to talk a little about your cooking style. For instance, do you like to try new recipes, do you need convenience? Talk in general, but then specifically about how you prepare vegetables, please.
Behavioral motivation	Finally, I'd like your thoughts on how to motivate people to choose more vegetables. Think as if you were selling vegetables in a grocery store or if you had a restaurant or cafeteria.

2.2. Data Analysis

Focus group conference calls were digitally recorded, and hand-written notes were taken. Recordings were transcribed verbatim. Two researchers met to review and agree on content analysis procedures [20] and inductive content development [21]. After 13 focus groups, transcriptions were evaluated independently by two researchers for codes, using content analysis. Researchers then met to discuss and agree on codes, and subsequently reviewed transcripts independently for themes. After meeting to discuss and reach agreement on themes, researchers independently reviewed the transcripts' saturation and outlying or emotionally presented responses. Concurrence was reached on these negative cases [22] for the final results as well as for saturation [23].

3. RESULTS

One-hundred forty-six people responded to recruitment efforts. Among them, 11 did not agree to participate before being screened, 34 did not respond to the invitation to schedule a focus group and six declined to participate after completing screening. After 13 focus groups with 54 participants, saturation of responses was achieved; thus, the remaining screened participants (n = 41) were not scheduled. Focus group sizes ranged from two to six participants. Smaller groups were not planned, but some scheduled participants did not answer when called or had to leave for an emergency. Participants were mostly female (n = 46), with only eight males included (Table 2). Four participants described themselves as vegetarians.

When asked to name their three favorite vegetables, 39 different vegetables were mentioned with broccoli, peppers (green or red) and asparagus with the highest frequencies. Participants primarily cited flavor as the reason for picking their favorite ve getable (n = 28), followed by versatility of a vegetable's use (n = 15). Most participants ate vegetables at home (n = 33), with family (n = 28) or as part of the evening meal (n = 41).

Table 2. Demographics of participants.

Demographic	Participants (n)
Gender	
Males	8
Females	46
Age	
18- to 39-years	26
40- to 59-years	26
over 60 years	2
Residence	
Illinois, USA	35
Other states	19

The majority (n = 37) prepared household meals themselves or with a family member. Most participants reported eating vegetables when at restaurants (n = 31), though eating in restaurants was an infrequent activity. However, a frequent comment was that restaurants did not have many vegetable options available.

Roasting was the most popular vegetable preparation method (n = 31), followed by steaming (n = 21) and sautéing (n = 20). Although many believed spices and herbs could enhance vegetables (n = 24), when asked about the herbs and spices they use with vegetables, salt and pepper was the predominant response (n = 19) "...most of my vegetables I do think I go simply with salt and pepper." A number of themes emerged from the focus groups and were categorized into two main ideas: motivations for eating vegetables (attitudes) and strategies to increase vegetable consumption (self-efficacy and behavioral control).

3.1. Motivations for Eating Vegetables

Family was the predominant thematic reason for eating vegetables. Health was identified as the main reason for consuming vegetables, both their own health as well as the health of their parents and children.

So for me it was more the way that I was raised and Mom now is... in a nursing home but Dad is still there so two or three times a week I'll go down and normally I'll make dinner and he'll do dishes. It's kind of a good trade off... And worrying about him eating enough vegetables and those kinds of things, I will always open up a jar of green beans or a jar of beets or something.

Children emerged as a motivator for serving vegetables at meals. Many stated their upbringing led them to believe that vegetables were a necessary part of the family's evening meal. Parents wanted to establish nutritious foods as part of their children's lives and form good dietary habits.

... if my children are there ... I feel it's always important to have that vegetable.

Um, and I as well try to offer two vegetables. That was done when I was a child and I still try to do it with my own kids.

Children were also important in deciding which vegetables to serve to the family. Many participants shared examples where their children determined which vegetables were served based on which ones they enjoyed eating. This illustrates how much participants value vegetable consumption as a part of their children's diets and lives.

Zucchini because my ... younger daughter who's 15 months now really enjoys it

... I have a 3 year old and that's what he'll eat

Seasonality was another strong motivator of vegetable consumption. Many participants had gardens at home, so harvest dictated which vegetables were consumed. Seasonality was mentioned regarding favorite vegetables, vegetable intake frequency, cooking methods and whether vegetables consumed at restaurants were the same as at home.

... And we got a big garden so it's pretty much seasonal what vegetables we have um, so this time of year it's a lot of salads um in the wintertime a lot of soup.

Additionally, some noted they used herbs because they were grown in their garden at home.

I use a lot of parsley, chives, and dill with my vegetables. Mainly because I grow it in my back yard.

Even those who did not garden mentioned they liked to eat vegetables that were in season. Seasonality was used in conjunction with freshness.

Actually my husband and I eat a lot of seasonal vegetables so we have our favorites for every season. Now that I'm shopping the farmer's market is open in my village I'm able to get fresh asparagus. We use it for all three meals.

... we use asparagus because it's just so delicious it makes you think of spring and I think we pretty much eat if it four times a week in the spring when it's fresh because it's just so good and so yummy.

3.2. Strategies to Increase Vegetable Consumption

The primary strategy suggested to increase vegetable consumption was to present new ideas or items to consumers. Ideas included sampling or tasting opportunities, providing use and meal pairing suggestions and presenting new preparation techniques. Participants noted that in-store tastings introduced them to now-favorite vegetables of which they were previously intimidated or unaware. Participants also commented that learning about the vegetable and how it could be used in family meals had helped make some vegetables more accessible.

... we were not a parsnip family until ... [named grocery store] had a parsnip sample thing and now we eat parsnips because they're really good.

Like, if I'm at the farmer's market and it's something that I'm not familiar with if they have samples and examples of how I can use this or how I can incorporate this um, that would help a lot.

Several participants shared stories about discovering they liked a vegetable when they tried it prepared in a novel way. They suggested trying new cooking methods would be a good strategy to increase vegetable consumption.

... she didn't like vegetables at all most of her life and then someone told her to try them roasted and she can't get enough and so she eats a lot of roasted vegetables and she's lost a lot of weight ...

When participants were asked which of a list of dish names would make a vegetable dish most appealing in a store or restaurant, the majority of participants preferred names describing the cooking method and ingredients used. Participants repeatedly explained that they liked to know what they were getting in a vegetable dish. Names that were less specific and instead tried to illustrate a general genre of taste, such as Italian, were less preferable.

It also gives you an expectation of what it will taste like. So yes, I think any and all of those descriptive names are good.

Respondents reported herbs and spices can add flavor to a vegetable dish, but no one mentioned using herbs and spices to increase vegetable consumption. However, many participants discussed eating vegetables in ethnic dishes, both at home and at restaurants, which typically use herbs and spices. Notably, Mexican ($n = 15$) and Asian ($n = 14$) dishes were popular choices.

... I tend to think a lot about ethnicities, you know, in terms of variety um, so say broccoli could be in an Asian stir fry or in Mexican fajitas, um, you know things like that just to [indecipherable] out how flavors have the different ingredients and foods go together along with the spices and flavors based on kind of more traditional cuisines.

Um, we also really like Thai food so, and there's a lot of like bamboo shoots or eggplant that I don't generally cook at home but I really like them in the Thai cooking or Indian food.

4. DISCUSSION

Food behavior is complex, with personal, interpersonal, and environmental factors influencing choice. In regards to vegetables, little is known except when vegetables and fruits are grouped together. For instance, a systematic review found that taste, availability, social influences, habits, marketing, time, and concerns about health may be factors influencing an individual's decision to consume fruits and vegetables [24]. However, choice about fruit may be quite different than choice concerning vegetables. For instance, the high cost of fruit is a perceived barrier, whereas poor appetite for vegetables was the reason for limited intake in a study of older adults [25].

While national surveys have reported that potatoes and tomatoes are the most frequently eaten vegetable by US adults [2] , little has been published about adults' reported favorite or most frequently eaten vegetable beyond these results. Published reports instead focus on the associations of total fruit and vegetable intake and chronic disease prevalence [26] , relationship to body weight [27] , or individual vegetables and chronic disease prevalence [28].

Broccoli, peppers (green or red) and asparagus were mentioned most often in this study as favorite vegetable. Although the script was pilot-tested and feedback obtained from an expert panel, there may have been ambiguity in using the term "favorite" without clarification as to frequency. With 39 vegetables recounted by participants as a favorite, it was not clear how often these favorites were consumed. This is a limitation of the study. Some were clearly a "special occasion" vegetable or a treat. Indeed, one participant said

You asked about favorite vegetables and It's not necessarily the ones we eat most often. I love artichokes... It's not really a normal everyday food.

Although this project includes no data to support this conclusion, national intake data support the idea that broccoli may be both a favorite and frequently eaten vegetable. Broccoli intake in the US has increased for both fresh and frozen from approximately 8.1 pounds/person in 2000 to 9.2 pounds/person in 2015. In comparison, intake of green beans was 6.4, carrots was 10.4 pounds/person and intake of asparagus was 1.7 pounds/person in 2015 [29]. However, it should be noted that the focus groups described in this study were conducted in the spring, when asparagus was in season, and many were gardeners themselves.

Reasons for increased frequency of intake included the versatility of the vegetable as well as the seasonality. Holmes and Gates [28] reported that seasonality was mentioned in regards to vegetable intake in interviews with older men and women, although not as often as was mentioned in regards to fruit. Neither of the reasons reported in these focus groups were previously included in the top twenty promoters of vegetable intake for urban, low income adults in 2008 [30].

Reports of eating at home more often than eating out was not a new finding, and the belief that vegetable options in restaurants are minimal has been reported by others [31]. Roasting and steaming vegetables were the most preferred preparation methods. Among six preparation methods for green beans and carrots, steamed was most preferred by children and young adults, although roasting was not a preparation method offered [32].

The concept of how participants were raised and how they want to raise their children were reasons for eating vegetables. These concepts have been reported by others, including Holmes and Gates [30] in a qualitative study with older men. Yeh et al. [33] also reported both childhood experiences with vegetables and desiring healthy meals for their current families among both older and younger focus group participants. The importance of family was also reported in a study investigating interventions to increase fruit and vegetable intake at the workplace or workplace plus home, concluding that the workplace plus home was more effective [34]. In addition, past experiences when eating vegetables may not need to be in a person's distance past in order to influence their vegetable eating habits. In a study of college students, Robinson, Blissett, and Higgs [35] found that the students recalled pleasant vegetable eating experiences and that this was related to higher vegetable liking and choice of larger vegetable portions. Therefore, connecting pleasant vegetable intake memories and current home practices may be valuable to increase intake.

Expectations of benefits on health have also been reported by others [31] [33]. However, providing more information about the health benefits of vegetables was mentioned by only a minority of focus group participants. Yeh et al. [33] noted that while most of their participants cited the health benefits of vegetables, and fruits, they also admitted they do not follow the guidelines for intake. Therefore, knowing the health benefits may not be a powerful motivator for improved intake.

The grocery store was the suggested place to learn about vegetables by participants advocating taste testing as a way to increase vegetable intake. In a study of child care providers, grocery stores were identified as the primary place they learned about vegetables (and fruits) and this information was then conveyed to the children [36]. None of the strategies for increasing vegetable and fruit intake outlined by the Center for Disease Control and Prevention (CDC) [37] were offered by participants as suggested strategies. This may be because the CDC's strategies had a broader viewpoint, but suggestions concerning access, cost and quality were not issues with which participants identified. However, although not suggested as a strategy, participants' home gardens clearly play a role in their personal vegetable intake and are part of the CDC's guidelines. Indeed, home gardens have been found to include a wide diversity of vegetables and can provide meaningful quantities of vegetables for home use [38].

Although participants reported growing herbs in addition to vegetables, reported combinations were mostly limited to tomatoes and basil. Indeed, most reported salt and pepper as the seasoning usually used in vegetable preparation, similar to the reported recipe analysis [12]. A lack of familiarity with vegetable seasoning could be overcome by tasting and recipe demonstrations.

While past research has explored strategies such as serving more of a vegetable [39] , serving a wider variety of vegetables [40] and offering a vegetable subsidy (price reduction) to make them more affordable [41] to increase vegetable intake, focus group findings here suggest that tasting or sampling opportunities as well as preparation suggestions and demonstrations may be effective means to introduce consumers to new vegetables. Additional research is needed to better explain the interaction of favorite vegetables and frequently eaten vegetables.

Although the use of herbs and spices was not explicitly listed by focus group participants as a strategy to increase vegetable consumption, herbs and spices could likely be incorporated in tasting or sampling demonstrations. In such demonstrations, it is important to use a descriptive and informative name for the vegetable. Focus groups revealed that consumers prefer descriptive, yet straightforward, names that explain what they are eating and how the food was prepared. For instance, a name like "Roasted zucchini with chili powder and lime" would be preferred to "Southwestern-style zucchini". Insights from this research can be used to tailor recipes and can be built into educational programs and campaigns designed to increase vegetable intake.

ACKNOWLEDGEMENTS

This research was supported by an unrestricted gift from the McCormick Science Institute. The funding agency had no role in the design of the study, collection or analysis of data, interpretation of results, or decision to submit the manuscript for publication.

REFERENCES

1. U.S. Department of Health and Human Services and U.S. Department of Agriculture. (2015) 2015-2020 Dietary Guidelines for Americans. 8th Edition. http://health.gov/dietaryguidelines/2015
2. Bowman, S.A., Clemens, J.C., Friday, J.E., Thoerig, R.C. and Moshfegh, A.J. (2014) Food Patterns Equivalents Intakes from Food: Mean Amounts Consumer per Individual. What We Eat in America, NHANES 2011-12. http://www.ars.usda.gov /Services/docs.htm?docid=23871
3. Moore, L.V. and Thompson, F.E. (2015) Adults Meeting Fruit and Vegetable Intake Recommendations—United States, 2013. Morbidity and Mortality Weekly Report, 64, 709-713.
4. Kongsbak, I., Skov, L.R., Nielsen, B.K., Ahlmann, F.K., Schaldemose, H., Atkinson, L., Wichmann, M. and Perez-Cueto, A. (2016) Increasing Fruit and Vegetable Intake among Male University Students in an Ad Libitum Buffet Setting: A Choice

Architectural Nudge Intervention. Food Quality and Preference, 49, 183-188. https:/doi.org/10.1016/j.foodqual.2015.12.006
5. Prochaska, J.O. and DiClemente, C.C. (1986) Toward a Comprehensive Model of Change. In: Miller, W.R. and Heather, N., Eds., Treating Addictive Behaviors: Processes of Change, Plenum Press, New York, 3-27. https:/doi.org/10.1007/978-1-4613-2191-0_1
6. Wiedemann, A.U., Lippke, S., Reuter, T., Schüz, B., Ziegelmann, J.P. and Schwarzer, R. (2009) Prediction of Stage Transitions in Fruit and Vegetable Intake. Health Education Research, 24, 596-607. https:/doi.org/10.1093/her/cyn061 [Citation Time(s):1]
7. Richert, J., Reuter, T., Wiedemann, A.U., Lippke, S., Ziegelmann, J. and Schwarzer, R. (2010) Differential Effects of Planning and Self-Efficacy on Fruit and Vegetable Consumption. Appetite, 54, 611-614. https:/doi.org/10.1016/j.appet.2010.03.006
8. Lhakhang, P., Godinho, C., Knoll, N. and Schwarzer, R. (2014) A Brief Intervention Increases Fruit and Vegetable Intake. A Comparison of Two Intervention Sequences. Appetite, 82, 103-110. https:/doi.org/10.1016/j.appet.2014.07.014
9. Ajzen, I. (1991) The Theory of Planned Behavior. Organizational Behavior and Human Decision Processes, 50, 179-211. https:/doi.org/10.1016/0749-5978(91)90020-T
10. Blanchard, C.M., Fisher, J., Sparling, P.B., Shanks, T.H., Nehl, E., Rhoades, R.E., et al. (2009) Understanding Adherence to 5 Servings of Fruits and Vegetables per Day: A Theory of Planned Behavior Perspective. Journal of Nutrition Education and Behavior, 41, 3-10. https:/doi.org/10.1016/j.jneb.2007.12.006
11. Poelman, A.A.M., Delahunty, C.M. and de Graaf, C. (2001) Vegetable Preparation Practices for 5-6 Years Old Australian Children as Reported by Their Parents; Relationships with Liking and Consumption. Food Quality and Preference, 42, 20-26. https:/doi.org/10.1016/j.foodqual.2015.01.005
12. Sherman, P.W. and Hash, G.A. (2001) Why Vegetable Recipes Are Not Very Spicy. Evolution and Human Behavior, 22, 147-163. https:/doi.org/10.1016/S1090-5138(00)00068-4
13. Pang, B., Zhou, Q., Zhao, T.Y., He, L., Guo, J., Chen, H., et al. (2015) Innovative Thoughts on Treating Diabetes from the Perspective of Traditional Chinese Medicine. Evidence-Based Complementary and Alternative Medicine, 2015, Article ID: 905432. https:/doi.org/10.1155/2015/905432
14. Rouhi-Boroujeni, H., Rouhi-Boroujeni, H., Heidarian, E., Mohammadizadeh, F. and Rafieian-Kopaei, M. (2015) Herbs with Anti-Lipid Effects and Their Interactions with Statins as a Chemical Anti-Hyperlipidemia Group Drugs: A Systematic Review. ARYA Atherosclerosis, 11, 244-251.
15. Hügel, H.M. (2015) Brain Food for Alzheimer-Free Ageing: Focus on Herbal Medicines. Advances in Experimental Medicine and Biology, 863, 95-116. https:/doi.org/10.1007/978-3-319-18365-7_5
16. Peters, J.C., Polsky, S., Stark, R., Zhaoxing, P. and Hill, J.O. (2014) The Influence of Herbs and Spices on Overall Liking of Reduced Fat Food. Appetite, 79, 183-188. https:/doi.org/10.1016/j.appet.2014.04.019
17. Ghawi, S.K., Rowland, I. and Methven, L. (2014) Enhancing Consumer Liking of Low Salt Tomato Soup over Repeated Exposure by Herb and Spice Seasonings. Appetite, 81, 20-29. https:/doi.org/10.1016/j.appet.2014.05.029
18. Gunn, C.A., Weber, J.L. and Kruger, M.C. (2013) Midlife Women, Bone Health, Vegetables, Herbs and Fruit Study. The Scarborough Fair Study Protocol. BMC Public Health, 13, 23. https:/doi.org/10.1186/1471-2458-13-23
19. Krueger R.A. and Casey, M.A. (2009) Focus Groups: A Practical Guide for Applied Research. 4th Edition, Sage Publications, Thousand Oaks.
20. Mayring, P. (2000) Qualitative Content Analysis, Forum: Qualitative Social Research, 1, Article Number: 20. http://www.qualitative-research.net/index.php/fqs/article/view/1089/2385
21. Lincoln, Y.S. and Guba, E.G. (1985) Naturalistic Inquiry. Sage, Thousand Oaks.

22. Brodsky, A. (2008) Negative Case Analysis. In: Given, L.M., Ed., The Sage Encyclopedia of Qualitative Research Methods, Sage Publications, Thousand Oaks, Vol. 2.
23. Strauss, A.L. and Corbin, J.M. (1990) Basics of Qualitative Research: Grounded Theory Procedures and Techniques. Sage Publications, Newbury Park.
24. Pollard, J., Kirk, S.F. and Cade, J.E. (2002) Factors Affecting Food Choice in Relation to Fruit and Vegetable Intake: A Review. Nutrition Research and Review, 15, 373-387. https:/doi.org/10.1079/NRR200244
25. Dijkstra, S.C., Neter, J.E., van Stralen, M.M., Knol, D.L., Brouwer, I.A., Huisman, M., et al. (2015) The Role of Perceived Barriers in Explaining Socio-Economic Status Differences in Adherence to the Fruit, Vegetable and Fish Guidelines in Older Adults: A Mediation Study. Public Health Nutrition, 18, 797-808. https:/doi.org/ 10.1017/S1368980014001487
26. Cavallo, D.N., Horino, M. and McCarthy, W.J. (2016) Adult Intake of Minimally Processed Fruits and Vegetables: Associations with Cardiometabolic Disease Risk Factors. Journal of the Academy of Nutrition and Dietetics, 116, 1387-1394. https:/doi.org/10.1016/j.jand.2016.03.019
27. Seguin, R.A., Aggarwal, A., Vermeylen, F. and Drewnowski, A. (2016) Consumption Frequency of Foods Away from Home Linked with Higher Body Mass Index and Lower Fruit and Vegetable Intake among Adults: A Cross-Sectional Study. Journal of Environmental and Public Health, 2016, Article ID: 3074241. https:/doi.org/10.1155/2016/3074241 [Citation Time(s):1]
28. Borgi, L., Rimm, E.B., Willett, W.C. and Forman, J.P. (2016) Potato Intake and Incidence of Hypertension: Results from Three Prospective US Cohort Studies. British Medical Journal, 353, i2351. https:/doi.org/10.1136/bmj.i2351
29. Economic Research Service (2016) 2016 Vegetables and Pulses Yearbook. http://www.ers.usda.gov/webdocs/DataFiles/Vegetable_and_Pulses_Yearbook_Tables__18035//37715_YRBK2016_Section%201_General.pdf
30. Holmes, T.S. and Gate, G.E. (2008) Influences on Fruit, Vegetable, and Grain Intake of Older Men. Journal of Nutrition for the Elderly, 22, 43-61. https:/doi.org/10.1300/J052v22n03_04
31. Lucan, S.C., Barg, F.K. and Long, J.A. (2010) Promoters and Barriers to Fruit, Vegetable, and Fast-Food Consumption among Urban, Low-Income African Americans—A Qualitative Approach. American Journal of Public Health, 100, 631-635. https:/doi.org/10.2105/AJPH.2009.172692
32. Zeinstra, G.G., Koelen, M.A., Kok, F.J. and de Graaf, C. (2010) The Influence of Preparation Method on Children's Liking for Vegetables. Food Quality and Preference, 21, 906-914. https:/doi.org/10.1016/j.foodqual.2009.12.006
33. Yeh, M.C., Ickes, S.B., Lowenstein, L.M., Shuval, K., Ammerman, A.S., Farris, R. and Katz, D.L. (2008) Understanding Barriers and Facilitators of Fruit and Vegetable Consumption among a Diverse Multi-Ethnic Population in the USA. Health Promotion International, 23, 42-51. https:/doi.org/10.1093/heapro/dam044
34. Sorensen, G., Stoddard, A., Peterson, K., Cohen, N., Stein, E., Palombo, R., et al. (1999) Increasing Fruit and Vegetable Consumption through Worksites and Families in the Treatwell 5-A-Day Study. American Journal of Public Health, 89, 54-70. https:/doi.org/10.2105/AJPH.89.1.54
35. Robinson, E., Blissett, J. and Higgs, S. (2011) Recall of Vegetable Eating affects Future Predicted Enjoyment and Choice of Vegetables in British University Undergraduate Students. Journal of the American Dietetic Association, 111, 1543-1548. https:/doi.org/10.1016/j.jada.2011.07.012
36. Fleischhacker, S., Cason, K. and Achterberg, C. (2007) Always a Vegetable at Dinner: A Fruit and Vegetable Qualitative Study with Primary Care Providers of Preschoolers Enrolled in an Inner-City, Head Start Childcare Center. Journal of Hunger and Environmental Nutrition, 1, 55-68. https:/doi.org/ 10.1300/J477v01n04_05

37. Centers for Disease Control and Prevention, US Department of Health and Human Services (2011) Strategies to Prevent Obesity and Other Chronic Diseases: The CDC Guide to Strategies to Increase the Consumption of Fruits and Vegetables. http://www.cdc.gov/obesity/downloads/fandv_2011_web_tag508.pdf
38. Conk, S.J. and Porter, C.M. (2016) Food Gardeners' Productivity in Laramie, Wyoming: More than a Hobby. American Journal of Public Health, 106, 854-856. https:/doi.org/10.2105/AJPH.2016.303108
39. Rolls, B.J., Roe, L.S. and Meengs, J.S. (2010) Portion Size Can Be Used Strategically to Increase Vegetable Consumption in Adults. American Journal of Clinical Nutrition, 91, 913-922. https:/doi.org/10.3945/ajcn.2009.28801
40. Meengs, J.S., Roe, L.S. and Rolls, B.J. (2012) Vegetable Variety: An Effective Strategy to Increase Vegetable Intake in Adults. Journal of the Academy of Nutrition and Dietetics, 112, 1211-1215. https:/doi.org/10.1016/j.jand.2012.05.013
41. Smith-Drelich, N. (2016) Buying Health: Assessing the Impact of a Consumer-Side Vegetable Subsidy on Purchasing, Consumption and Waste. Public Health Nutrition, 19, 520-529.

CHAPTER 15

Proximate Composition, Mineral and Vitamin Content of Some Wild Plants Used as Spices in Cameroon

Armand Abdou Bouba[1], Nicolas Yanou Njintang[2,3], Harquin Simplice Foyet[1], Joel Scher[4], Didier Montet[5], Carl Moses F. Mbofung[3]

[1]Department of Agriculture, Livestock and By-Products, University of Maroua, Maroua, Cameroon; [2]Department of Biological Sciences, Faculty of Science, University of Ngaoundere, Ngaoundere, Cameroon; [3]Department of Food Science and Nutrition, University of Ngaoundere, Ngaoundere, Cameroon; [4]Laboratoire d'Ingénierie et des Biomolécules (LiBio), Nancy-Université, Nancy, France; [5]CIRAD, Montpellier, France.

ABSTRACT

The aim of this work was to contribute to our knowledge of the proximate composition, mineral and vitamin content of 20 edible wild plants used as spices in Cameroon. The plant species were collected from 3 different markets in the West Region of Cameroon and analysed for their content of crude proteins, and lipid, ash, moisture, available sugars, total phenols, carotenoids, minerals (Ca, Zn, K, Na, Mg, Al, Mn, Cu and Se), and vitamins (A, E and C) as well as for their pH and colour. Results revealed that all the plants were low in moisture (7.7 to 10.5 g/100g) but high in ash content (7.7 to 10.5 g/100g). Hua gabonii (bark) (1594.5 mg/100g) was relative source of calcium, Echinops giganteus (206.4 mg/100g) exhibited the highest level of iron and Scorodophleus zenkeri (310.0 μg/100g) the highest level of selenium. Generally all the plants were found to contain low levels of Zn, Cu and Mg. Wide variations were observed for the proteins and available sugars among the samples. The lipid content of some of the plants were surprisingly relatively high as was the case with Monodora myristica (53.4 g/100g), Xylopia aethiopica (33.7 g/100g), Fagara leprieuri (32.1 g/100g), and Aframomum daniellii (23.1 g/100g). All the plants were rich in phenols, carotenoids, vitamin E and C. They are dark in colour and in solution they tended to provoke a fair acidification.

Keywords: Spices; Proximate Composition; Minerals; Vitamins

1. INTRODUCTION

Spices refer to all of the edible parts of a plant used for flavoring or coloring foods, including fruit, seed, root, bark or vegetable substance [1]. Some are often used as preservative against the action of harmful bacteria or prevents their growth [2]. Actually the specific uses of spices tend to vary considerably among cultures and countries: medicine, religious rituals, cosmetics, perfumery, and foods. As food, they have been shown to play an important role in health partially as sources of nutrients [3,4]. From a nutritional view point research attention on spices conducted mostly but not exclusively in India and neighbouring countries has given rise to our knowledge of the volatile oils and vitamin and mineral contents [5,6]. Of interest also has been to report of some spices as sources of polyunsaturated fatty acids as is the case with Ricinodendron heudoleutii [7] or Black curmin oil which is subject to commercial transaction and scientific research [4].

The forest zone of Central Africa region is a rich source of a wide variety of spices but most of which has hardly been studied from the view point of its proximate composition and nutritional profile. Previous works on this aspect have revealed the potential of some spices as lipid, proteins or minerals sources [7,8]. However these studies were limited in terms of the types of spices studied. In the present study a total of 20 spices identified as common ingredients in the preparation of some common meals in Cameroon earlier shown to be potential sources of antioxidants and health useful polyphenols [9] were analysed to bridge the wide knowledge gap on their nutritional potential especially with regards their proximate composition, vitamin (carotenoids, Vitamin C and E) and mineral content.

2. MATERIALS AND METHODS

2.1. Selection of Sample

Samples were purchased from specific markets in Bafoussam, Douala and Bangangte—Cameroon, famous for their trade in wild plants. The common scientific and family names of plants species are indicated in **Table 1**. Each spice sample was cleaned of debris, sorted, graded, trimmed and ground to a fine powder using a culatti grinder (Polymix, France) fitted with a 500 μm mesh. Each ground sample was then sealed in polyethylene plastic and stored at 4°C in a refrigerator until needed for analysis.

2.2. Proximate Analysis

Analyses of spices for crude protein, non protein nitrogen, ash and moisture contents were carried out essentially according to the standard AOAC methods [10]. Flour samples were acid-hydrolyzed, and the reducing sugar known as available carbohydrates was determined by the dinitrosalicylic acid (DNS) method as described by Fisher and Stein [11].

For the determination of lipid content, the cold extraction method was used [12]. Following this procedure, 50 g of spice powder samples were introduced in volumetric flacks containing 300 mL of a mixture of chloroform/ methanol (200 mL/100mL) and mixed for 20 min using an agitator. The mixture was filtered under N_2 and the residue re-extracted in 200 mL of the same solvent and filtered. The extracts were then mixed and allowed to separate after the addition of 0.2 mL of NaCl solution 0.7 g/100mL. The lipid in the lower phase solution was flushed out and recovered by rotary evaporation at 50°C under liquid Nitrogen. The flasks containing the extracted oil were weighed and the difference in weight expressed as percentage oil content.

2.3. Determination of Minerals Content

Samples of the spices powders (2.5 g) were weighed out into beakers and ashed at 550°C overnight. The resulting white-ash residue was dissolved in 4 mL of concentrated hydrochloric acid and filtered into a 10 mL volumetric flask and the volume made up with distilled water. The resulting extract was then subjected for the analysis of their content of specific major minerals. Analyses were performed in duplicate. Iron (Fe) was determined with an ICP-OES [13]. The experimental conditions of the ICP-OES were: instrument ICP-OES (Varian-Vista), RF

Table 1. Name and parts of the spices used in the study.

Botanical name	Common name	General French name	Family	Used parts
Aframomum daniellii	Bastered melegueta	Maniguette sucrée	Zingiberaceae	Fruit
Capsicum frutescens	Bird pepper	Petit pigment rouge	Solanaceae	Fruit
Dichrostachys glomerata	Sickle bush	oreilles de souris	Mimosaceae	Fruit
Dorstenia psilirus		Remèdes des serpents	Moraceae	Root
Echinops giganteus	Giant Japanese butterbur	Racines tubéreuse	Asteraceae	Root
Fagara leprieuri	Prickly ash	Grappe odoriférante	Rutaceae	Fruit
Fagara xanthoxyloïdes		Bouche béante	Rutaceae	Fruit
Hua gabonii	Garlic tree	Fruit de l'arbre de l'ail	Huacaceae	Fruit
Hua gabonii	Garlic tree	Ecorce de l'arbre de l'ail	Huacaceae	Bark
Mondia whitei	White ginger	Racine sucrée	Periplocaceae	Root
Monodora myristica	Calabash nutmeg	Fausse noix de muscade	Annonaceae	Almond
Pentadiplandra brazzeana	Joy perfume tree	Liane blanche	Pentadipandraceae	Root
Piper guineense	Black pepper (Ashandi)	Poivre sauvage	Piperaceae	Fruit
Piper umbellatum	Cordoncillo		Piperaceae	Flower
Scleria striatinux		Racine de chaume	Cyperaceae	Fruit
Scorodophleus zenkeri	Divida (African)	Fruit de l'arbre à l'ail	Caesalpinaceae	Fruit
Scorodophleus zenkeri	Divida (African)	Ecorce de l'arbre à l'ail	Caesalpinaceae	Bark
Solanum melongena	Aubergine	Aubergine	Solonaceae	Fruit
Tetrapleura tetraptera	Aidan tree	Fruit à 4 ailes	Mimosaceae	Fruit
Xylopia aethiopica	Ethiopian pepper (African)	Poivre d'Éthiopie	Annonaceae	Fruit

Power 0.7 - 1.5 kw (1.2 - 1.3 kw for Axial), Plazma gas flow rate (Ar) 10.5 - 15 L/min (radial) 15 L/min (axial), auxiliary gas flow rate (Ar) b1.5 L/min, Viewing height 5 - 12 mm, copy and reading time 1 - 5 s (max 60 s), copy time 3 s (max 100 s). The other elements Na, K, Mn, Mg, Ca, Zn and Cu were determined by atomic absorption spectrophotometer (Pharmacia biotech Ultraspec 4000) [14].

For the determination of selenium content, the mineralization of the powder (1g) in closed system was carried out by acid digestion [2 mL, HNO_3 + HCl (1/1)] in a furnace microwave and the solutions obtained were analyzed for selenium content with a ICP-CCC equipment (Varian vista) equipped with the Coupled Charge Device detector [13]. Standard curves made from known concentrations of selenium were used for this purpose.

2.4. Vitamin and Phenols Analysis, pH-Buffering Capacity and Colour Measurement

Vitamin A precursor and vitamin E were evaluated on oil extracts by spectrophotometry as previously described with some modifications [15,16]. In this respect, 0.1 g of oil extracted from each powdered sample as described above, was diluted in an acetone/hexane mixture (70/30; v/v) and total carotenoid and vitamin E determination carried out on an aliquot by measuring absorbance at 470 nm and 270 nm respectively, in a Genesis-5 Spectronic spectrophotometer (Rochester, NY) against a blank sample (solvent). Standard curves made with pure vitamin A and vitamin E were used for this purpose and the results expressed as mg vitamin A equivalent or mg vitamin E equivalent per 100 g dried powder sample.

Vitamin C was analyzed by titration as previously described with some modifications [17]. In this procedure, 2 g of each sample was dissolved in a 25 mL volumetric flask containing 20 mL of distilled water. After mixing for 10 min using an agitator, the mixture was titrated with 2,6 dichloro phenol indophenol solution using phenolphthalein as indicator. A standard sample of ascorbic acid (0.1 mg/100mL) was used as reference and the results expressed in mg vitamin C/100g of dry powder.

Total phenolic content was determined following the Folin-Ciocalteu method using gallic acid as standard [18]. The pH-buffering capacity was evaluated for different water suspension of powders. In this respect 1 g of each individual sample was dissolved in 50 mL of deionised water and homogenised using an agitator for 10 min. The pH of the solution was then read with a pH-meter (Checker pH Tester, Hanna Instruments).

Colour measurements of flour samples were carried out using a portable tintometer (Lovibond RT Colour Measurement Kit V2.28) with a 10° observer window and a D-65 light source. The color values generated by the equipment were expressed as L—(whiteness or darkness), a—(redness/greenness) and b—(blueness/yellowness).

2.5. Statistical Analysis

Analyses were performed in triplicate and variation in the samples was routinely less than 10%. Mean values and standard deviations were calculated for each plant using the values obtained from three locations. Data obtained were subjected to one way analysis of variance. Differences between means were verified using the Duncan multiple range test at significant level P = 0.05 using Statgraphics 3.0 (Manugistics, Rockville, Maryland, USA) statistical software.

3. RESULTS AND DISCUSSION

3.1. Proximate Composition

The proximate compositions of plants samples are shown in **Table 2**. The moisture content of plants was found to range from 7.7 g/100g for M. myristica to 10.5 g/100g for M. Whitei, which makes them more stable during storage and packaging. Similar low moisture contents (8 - 14 g/100g) have been reported on some of these spices [8]. The ash content oscillated between 7.7 g/100g and 10.5 g/100g—a range that is much higher than that (2 - 9 g/100g) reported earlier [8]. These differences may probably reflect the difference in the origin and varieties of samples.

Compared to other foodstuffs, the edible plants analyzed could be considered all as good sources of minerals. In fact the values are high compared to data reported in the literature for leguminous seeds (3.7 - 4.4 g/100g) [19], for marama bean (2.8 - 4.6 g/100g) [20], for lima beans (2.78 g/100g) from Nigeria [21] and for eight varieties of tomatoes (0.74% - 1.41%) [22], but were in the range (4.21 - 26.0 g/100g) reported for leafy vegetables [23]. The available carbohydrate values varied significantly and reflected the nature of each plant. Whearas E. giganteus (33.1 g/100g), P. brazeana (22.2 g/100g) and Hua gabonii fruit (20.4 g/100g) can be considered as sources of available carbohydrates, X. aethiopica (0.4 g/100g) and M. myristica (0.4 g/100g), S. zenkeri root (1.5 g/100g) and fruit (3.2 g/100g), and F. Leprieurii (3.4 g/100g) had low levels of available carbohydrates. Results showed that E. giganteus and P. brazzeana can accumulate sugar in roots; S. Zenkeri on the other hand did not seem to accumulate sugar either in root or in the fruits. Actually, although some of these spices contain some dietary sugars, they cannot be considered as carbohydrate sources as compared to tubers and cereals which are spread throughout the world [24].

For the organic nitrogen content, expressed in terms of proteins (Nitrogen × 6.25), the highest values were found in S. zenkeri fruit (14.0 g/100g) and root (12.2 g/100g) and M. myristica (13.6 g/100g), while the lowest were in M. whitei (4.7 g/100g), S. striatinus (4.9 g/100g), D. psilurus (4.8%) and T. tetraptera (5.0 g/100g). These observations suggest the relative dietary importance of S. zenkeri fruit and root, or M. myristica to the improvement of protein content of foods made from them. It is worth noting that our results for S. Zenkeri, M. myristica, M. whitei and T. tetraptera are fairly higher than those reported by Tchiegang and Mbougueng [8] on these

Table 2. Proximate composition of the selected wild plants.

Spice names	*Moisture* (g/100g)	*Ash* (g/100g)	*Proteins (N × 6.25)* (g/100g)	*Non proteins Nitrogen* (g/100g)	*Fats* (g/100g)	*Available carbohydrate* (g/100g)
A. daniellii	10.4 ± 0.1^{i}	9.3 ± 0.1^{cd}	8.5 ± 1.0^{cd}	0.31 ± 0.07^{h}	23.1 ± 0.6^{l}	11.9 ± 0.1^{h}
C. frutescens	9.3 ± 0.1^{cde}	9.4 ± 0.2^{de}	9.4 ± 0.8^{def}	0.41 ± 0.03^{j}	11.1 ± 0.2^{e}	7.0 ± 0.1^{e}
D. glomerata	9.0 ± 0.1^{c}	9.8 ± 0.1^{fg}	8.6 ± 0.4^{cde}	0.43 ± 0.02^{i}	11.2 ± 0.6^{e}	17.8 ± 0.1^{k}
D. psilirus	9.7 ± 0.1^{fgh}	9.5 ± 0.1^{def}	4.8 ± 0.2^{a}	0.22 ± 0.03^{a}	15.3 ± 0.9^{h}	8.9 ± 0.1^{f}
E. giganteus	9.3 ± 0.1^{cde}	9.0 ± 0.1^{bc}	5.9 ± 0.3^{b}	0.35 ± 0.04^{e}	12.2 ± 0.4^{f}	33.1 ± 0.1^{o}
F. leprieuri	9.5 ± 0.1^{defg}	9.7 ± 0.1^{efg}	6.4 ± 0.2^{b}	0.40 ± 0.02^{f}	32.1 ± 0.2^{m}	3.4 ± 0.1^{c}
F. xanthoxyloïdes	9.6 ± 0.1^{efgh}	9.5 ± 0.1^{def}	8.0 ± 0.1^{c}	0.44 ± 0.02^{h}	19.3 ± 0.9^{j}	6.0 ± 0.1^{d}
H. gabonii (fruit)	9.8 ± 0.1^{gh}	9.6 ± 0.1^{defg}	6.2 ± 0.7^{b}	0.13 ± 0.03^{bc}	17.0 ± 0.3^{i}	12.1 ± 0.1^{h}
H. gabonii (bark)	9.7 ± 0.1^{fgh}	9.7 ± 0.1^{efg}	10.2 ± 0.4^{f}	0.46 ± 0.04^{k}	2.2 ± 0.3^{a}	20.5 ± 0.1^{l}
M. myristica	7.7 ± 0.1^{a}	9.6 ± 0.1^{defg}	13.6 ± 0.2^{h}	0.15 ± 0.03^{j}	53.4 ± 0.2^{o}	0.4 ± 0.1^{a}
M. whitei	10.5 ± 0.1^{i}	8.7 ± 0.1^{b}	4.7 ± 0.1^{a}	0.47 ± 0.01^{de}	4.2 ± 0.3^{b}	21.4 ± 0.1^{m}
P. brazzeana	9.5 ± 0.1^{defg}	9.6 ± 0.1^{b}	11.3 ± 0.1^{g}	0.65 ± 0.04^{m}	6.2 ± 0.2^{c}	22.2 ± 0.1^{n}
P. guineense	9.2 ± 0.1^{cd}	7.7 ± 0.1^{defg}	8.5 ± 0.2^{cd}	0.49 ± 0.03^{ij}	20.3 ± 0.7^{k}	9.6 ± 0.1^{g}
P. umbellatum	8.7 ± 0.1^{b}	10.4 ± 0.1^{a}	11.3 ± 0.1^{g}	0.35 ± 0.01^{k}	6.0 ± 0.4^{c}	9.0 ± 0.1^{f}
S. melongena	9.4 ± 0.2^{def}	9.3 ± 0.1^{h}	9.5 ± 0.2^{ef}	0.35 ± 0.02^{ij}	17.4 ± 1.1^{i}	12.3 ± 0.1^{h}
S. striatinux	9.6 ± 0.1^{efgh}	9.2 ± 0.1^{cd}	4.9 ± 0.2^{a}	0.24 ± 0.03^{ab}	7.1 ± 0.5^{d}	16.3 ± 0.2^{j}
S.zenkeri (fruit)	9.5 ± 0.1^{defg}	9.9 ± 0.1^{g}	12.2 ± 0.7^{g}	0.39 ± 0.08^{i}	2.0 ± 0.3^{a}	1.5 ± 0.1^{b}
S.zenkeri (bark)	9.6 ± 0.1^{efgh}	9.6 ± 0.1^{defg}	14.0 ± 0.1^{h}	1.07 ± 0.02^{n}	3.6 ± 0.3^{b}	3.2 ± 0.1^{c}
T. tetraptera	9.9 ± 0.1^{h}	10.5 ± 0.1^{h}	5.0 ± 0.3^{a}	0.36 ± 0.07^{cd}	5.6 ± 0.5^{c}	13.8 ± 0.2^{i}
X. aethiopica	9.6 ± 0.1^{efgh}	9.5 ± 0.1^{def}	7.9 ± 0.3^{c}	0.29 ± 0.02^{g}	33.7 ± 1.0^{n}	0.4 ± 0.1^{a}

N = 6; means ± sd; values with different letters within a column are significantly different at 5% level.

spices and nearly similar to those obtained for Opuntia spp (6.7 - 19 g/100g) [25], while values for S. striatinus and D. psilurus were lower. This may be due to the differences in the analytical methods used which, in their case was a spectrophotometric evaluation of ammonium [26]. In addition this could also be due to the inorganic nitrogen content which had not been taken into consideration by these authors, and which in the present work varied significantly amongst the samples (**Table 2**). In fact, non protein nitrogen were found in relatively large amounts in S. zenkeri fruit (1.07 g/100g) while low amounts were found in M. myristica (0.15 g/100g) and H. gabonii fruit (0.13 g/100g).

The most significant variation observed on the proximate composition of the spices under study was in their fatcontent. Generally the fat content of the different spice samples varied from 2.0 g/100g (S. zenkeri fruit) to 53.4 g/100g (M. myristica). Similar high levels of fats (47.5 g/100g), but lower than our results have been reported for M. Myristica [8] and for marama bean (41.9 g/100g) [20]. It was equally found that F. leprieuri (32.1 g/100g) in concordance with literature and X. aethiopica (33.7 g/100g) possessed high fat contents. Comparatively the low fat content reported in literature [8] for X. aethiopica (20.2 g/100g) may be attributed to the difference in the extraction method which in the case of these authors, Soxhlet, is known to be destructive. Compared to other oils sources (olive, palm, soy, and groundnuts), M. myristica, F. leprieuri and X. aethiopica can be considered as good sources of fat.

Since vegetable oils are usually associated with high contents of unsaturated fatty acids [27], the likelihood of the oils in the spices analysed being of good nutritional and even commercial value are high but this needs to be verified using appropriate analytical methods.

3.2. Minerals

The mineral content of plants is given in **Table 3**.

Potassium content was relatively high in all cases except in two samples (F. leprieuri, 31.9 mg/100g and S. striatinus, 69.9 mg/100g). The highest value was observed for M. whitei with an average value of 592.7 mg/100g. At the best of our knowledge, this is the first time data is available in the literature on the potassium level of these spices. Compared to other vegetables, the range of values reported for the spices (280 - 670 mg/ 100g) [28], are significantly lower than those for leguminous seeds (776 - 1240 mg/100g) [19], turkey spices (591 - 2481 mg/100g) [28], medicinal and aromatic plants used as spices, condiments and herbal teas (357 - 2767 mg/100g) [30] and for lima beans (2.78 g/100g) from Nigeria [21].

Calcium content was found in low levels in most of the plants analyzed, especially in A. danielli (62 mg/100g) and H. gabonii fruit (74 mg/100g) whereas H. gabonii bark (1594 mg/100g) and M. whitei (1060 mg/100g) exhibited higher calcium content. For H. gabonii bark the value observed was lower than the value of 2571 mg/100g previously reported [8]. Moreover these authors reported high calcium content for S. zenkeri bark (2789 mg/100g) as compared to significantly lower data (224 mg/100g) obtained in this study. Compared to other foodstuffs, the values fall in the range reported for vegetables (17 - 2596 mg/100g) [23] and for marama bean (937 - 2038 mg/100g) [20], but covers those reported for Turkey spices (633 - 1236 mg/100g) [28], root vegetables (37 - 431 mg/100g) [31] and leguminous seeds (78 - 152 mg/100g) [19].

Table 3. Mineral composition of selected wild plant samples expressed in mg/100g dry weight basis.

Plants	Minerals								
	Ca (mg/100g)	Mg (mg/100g)	Na (mg/100g)	K (mg/100g)	Fe (mg/100g)	Se (µg/100g)	Mn (mg/100g)	Zn (mg/100g)	Cu (mg/100g)
A. daniellii	62 ± 3^{a}	125 ± 5^{de}	7.7 ± 0.2^{abcd}	432 ± 22^{i}	27.6 ± 1.2^{defg}	70.0 ± 3.5^{cd}	15.4 ± 0.7^{k}	6.7 ± 0.2^{f}	0.20 ± 0.01^{g}
C. frutescens	173 ± 8^{e}	254 ± 10^{k}	8.0 ± 0.3^{abcd}	206 ± 10^{bc}	6.4 ± 0.2^{a}	80.0 ± 4^{de}	9.1 ± 0.4^{h}	3.7 ± 0.1^{d}	0.20 ± 0.01^{g}
D. glomerata	218 ± 11^{e}	162 ± 7^{ghi}	12.9 ± 0.5^{bcde}	404 ± 19^{hi}	16.9 ± 0.6^{abcd}	110.0 ± 5.5^{g}	2.20 ± 0.07^{cde}	0.8 ± 0.0^{a}	0.20 ± 0.01^{g}
D. psilirus	233 ± 9^{e}	116 ± 6^{cde}	9.7 ± 0.4^{abcd}	340 ± 13^{fg}	60.4 ± 2.8^{j}	50.0 ± 2.5^{b}	0.70 ± 0.03^{a}	5.8 ± 0.2^{e}	0.03 ± 0.01^{b}
E. giganteus	176 ± 7^{e}	89 ± 4^{bc}	6.6 ± 0.2^{abc}	368 ± 16^{gh}	206.4 ± 10.7^{l}	0.0^{a}	13.0 ± 0.6^{j}	1.1 ± 0.1^{g}	0.10 ± 0.01^{d}
F. leprieri	182 ± 10^{e}	212 ± 10^{j}	0.0^{a}	32 ± 1^{a}	18.3 ± 0.8^{abcde}	0.0^{a}	2.7 ± 0.1^{de}	1.5 ± 0.1^{b}	0.25 ± 0.01^{i}
F.xanthoxyloides	188 ± 6^{e}	143 ± 6^{fgh}	13.4 ± 0.6^{bcde}	518 ± 26^{j}	44.7 ± 2.7^{j}	50.0 ± 2.5^{b}	1.4 ± 0.03^{bcd}	0.7 ± 0.0^{a}	0.14 ± 0.01^{e}
H.gabonii (fruit)	74 ± 3^{a}	118 ± 6^{cde}	5.6 ± 0.2^{ab}	340 ± 13^{fg}	13.0 ± 0.5^{abc}	160.0 ± 8.0^{h}	6.3 ± 0.3^{g}	2.7 ± 0.1^{c}	0.22 ± 0.01^{h}
H.gabonii (bark)	1594 ± 80^{j}	72 ± 3^{b}	0.5 ± 0.1^{a}	371 ± 18^{efg}	8.5 ± 0.4^{ab}	90.0 ± 4.5^{ef}	2.9 ± 0.1^{e}	0.8 ± 0.1^{a}	0.19 ± 0.01^{f}
M. whitei	1060 ± 53^{i}	532 ± 27^{l}	21.4 ± 0.9^{f}	593 ± 23^{k}	102.1 ± 5.3^{k}	00^{a}	6.6 ± 0.3^{g}	3.9 ± 0.1^{d}	0.16 ± 0.01^{f}
M. myristica	375 ± 14^{f}	132 ± 6^{ef}	9.1 ± 0.4^{abcd}	432 ± 18^{i}	36.7 ± 1.4^{gh}	80.0 ± 4^{de}	10.9 ± 0.2^{i}	1.4 ± 0.05^{b}	0.2 ± 0.01^{g}
P. brazzeana	156 ± 6^{cde}	250 ± 12^{k}	9.9 ± 0.3^{abcd}	188 ± 7^{b}	22.7 ± 0.9^{cdef}	80.0 ± 4^{de}	2.8 ± 0.1^{e}	0.7 ± 0.01^{a}	0.01 ± 0.00^{a}
P. guineense	466 ± 20^{g}	296 ± 13^{l}	12.9 ± 0.6^{bcde}	317 ± 14^{efg}	21.8 ± 1.1^{bcdef}	50.0 ± 2.5^{b}	1.8 ± 0.1^{bcde}	5.2 ± 0.2^{e}	0.24 ± 0.01^{i}
P. umbellatum	845 ± 33^{h}	490 ± 23^{l}	28.1 ± 1.3^{h}	350 ± 13.2^{a}	97.8 ± 4.7^{k}	60.0 ± 3^{bc}	4.4 ± 0.2^{f}	2.4 ± 0.1^{c}	0.13 ± 0.01^{e}
S. striatinux	82 ± 3^{bc}	175 ± 8^{i}	18.5 ± 0.7^{de}	70 ± 2^{a}	28.7 ± 1.3^{defg}	110.0 ± 5.5^{g}	6.5 ± 0.3^{g}	1.5 ± 0.06^{b}	0.06 ± 0.00^{c}
S. zenkeri (fruit)	87 ± 3^{e}	145 ± 7^{fghi}	31.1 ± 1.4^{g}	289 ± 11^{def}	44.8 ± 2.1^{j}	310 ± 16^{i}	10.1 ± 0.5^{hi}	6.7 ± 0.3^{f}	0.25 ± 0.01^{i}
S. zenkeri (bark)	224 ± 10^{bcd}	22 ± 1^{a}	6.5 ± 0.2^{abc}	368 ± 15^{gh}	5.9 ± 0.3^{a}	110.0 ± 5.5^{g}	20.5 ± 1.1^{l}	0.8 ± 0.05^{a}	0.07 ± 0.00^{c}
S. melongena	156 ± 6^{cde}	250 ± 12^{k}	0.0^{a}	319 ± 12^{d}	38.2 ± 1.6^{hi}	00^{a}	0.9 ± 0.03^{bc}	5.5 ± 0.2^{e}	0.33 ± 0.01^{j}
T. tetraptera	167 ± 6^{de}	96 ± 4^{bcd}	10.1 ± 0.4^{abcd}	255 ± 8^{cd}	30.9 ± 0.9^{efg}	50.0 ± 2.5^{b}	1.8 ± 0.01^{bcde}	0.8 ± 0.1^{efg}	0.09 ± 0.00^{d}
X. aethiopica	228 ± 11^{e}	172 ± 7^{hi}	17.1 ± 0.7^{cde}	277 ± 9^{de}	34.7 ± 1.4^{fgh}	60.0 ± 3^{bc}	11.0 ± 0.7^{i}	3.8 ± 0.2^{fgh}	1.37 ± 0.03^{k}

N = 3; mean ± sd; values with different letters within a column are significantly different at 5% level.

Compared to other major elements evaluated in this study, Na was found in very low amounts (i.e. 40 times lower than K) in spice samples with values ranging from 0.0 mg/100g (F. leprieuri and S. melongena) to 28.1 mg/100g (P. umbellatum). Similar low amounts of Na were reported for wild leguminous seeds (3.7 - 11.8 mg/100g), and a processed spice made of Hisbiscus sabdariffa and Parkia biglobosa (2.2 - 18.2 mg/100g) and condiments [32]. Mg content oscillated between 22.5 mg/100g (S. zenkeri) and 531.6 mg/100g (M. whitei). Our results are much lower than those reported in literature [8] and cover the range value reported for edible wild leaves (32 - 160 mg/100 mg) [28]. Cu was found in small percentages in all the species analyzed (0.03 - 1.37 mg/100g), with X. aethiopica (1.37 mg/100g) exhibiting the highest value and D. psilirus (0.03 mg/100g) the lowest. Our data were much higher than those reported on wild leaves (0.12 - 0.33 mg/100g) [28]. Zn content ranged from 0.7 mg/100g (F. xanthoxyloides and P. brazzeana) to 6.7 mg/100g (A. Daniellii and S. Zenkeri bark). These values were significantly higher than those reported for wild leaves (0.41 - 1.20 mg/100g) [28] and lower than those found in marama bean (31 - 39 mg/100g) [20]. Mn was found to be high (range 0.7 to 20.5 mg/100g) in our samples than those reported on other plants (0.29 - 1.0 mg/100g) [28]. Fe content largely varied ($p < 0.001$) between plants and ranged from 5.9 mg/100g for Scorodophleus zenkeri (bark) to 206.4 mg/100g for Echinops giganteus. Our results were in the range (4.6 - 123.0 mg/100g) reported for some plants used as condiments [29], but higher than those reported in literature for similar wild plants (0.4 - 41.9 mg/100g) [8].

Selenium values ranged from 0.0 µg/100g to 310.0 µg/100g. To the best of our knowledge, this is the first reported data on Se content of these spices. Various values of Se were reported in literature for various foodstuffs: 4.6 - 5.4 µg/100g for rice, 5.8 - 23.0 µg/100g of dry legumes and seeds, 0.1 - 12.7 µg/100g for various vegetables [33] and 21.6 µg/100g of Korean diet [34]. The low level of Se in foods has been ascribed to the low soil content in selenium, such as in volcanic regions [35]. In addition it has been demonstrated that some plants, known as Se-accumulator, highly absorb Se compared to non Se-accumulator plants. The high Se content of some wild plants offers dietary Se sources to consumers for the management of several chronic diseases such as cardiovascular diseases, cancers and HIV [33].

3.3. pH and Colour Characteristics L, a*, b* of the Spices

Spice powders induced some acidification of the solution in which they were introduced. Generally the powdered samples of S. zinkeri "root", M. myristica, and A. Daniellii have low pH values (3 - 4), while P. brazzeana, C. frutescens and H. gabonii "root", have pH of fairly neutral (**Table 4**). These results suggest that the spices used in this study contain some acidic molecules. Acidification is well established as a mechanism of food preservation. From this perspective, some of the spices analysed may be useful in the preservation of foods through this mechanism, but this needs to be investigated.

Conversely, all the spices exhibited a characteristic dark colour but this varied significantly from one plant species to another. In fact values for lightness, L, varied from 20.6 (less light, Xylopia aethiopica, Fagara leprieuri) to 78.6 (slightly light, M. myristica).

The a* values were higher than the white standard (a = 0.19) varying from 0.9 (S. striatinux) to 22.7 (C. frustescens). The b* coordinate of the colour also varied significantly among samples with X. aethiopica exhibiting the lower level (7.6) and S. zenkeri bark exhibiting the highest (35.2). Compared to the literature our results are nearly similar with those found in the dry bean varieties grown in Ethiopia [36] and when we compared to a white standard, these results showed that all the samples were less light, redder and more yellow.

Table 4. Colour characteristics and pH of selected wild plants.

Spice Name	Colour Parameters			pH
	L*	a*	b*	
A. daniellii	51.7 ± 1.4^{e}	7.2 ± 0.7^{cde}	20.1 ± 0.2^{cd}	3.25 ± 0.01^{a}
C. frustescens	39.9 ± 0.1^{c}	22.7 ± 0.1^{i}	31.0 ± 0.1^{g}	5.36 ± 0.01^{b}
D. glomerata	46.7 ± 0.6^{d}	7.0 ± 0.1^{cd}	19.4 ± 0.8^{bc}	6.16 ± 0.0^{b}
D. psilurus	40.8 ± 1.3^{c}	10.1 ± 0.1^{g}	24.1 ± 1.0	5.42 ± 0.01^{b}
E. giganteus	46.6 ± 0.6^{d}	8.1 ± 0.1^{ef}	25.2 ± 0.2^{ef}	6.02 ± 0.01^{b}
F. leprierui	26.4 ± 3.9^{b}	8.0 ± 0.7^{cde}	16.3 ± 3.3^{b}	5.70 ± 0.01^{b}
F. xantholozoïde	40.8 ± 1.1^{c}	14.2 ± 0.1^{h}	21.4 ± 0.4^{de}	5.32 ± 0.03^{b}
H. gabonii (fruit)	76.6 ± 0.7^{i}	3.7 ± 0.1^{b}	19.4 ± 0.3^{cd}	5.60 ± 0.03^{b}
H. gabonii (root)	39.0 ± 0.1^{c}	13.5 ± 0.2^{h}	24.8 ± 0.2^{ef}	5.35 ± 0.07^{b}
M. myristica	78.6 ± 1.1^{i}	4.7 ± 0.3^{bc}	21.0 ± 0.5^{d}	3.87 ± 0.01^{a}
M. whitei	73.4 ± 0.1^{h}	5.8 ± 0.1^{bc}	20.0 ± 0.1^{c}	5.09 ± 0.01^{b}
P. brazzeana	51.0 ± 0.5^{e}	8.4 ± 0.1^{f}	25.9 ± 0.2^{ef}	5.12 ± 0.01^{b}
P. guineense	45.7 ± 0.8^{d}	7.6 ± 0.1^{cde}	23.2 ± 0.7^{ef}	5.17 ± 0.01^{a}
P. umbellatum	53.2 ± 1.4^{e}	2.9 ± 0.5^{b}	16.3 ± 1.8^{b}	5.51 ± 0.01
S. melongena	56.5 ± 2.2^{f}	7.6 ± 0.1^{cde}	27.4 ± 0.1^{f}	5.58 ± 0.01^{b}
S. striatinux	66.1 ± 1.5^{gh}	0.9 ± 0.1^{a}	20.7 ± 0.9^{d}	5.40 ± 0.01^{b}
S. zinkeri (fruit)	70.5 ± 3.0^{h}	6.6 ± 0.6^{bc}	35.2 ± 0.8^{h}	5.79 ± 0.01^{b}
S. zinkeri (bark)	60.2 ± 0.5^{g}	8.6 ± 0.7^{f}	25.9 ± 0.9^{ef}	5.75 ± 0.01^{b}
T. tetraptera	40.2 ± 0.2^{c}	8.6 ± 0.5^{f}	18.0 ± 0.7^{bc}	5.74 ± 0.01^{b}
X. aethiopica	20.6 ± 0.3^{a}	5.1 ± 0.6^{bc}	7.6 ± 0.8^{a}	5.31 ± 0.01^{b}

N = 6; means ± sd; values with different letters within a column are significantly different at 5% level.

3.4. Phenols and Vitamins

Phenolic and vitamin content of foods are known to have positive effects on health. Epidemiological studies have shown high intakes of phenols, vit C, vit E and carotenoid-rich vegetables to be associated with decreased incidence of some cancers and cardiovascular diseases [37]. The mean content of total phenols, β-carotene, vit C and E of the wild plants analysed are given in **Table 5**. The total phenolic content, obtained varied between 1.05 g GAE/100g for S. melongena and 38.80 g GAE/100g for D. glomerata. From a comparative point of view, the values found here are lower than

Table 5. Total phenol, total carotenoid, vit C and E contents of selected wild plants.

Spice names	Total phenols (g GA eq/100g)	Total carotenoids (g vit A eq/100g)	Vit E (g α-tocopherol eq/100g)	Vit C (g/100g)
A. daniellii	9.0 ± 0.0^{h}	11.0 ± 0.1^{j}	3.6 ± 0.1^{i}	0.9 ± 0.0^{bc}
C. frustescens	6.4 ± 0.1^{g}	73.8 ± 0.0^{s}	6.2 ± 0.1^{k}	1.4 ± 0.1^{efgh}
D. glomerata	38.8 ± 0.1^{n}	10.0 ± 0.1^{i}	2.0 ± 0.0^{e}	2.8 ± 0.2^{j}
D. psilurus	8.5 ± 0.1^{h}	12.0 ± 0.1^{l}	4.0 ± 0.0^{j}	1.3 ± 0.0^{ef}
E. giganteus	14.9 ± 1.0^{k}	8.7 ± 0.0^{g}	2.9 ± 0.0^{h}	1.3 ± 0.0^{cdef}
F. leprierui	34.6 ± 0.2^{m}	7.9 ± 0.1^{e}	2.4 ± 0.1^{f}	1.2 ± 0.0^{bcde}
F.xantholozoïde	5.8 ± 0.2^{f}	26.7 ± 0.1^{o}	10.0 ± 0.1^{o}	3.3 ± 0.1^{k}
H. gabonii (fruit)	1.6 ± 0.0^{bc}	27.5 ± 0.1^{p}	8.1 ± 0.0^{m}	0.2 ± 0.0^{a}
H. gabonii (root)	1.3 ± 0.0^{ab}	13.4 ± 0.0^{n}	1.6 ± 0.0^{d}	1.6 ± 0.1^{fgh}
M. myristica	6.4 ± 0.1^{g}	2.9 ± 0.0^{b}	1.2 ± 0.0^{c}	1.0 ± 0.1^{bcd}
M. whitei	1.9 ± 0.0^{bc}	8.4 ± 0.1^{f}	3.6 ± 0.1^{i}	0.4 ± 0.0^{a}
P. brazzeana	2.2 ± 0.1^{cd}	31.0 ± 0.2^{q}	8.1 ± 0.1^{m}	0.9 ± 0.1^{b}
P. guineense	4.4 ± 0.1^{e}	9.3 ± 0.2^{h}	2.7 ± 0.1^{g}	0.9 ± 0.0^{b}
P. umbellatum	2.5 ± 0.27^{d}	38.7 ± 0.1^{r}	7.1 ± 0.1^{l}	1.7 ± 0.1^{h}
S. melongena	6.9 ± 0.1^{g}	12.6 ± 0.1^{m}	8.4 ± 0.1^{m}	0.9 ± 0.0^{b}
S. striatinux	4.5 ± 0.10^{e}	11.4 ± 0.0^{k}	1.6 ± 0.1^{d}	1.4 ± 0.2^{efg}
S. zinkeri (fruit)	13.4 ± 0.14^{j}	4.6 ± 0.1^{a}	0.5 ± 0.0^{a}	2.4 ± 0.2^{i}
S. zinkeri (bark)	1.05 ± 0.01^{a}	2.1 ± 0.0^{c}	0.4 ± 0.0^{a}	3.8 ± 0.5^{l}
T. tetraptera	22.8 ± 0.3^{l}	11.0 ± 0.1^{j}	7.0 ± 0.0^{k}	1.7 ± 0.1^{gh}
X. aethiopica	12.7 ± 0.2^{i}	5.7 ± 0.0^{d}	0.8 ± 0.0^{b}	1.3 ± 0.0^{def}

N = 6; means ± sd; values with different letters within a column are significantly different at 5% level.

those reported by Agbor et al. for P. guineense, P. umbellatum, S. zenkeri (fruit), and S. zenkeri (bark) with respective mean values of 20.9, 19.0, 18.4 and 15.1 mg equivalent catechin/g or 990.9, 899.1, 339.0, and 480.8 mg equivalent vit E/g [38]. These differences in the phenolic content may be due to difference in the standard used. When compared to other foodstuffs, similar high levels of antioxidant was observed for Arctostaphylos uva-ursi, (31.2 g GAE/100g), the aerial parts of Horsetail, Equisetum spp. (21.6 g GAE/100g) [39] and for tea and herbal infusions (44 - 209 mg GAE/g) from Poland [40].

Vit E contents in samples varied between 0.39 and 10.02 mg eq α tocopherol/100g dwb. F. Xanthoxyloides, H. gabonii fruit, P. Brazzeana, and S. striatinux possessed the highest vit E content while the lowest were found in S. Melongena, S. zenkeri bark, and X. Aethiopica. Compared to other food products, our results are in the range found in the literature: 1.3 mg eq α tocopherol /100 dwb in spirulina, 132.2 mg eq α tocopherol/100g fresh in bay leaf, and 20.61 mg/100g fresh sample in thyme [41]. Vit C content of plants varied significantly from 0.22 to 3.75 g/100g. The highest levels were found in S. melongena and F. xanthoxyloides while the lowest levels were observed in H. gabonii fruit and M. myristica. These values are high when compared to those reported for some fruits known as sources of vitamin C such as apples (1.1 - 3.5 mg/100g fresh samples), grapefruits (28.5 - 52.0 mg/100g fresh samples), mangoes (9.1 - 18.6 mg/100g fresh samples) and oranges (42.1 - 62.4 mg/ 100g) [42].

4. CONCLUSION

This study has revealed a wide variation wide variation in the carbohydrate, crude fat, crude protein, total phenols, vitamins and high levels in ash content of the spices analysed. C. frutescens is particularly high in provitamin A, just as is F. xanthoxyloides in its vit E and β carotene content, and M. myristica, F. leprieuri and X. aethiopica in their overall lipid content. These spices have the potential of contributing tothe nutritional and health needs of their consumers.

5. ACKNOWLEDGEMENTS

We wish to thank the International Foundation for Sci-Armand Abdou Bouba, Nicolas Yanou Njintang, Harquin Simplice Foyet, Joel Scher, Didier Montet, Carl Moses F. Mbofung ences, Stockholm, Sweden, for the financial support and the French Cooperation in Cameroon for the additional financial assistance.

REFERENCES

1. U. R. Susheela, "Handbook of Spices, Seasoning, and Flavorings," TECHNOMIC Publishing Co., Inc., Lancaster, 2000, p. 329.
2. J. Bitting and P. W. Sherman, "Antimicrobial Functions of Spices—Why Some Like It Hot," Quarterly Review of Biology, Vol. 73, No. 1, 1998, pp. 3-49. doi:10.1086/420058
3. A. Barakat, A. O. Maslat and M. M. AL-Kofahi, "Element Analysis and Biological Studies on Ten Oriental Spices Using XRF and Ames Test," Journal of Trace Element Medicine Biology, Vol. 17, No. 2, 2003, pp. 85-90. doi:10.1016/S0946-672X(03)80003-2
4. H. R. H. Takruri and A. F. M. Dameh, "Study of the Nutritional Value of Black Cumin Seeds (Nigella sativa L)," Journal of the Science of Food and Agriculture, Vol. 76, No. 3, 1998, pp. 404-410. doi:10.1002/(SICI)1097-0010(199803)76:3<404::AID-JSFA964>3.0.CO;2-L
5. C. Gopalan, B. V. Ramasastri, S. C. Balasubramanian, B. S. Narsinagarao, Y. G. Deosthale and K. C. Pant, "Nutritive Value of Indian Foods," India National Institute of Nutrition, Hyderabad,1999.
6. J. S. Pruthi, "Quality Assurance in Spices and Spice Products, Modern Methods of Analysis," Allied Publishers Ltd, New Delhi, 1999.

7. C. M. F. Mbofung, J. M. Gee and J. D. Knight, "Fatty Acid Profile of Some Cameroonian Spices," Journal of the Science of Food and Agriculture, Vol. 66, No. 2, 1994, pp. 213-216. doi:10.1002/jsfa.2740660216
8. C. Tchiegang and D. Mbougueng, "Composition Chimique des Epices Utilisées dans la Préparation du na'a poh et du kui de l'Ouest Cameroun," Tropicultura, Vol. 23, No. 4, 2005, pp. 193-200.
9. A Abdou Bouba, N. Y. Njintang, J. Scher and C. M. F. Mbofung, "Phenolic Compounds and Radical Scavenging Potential of Twenty Cameroonian Spices," Agriculture and Biology Journal of North America, Vol. 1, No. 3, 2010, pp. 213-224. doi:10.5251/abjna.2010.1.3.213.224
10. AOAC, Association of Official Analytical Chemists, Washington DC, 1990.
11. Y. N. Njintang, C. M. F. Mbofung, F. Balaam, P. Kitissou and J. Scher, "Effect of Taro (Colocasia esculenta) Flour Addition on the Functional and Alveographic Properties of Wheat Flour and Dough," Journal of the Science of Food and Agriculture, Vol. 88, No. 2, 2008, pp. 273-279. doi:10.1002/jsfa.3085
12. J. Folch, M. Lees and G. S. H. Sloane, "A Simple Method for the Isolation and Purification of Total Lipids from Animal Tissue," Journal of Biolology and Chemistry, Vol. 226, 1957, pp. 497-509.
13. S. Skunjins, "Handbook for ICP-AES (Varian-Vista), A Short Guide to Vista Series ICP-AES Operation," Version 1.0., Varian International AG, 1998.
14. AACC, "Approved Methods of the American Association of Cereal Chemists," The Association, St. Paul, 1995, pp. 40-71.
15. D. B. Milne and J. Botnen, "Retinol, α-Tocopherol, Lycopene, and α- and β-Carotene Simultaneously Determined in Plasma by Isocratic Liquid Chromatography," Clinical Chemistry, Vol. 32, 1986, pp. 874-876.
16. P. De Leenheer, H. J. Nelis, W. E. Lambert and R. M. Bauwen, "Chromatography of Fat-Soluble Vitamins in Clinical Chemistry," Journal of Chromatography, Vol. 429, 1988, pp. 3-58.
17. V. Raghu, P. Kalpana and K. Srinivasan, "Comparison of Ascorbic Acid Content of Emblica officinalis Fruits Determined by Different Analytical Methods," Journal of Food Composition and Analysis, Vol. 20, No. 6, 2007, pp. 529-533. doi:10.1016/j.jfca.2007.02.006
18. X. Gao, M. Ohlander, N. Jeppsson, L. Björk and V. Trajkovski, "Changes in Antioxidant Effects and Their Relationship to Phytonutrients in Fruits of Sea Buckthorn (Hippophae rhamnoides L.) during Maturation," Journal of Agriculture and Food Chemistry, Vol. 48, No. 5, 2000, pp. 1485-1490. doi:10.1021/jf991072g
19. J. O. Amarteifio and D. Moholo, "The Chemical Composition of Four Legumes Consumed in Botswana," Journal of Food Composition and Analysis, Vol. 11, No. 4, 1998, pp. 329-332. doi:10.1006/jfca.1998.0595
20. M. Holse, S. Husted and A. Hansen, "Chemical Composition of Marama Bean (Tylosema esculentum) A Wild African Bean with Unexploited Potential," Journal of Food Composition and Analysis, Vol. 23, No. 6, 2010, pp. 648-657. doi:10.1016/j.jfca.2010.03.006
21. E. Ezeagu and M. D. Ibegbu, "Biochemical and Nutritional Potential of Ukpa: A Variety of Tropical Lima Beans (Phaseolus lunatus) from Nigeria—A Short Report," Poland Journal of Food Nutrition Science, Vol. 60, No. 3, 2010, pp. 231-235.
22. L. Guil-Guerrero and M. M. Rebolloso-Fuentes, "Nutrient Composition and Antioxidant Activity of Eight Tomato (Lycopersicon esculentum) Varieties," Journal of Food Composition and Analysis, Vol. 22, No. 2, 2009, pp. 123-129. doi:10.1016/j.jfca.2008.10.012
23. O. Hels, T. Larsen, L. P. Christensen, U. Kidmose, N. Hassan and S. H. Thilsted, "Contents of Iron, Calcium, Zinc and b-Carotene in Commonly Consumed Vegetables in Bangladesh," Journal of Food Composition and Analysis, Vol. 17, No. 5, 2004, pp. 587-595. doi:10.1016/j.jfca.2003.08.007
24. Jayakody, R. Hoover, Q. Liu and E. Weber, "Studies on Tuber and Root Starches. I. Structure and Physicochemical Properties of Innala (Solenostemon rotundifolius)

Starches Grown in Sri Lanka," Food Research International, Vol. 38, No. 6, 2005, pp. 615-629. doi:10.1016/j.foodres.2004.11.015
25. T. Guevara-Figueroa., H. Jimenez-Islas., M. L. ReyesEscogido, A. G. Mortensen, B. B. Laursen., L-W. Lin, A. De Leon-Rodrıguez, I. S. Fomsgaard and A. P. Barba de la Rosa, "Proximate Composition, Phenolic Acids, and Flavonoids Characterization of Commercial and Wild Nopal (Opuntia spp)," Journal of Food Composition and Analysis, Vol. 23, 2010, pp. 525-532.
26. B. Devani, I. Shishoo, S. A. Shah and B. N. Suhagia, "Spectrophotometric Method for the Micro Determination of Nitrogen in Kjeldahl Digest," Journal of Association Official Analytical Chemists, Vol. 72, No. 6, 1989, pp. 953-956.
27. K. Huang, C. C. Akoh and C. M. Erickson, "Enzymatic Modification of Melon Seed Oil: Incorporation of Eicosapentaenoic Acid," Journal of Agriculture and Food Chemistry, Vol. 42, No. 11, 1994, pp. 2646-2648. doi:10.1021/jf00047a050
28. J. L. Guil, J. J. G. Martinez and M. E. Isasa, "Mineral Nutrient Composition of Edible Wild Plants," Journal of Food Composition and Analysis, Vol. 11, No. 4, 1998, pp. 322-328. doi:10.1006/jfca.1998.0594
29. Özcan, "Mineral Contents of Some Plants Used as Condiments in Turkey," Food Chemistry, Vol. 84, No. 3, 2004, pp. 437-440. doi:10.1016/S0308-8146(03)00263-2
30. M. Ozcan and M. Akbulut, "Estimation of Minerals, Nitrate and Nitrite Contents of Medicinal and Aromatic Plants Used as Spices, Condiments and Herbal Tea," Food Chemistry, Vol. 106, No. 2, 2007, pp. 852-858. doi:10.1016/j.foodchem.2007.06.045
31. Ekholm, H. Reinivuo, P. Mattila, H. Pakkala, J. Koponen, A. Happonen, J. Hellström and M. Ovaskainen, "Changes in the Mineral and Trace Element Contents of Cereals, Fruits and Vegetables in Finland," Journal of Food Composition and Analysis, Vol. 20, No. 6, 2007, pp. 487-495. doi:10.1016/j.jfca.2007.02.007
32. D.A. Dashak, M. L. Dawang and N. B. Lucas, "An Assessment of the Proximate Chemical Composition of Locally Produced Spices Known as Dadawa Basso and Dadawa Kalwa from Three Markets in Plateau State of Nigeria," Food Chemistry, Vol. 75, No. 2, 2001, pp. 231-235. doi:10.1016/S0308-8146(01)00205-9
33. Sirichakwal, P. Puwastien, J. Polngam and R. Kongkachuichai, "Selenium Content of Thai Foods," Journal of Food Composition and Analysis, Vol. 18, No. 1, 2005, pp. 47-59. doi:10.1016/j.jfca.2003.10.010
34. K. Hwang, H. K. Park and C.-H. Ohe, "Selenium Content in Representative Korean Foods," Journal of Food Composition and Analysis, Vol. 22, No. 2, 2009, pp. 117-122. doi:10.1016/j.jfca.2008.11.009
35. M. Rayman, "The Importance of Selenium to Human Health," Lancet, Vol. 356, No. 9225, 2000, pp. 233-241. doi:10.1016/S0140-6736(00)02490-9
36. E. A. Shimelis and S. K. Rakshit, "Proximate Composition and Physico-Chemical Properties of Improved Dry Bean (Phaseolus vulgaris L.) Varieties Grown in Ethiopia," LWT—Food Science and Technology, Vol. 38, No. 4, 2005, pp. 331-338. doi:10.1016/j.lwt.2004.07.002
37. M. L. Slattery, J. Benson, K. Ma, K. N. Curtin, D. Schaeffer and J. D. Potter, "Carotenoids and Colon Cancer," American Journal of Clinical Nutrition, Vol. 71, 2000, pp. 575-582.
38. G. A. Agbor, J. E. Oben, J. Y. Ngogang, X. X. Cai and A. J. Vinson, "Antioxidant Capacity of Some Herbs/Spices from Cameroon: A Comparative Study of Two Methods," Journal of Agriculture and Food Chemistry, Vol. 53, No. 17, 2005, pp. 6819-6824. doi:10.1021/jf050445c
39. Amarowicz, R. B. Pegg, P. Rahimi-Moghaddam, B. Barld and J. A. Weil, "Free-Radical Scavenging Capacity and Antioxidant Activity of Selected Plant Species from the Canadian Prairies," Food Chemistry, Vol. 84, No. 4, 2004, pp. 551-562. doi:10.1016/S0308-8146(03)00278-4
40. Rusaczonek, F. Świderski and B. Waszkiewicz-Robak, "Antioxidant Properties of Tea and Herbal Infusions," Poland Journal of Food Nutrition Science, Vol. 60, No. 1, 2010, pp. 33-35.

41. J. M. D. Gomez-Coronadoa, E. Ibanez, J. F. Rupereza and C. Barbasa, "Tocopherol Measurement in Edible Products of Vegetable Origin," Journal of Chromatography Analysis, Vol. 1054, 2004, pp. 227-233.
42. A. Franke, J. L. Custer, C. Arakaki and P. S. Murphy, "Vitamin C and Flavonoid Levels of Fruits and Vegetables Consumed in Hawaii," Journal of Food Composition and Analysis, Vol. 17, No. 1, 2004, pp. 1-35. doi:10.1016/S0889-1575(03)00066-8

CHAPTER 16

Evaluation of the Antifungal Activity of Aqueous and Alcoholic Extracts of Six Spices

Moses Ikegbunam[1], Maduagwu Ukamaka[1], Osazuwa Emmanuel[2]

[1]*Department of Pharmaceutical Microbiology and Biotechnology, Nnamdi Azikiwe University, Awka, Nigeria*

[2]*Department of Pharmaceutical Microbiology, University of Benin, Benin, Nigeria*

ABSTRACT

Background: Nigeria is plagued with a variety of socioeconomic problems mainly poverty, poor access to quality healthcare and poor hygienic conditions resulting in the myriad of fungal infections that are frequently encountered in clinical practice. Method: The antifungal activity of aqueous, methanolic, propanolic and benzyl alcohol extracts of Capsicum annuum fruits and seeds, Capsicum chinense fruits and seeds, Aframomum melegueta pods and seeds, Allium sativum bulbs, Allium cepa bulbs and Zingiber officinale rhizomes on Candida albicans (yeast), Aspergillus niger (mould) and Trichophyton rubrum (dermatophyte) were evaluated by the agar well diffusion method. The aqueous and alcoholic filtered plant extracts were obtained by maceration, and also methanolic extracts were also obtained by Soxhlet extraction. The results were obtained by measuring the inhibition zone diameter in millimeters and were presented by subtracting the activity of the control. Results: Aqueous and methanolic extracts of Allium sativum gave the highest inhibition of the growth of

Candida albicans (22 mm), followed closely by its propanolic extract with inhibition zone diameter of 15 mm, and also, propanolic extracts of Aframomum melegueta and Allium cepa gave inhibition zone diameters of 12 mm each. Soxhlet methanolic extract of Allium sativum had the highest inhibition of the growth of Aspergillus niger with an inhibition zone diameter of 25 mm, followed closely by Zingiber officinale Soxhlet methanolic extract with an inhibition zone diameter of 22 mm, also, the propanolic extract of Allium sativum gave an inhibition zone diameter of 21 mm, whereas Soxhlet methanolic extracts of Aframomum melegueta and Allium cepa gave an inhibition zone diameter of 19 mm each. The highest activity against Trichophyton rubrum was obtained with the Soxhlet methanolic extract of Allium sativum (39 mm), followed closely by its propanolic extract with an inhibition zone diameter of 27 mm. An inhibition zone diameter of 22 mm was recorded with the benzyl alcohol extract of Allium cepa, 22 mm with the Soxhlet methanolic extract of Aframomum melegueta and 19mm with the aqueous extract of Capsicum chinense seeds. Conclusion: The in-vitro inhibitory effects of these spice extracts indicated that the test spices could serve as potential candidates for developing new systemic and topical antifungal drugs against the wide range of pathogenic fungal strains, and they could also serve as natural prophylaxis against the fungal infections.

Keywords: Spices, Antifungal Activity, Agar Well Diffusion

1. INTRODUCTION

People all over the world are still affected by quite a large number of microbial infections with fungi causing a good number of them. It has since been discovered that active medicinal substances are present in plants and this has encouraged the inclusion of herbal remedies in the delivery of health care [1] [2].

Spices are aromatic or pungent plant parts used for enhancing the taste of foods. Although spices are commonly used to improve the taste of foods, they have also been exploited for their medicinal as well as their antimicrobial activities [3] [4].

Capsicum annuum (red pepper) is a fruit spice of the Solanaceae family, rich in proteins, lipids, vitamins, carbohydrates, and health phytochemicals such as carotenoids, flavonoids and capsaicinoids known to prevent diseases such as asthma, coughs, sore throats etc. [5] [6]. Capsicum chinense (Cameroon pepper) is also from the family Solanaceae. It is very popular in the Nigerian market and is used by so many households and individuals.

Aframomum melegueta (Alligator pepper) is from the family Zingiberaceae together with Zingiber officinale (ginger). Alligator pepper seeds have been shown to possess phytoconstituents that have potent antimicrobial effect [7]. Alligator pepper have been found to contain the phytoconstituents; gingerol, methyl-6-gingero, shogaol and paradol that contribute to its antimicrobial properties [8]. Ginger has strong aromatic and medicinal properties [9]. It has been utilized for the treatment of infectious diseases in many countries [10] [11]. The pungent taste of ginger has been attributed to gingerols, shagaols, and zingerone.

Allium cepa (onions) is from the family Amaryllidaceae and contains water, sugar, protein, fat and fibre [12]. Flavonoids and organosulphur compounds are the two major classes of phytochemicals present in onions believed to provide its beneficial health effects [13]. Allium sativum (garlic) also from the Amaryllidaceae family has been used both for its culinary and medicinal purposes [14]. The sulphur containing compounds give them the characteristic flavor and exhibit potent antifungal properties [15] together with the non-sulphur compounds [16] -[18].

Fungal pathogens of humans are grouped as: Yeasts (Candida albicans), Moulds (Aspergillus niger), and Dermatophytes (Trichophyton rubrum). C. albicans is responsible for a wide range of superficial and systemic infections. 75% of women are affected with genital Candidiasis in their lifetime while in men, alcoholics and diabetics are prone to genital Candidiasis [19] [20]. Aspergillus niger is ubiquitous in nature, exposure is common but disease is rare. Aspergillosis occurs in immunocompromised individuals. A. niger has been implicated in otomycosis in healthy persons. Massive inhalation of A. niger spores can cause allergic reactions such as asthma and pneumonitis. Dermatophytes are fungi that are capable of colonizing the skin, nails, or hair. Trichophyton rubrum is an anthropophilic saprotroph, it colonizes the upper layers of dead skin and it is the most common cause of Tinea pedis, Tinea unguuim, Tinea manuum, Tinea cruris, and Tinea corporis. Infection can be avoided by lifestyle and hygiene modification [19].

The aim of this study is to investigate the inhibitory effects of six commonly used spices on the growth of three pathogenic fungi to determine whether they can serve as natural prophylactics and the possibility of developing systemic and topical antifungal agents with them using the agar well diffusion technique.

2. METHODS

2.1. Test Micro-Organisms

Three microorganisms were tested in this study: Candida albicans, Trichophyton rubrum, Aspergillus niger. They were characterized strains obtained from the Laboratory of Pharmaceutical Microbiology and Biotechnology, Nnamdi Azikiwe University, Awka, Nigeria.

2.2. Plant Materials

In this study, six commonly used spices were selected to analyze their antifungal activity based on previous literature and their popularity in the Nigerian markets. The fresh bulbs of A. cepa and A. sativum, fresh rhizomes of Z. officinale, fresh fruits of C. annuum, dried fruits of C. chinense and pods and seeds of A. melegueta were purchased from Eke Awka and Ose market in Onitsha, Anambra State in the month of April. They were identified and authenticated by Mrs Oduche Anthonia in the Department of Pharmacognosy and Traditional medicine, Faculty of Pharmaceutical Sciences, Nnamdi Azikiwe University Awka.

2.3. Extraction of Crude Drugs

The fresh fruits of C. annuum were dried in a hot air oven at 60°C for 8 hours. Seeds from some of the dried fruits of C. annuum and C. chinense were separated, along with the seeds of A. melegueta and were powdered separately. The dried materials including the whole fruits of C. annuum and C. chinense and pods of A. melegueta and the separated seeds were reduced to fine powder with a mechanical grinder. The bulbs of A. cepa and A. sativum, and the Z. officinale rhizome were washed thoroughly using tap water, a clean kitchen knife was used to manually peel the outer coverings, and then the fleshy parts were washed and rinsed with distilled water again. A sterile ceramic mortar and pestle was used to crush the parts into smaller marshy parts. Four solvents were used for the preparation of the extracts namely: cold water, methanol, propanol and benzyl alcohol.

The aqueous extracts were prepared by weighing out 2 g of the powdered seeds and whole fruits of C. annuum, C. chinense and A. melegueta, and 5 g of the crushed bulbs of A. cepa and A. sativum and rhizome of Z. officinale, and macerating in 40 ml of cold water in a glass jar. The combination was allowed to stand for 24 hours at room temperature (32°C ± 2°C) with occasional agitation. The alcoholic extracts; methanol 99.5%, propanol 99.5% and benzyl alcohol 99% were obtained by weighing out same fraction (2 g) of the powdered whole fruits and separated seeds of C. annum, C. chinense and A. melegueta, and 5 g of the crushed bulbs of A. cepa and A. sativum and rhizome of Z. officinale and macerating in 40 ml of the 99.5% methanol, 99.5% propanol and 99% benzyl alcohol in a covered glass jar. The combination was allowed to stand for 24 hours with occasional agitation at room temperature (32°C ± 2°C). The extracts were then filtered using a Whatman no. 1 filter paper into plastic sample bottles and stored in a refrigerator at 2°C - 8°C. There was altogether nine aqueous extracts, nine methanolic, nine propanolic and nine benzyl alcohol extracts.

The Soxhlet extractor was also used to afford nine different methanolic extracts of the whole fruits and separated seeds of C.annum, A. melegueta and C. chinense, and the bulbs of A. cepa and A. sativum and rhizome of Z. officinale. The extracts were then stored in a refrigerator at 2°C - 8°C. Altogether, 45 extracts was tested.

2.4. Identification of Organisms

Three test tubes of 5 ml sterile Sabouraud dextrose broth was prepared and labeled appropriately and loopfuls of organisms were collected from the stock cultures of C. albicans, A. niger and T. rubrum and placed in the individual test tubes. The organisms were identified using morphological and cultural characteristics.

2.5. Fungal Identification

The morphological and cultural characteristics of the fungi isolates were used to identify them. The method that was used is the direct observation of plates. Colony morphology includes the type of pigment (if present), size of colony, texture (opaque, translucent, or transparent), adherence to agar and undulating/round/dentate edge. The plates were observed daily for the rate of growth of each of the isolates. The colour and morphology of the colonies were noted. The base of the plate, odour, were noted.

Standard suspensions containing 10^5 cfu/ml of the test organism was made by transferring colonies from the subculture into 5 mls of sterile water and then adjusting and comparing with McFarland's 0.5 standard.

2.6. Antifungal Activity of the Extracts

The antifungal activity of these crude extracts was determined against a yeast (C. albicans), mold (A. niger) and dermatophyte (T. rubrum). Agar well diffusion technique was used to determine the antifungal activities of the extracts. 25 mls of molten sterilized Sabouraud's dextrose agar, fortified with 0.05 mg/ml of chloramphenicol in a McCartney bottle was seeded with 0.2 mls of 0.5 McFarland standard suspension of each fungal isolate and poured into appropriately labeled sterile Petri dishes aseptically. The Petri dishes were rotated gently to achieve a uniform distribution of the fungal isolate and was then allowed to set. A standard sterile cork borer of 8 mm diameter was used to cut uniform wells on the surface of the solidified seeded agar. The wells were then filled with 0.1 ml of each extract with the aid of a sterile syringe and needle. One of the wells in each SDA agar plate was filled with 0.1 mls of sterile water, 99.5% methanol, 99.5% propanol or 99% benzyl alcohol as a control. The plates were then incubated at room temperature (32°C ± 2°C) on the laboratory bench for 24 hours (C. albicans), 48 hours (A. niger), and 72 hours (T. rubrum) respectively and observed for zones of inhibition. A zone of clearance round each well signifies inhibition and the diameter of the zones were measured in millimeter (mm) and presented by subtracting the activity of the control. This was performed in triplicates for each fungal isolate and extracts and the average of the three readings was taken.

3. RESULTS AND DISCUSSIONS

The fungal isolates were identified by their morphological characteristics (Table 1).

The spice extracts differed significantly in their potential to inhibit the growth of C. albicans, A. niger and T. rubrum. Even some of the aqueous extracts had inhibition of the growth of the fungal isolates. The fruits of C. annuum and C. chinense had ten (10) seeds each. This seeds were separated, milled and tested alone for their inhibitory effects on the test fungi. The seeds of A. melegueta were also purchased from the market, different from the pods, milled, and also tested for antifungal activity.

The highest activity on C. albicans was obtained with the methanolic and propanolic extracts, moderate activity was obtained with the aqueous extracts and the lowest activity was with the benzyl alcohol and Soxhlet methanolic extraction extracts. There was no growth on the C. albicans agar plates filled with the propanol and benzyl alcohol extracts for three and four days respectively. The table below shows the IZD of tested spice extracts on Candida albicans.

Soxhlet methanol extraction gave the best activity on A. niger, propanol gave moderate activity while aqueous and benzyl alcohol extracts gave the lowest activity. There was no growth on the A. niger agar plates filled with the propanol and benzyl alcohol extracts for four days. The table below shows the inhibition zone diameter of the tested spice extracts on Aspergillus niger.

Methanolic extracts gave the highest activity against T. rubrum; aqueous extracts gave intermediate activity, while the lowest activity was obtained with the propanolic extracts. There was no growth on the T. rubrum agar plates filled with the propanol and benzyl alcohol extracts for seven days. The table below shows the inhibition zone diameter of the tested spice extracts against T. rubrum.

Significant antifungal activities were evident with extracts of selected spices from members of Solanaceae, Zingiberaceae and Amaryllidaceae families. The result obtained from this study indicated that aqueous or methanolic, propanolic or even benzyl alcohol extracts of dried fruits of C. annuum, its seeds, C. chinense, its seeds, A. melegueta pods and seeds, A. cepa and A. sativum bulbs and Z. officinale rhizomes inhibited the growth of C. albicans, A. niger and T. rubrum.

From the Solanaceae family, C. annuum and C. chinense dried fruits and seeds were studied. Only the aqueous extracts of the seeds of C. annuum and C. chinense inhibited the growth of C. albicans. Only the aqueous extract of C. annuum seeds inhibited the growth of A. niger whereas the aqueous extract of C. annum fruit and C. chinense seeds inhibited the growth of T. rubrum. This indicates that these peppers have active constituents that inhibit the growth of these fungi. Benzyl alcohol extracts of C. annuum seeds and fruits had no activity on C. albicans indicating that the active constituents might be insoluble in the alcohol. Methanolic and benzyl alcohol extracts of C. annuum and C. chinense fruits and seeds had no activity on A. niger but the propanol extracts did indicating that their active constituents for antifungal activity are best extracted with propanol. The aqueous and propanol extracts of C. annuum seeds had the same inhibition zone diameter (15 mm) on A. niger. The aqueous and methanolic extracts of C. chinense seeds inhibited the growth of T. rubrum with IZD of 19 mm and 16 mm respectively. The aqueous extract had more activity than its methanol and benzyl alcohol extracts on T. rubrum. The propanol extract had no activity.

Table 1. Colony morphology of fungal isolates.

Isolate Identification Code	Cultural Characteristics	Organism
A	White to cream coloured opaque colonies, with fast rate of growth, produced in less than 24 hours.	*Candida albicans*
B	Fast growing, white, yellow, yellow-brown to black opaque colonies, arranged in whorls	*Aspergillus niger*
C	The colonies are opaque, flat to slightly raised, white to cream with a pale brown reverse pigment and a slow rate of growth.	*Trichophyton rubrum*

In the Zingiberaceae family, A. melegueta seeds and pods, and fresh rhizomes of Z. officinale was studied. Their aqueous extracts had no activity on C. albicans (Table 2) and A. niger (Table 3) The aqueous extract of A. melegueta pods had activity (18 mm) on T. rubrum. The methanolic extract of A. melegueta pods and seeds and ginger had activity on A. niger and T. rubrum. A. melegueta had more activity on the three test fungi than Z. officinale. In a study conducted by Odetunde et al., 2015, A. melegueta leaves aqueous and methanolic extracts inhibited the growth of C. albicans and A. niger with 1ZD up to 40 mm in both cases indicating that A. melegueta possess phytochemicals with potent antifungal activity [21]. The antifungal activities of A. melegueta have been attributed to phenolic compounds such as gingerol, shagaol, and paradol present in the plant parts [22]. The phytochemical screening of A. melegueta extracts indicated that it contains high doses of tannins, saponins, glycosides and polyphenols than many other weed plants [23].

Only the methanolic extract of ginger had activity on T. rubrum. Also, only the methanolic and benzyl alcohol extract of ginger had activity on A. niger. An inhibition zone diameter of 11 mm was obtained with ethanolic extracts of ginger against C. albicans in a study conducted by Atai et al., 2009. Similarly, an IZD of 16 mm was obtained with ethanolic extracts of ginger against C. albicans in a study conducted by Supreetha et al., 2011 [24]. This indicates that active constituents responsible for its activity against C. albicans are more soluble in ethanol as methanolic extracts used in this study only gave IZD of 9 mm against C .albicans. This study also suggests that the different antifungal agents present in ginger are insoluble in water as none of the aqueous extracts of ginger inhibited the growth of the three test fungi. Gingerols were identified as the major active component in the fresh ginger rhizome [10] indicating that it potential as an antifungal agent. Thus, methanol is a good solvent for obtaining A. melegueta extracts whereas ethanol is better for Z. officinale extracts.

Table 2. Inhibition zone diameter of tested spice extracts on Candida albicans.

Spices	Aqueous Extracts (mm)	Methanolic Extracts (mm)	Soxhlet Methanolic Extracts (mm)	Propanol Extracts (mm)	Benzyl Alcohol Extracts (mm)
C. annuum	-	8 mm	-	-	-
C. annuum seeds	5 mm	5 mm	-	3 mm	
C. chinense	-	8 mm	-	3 mm	7 mm
C. chinense seeds	6 mm	6 mm	7 mm	4 mm	6 mm
A.melegueta	-	-	10 mm	12 mm	1 mm
A.melegueta seeds	-	8 mm		2 mm	
A .cepa	-	10 mm		12 mm	
Z. officinale	-	9 mm	1 mm	5 mm	
A. Sativum	22 mm	22 mm	17 mm	15 mm	4 mm

Table 3. Inhibition zone diameter of tested spice extracts on Aspergillus niger.

Spices	Aqueous Extracts (mm)	Methanolic Extracts (mm)	Soxhlet Methanolic Extracts (mm)	Propanol Extracts (mm)	Benzyl Alcohol Extracts (mm)
C. annuum	-	-	-	10 mm	-
C. annuum seeds	15 mm	-	-	15 mm	-
C. chinense	-	-	14 mm	12 mm	-
C. chinense seeds	-	-	17 mm	6 mm	-
A. melegueta	-	15 mm	19 mm	13 mm	-
A. melegueta seeds	-	10 mm	12 mm	7 mm	-
A. cepa	-	11 mm	19 mm	-	11 mm
Z. officinale	-	5 mm	22 mm	-	7 mm
A. sativum	20 mm	14 mm	25 mm	21 mm	7 mm

A. sativum and A. cepa are members of the Amaryllidaceae family studied. The aqueous and methanolic extracts of garlic had the same activity (22 mm) on C. albicans. Similarly, in a study conducted by Iwalokun et al., 2004 [17] , Suleiman and Abdallah, 2014 [25] , an inhibition zone diameter of 27 mm and 16 mm respectively were obtained with aqueous extracts of garlic against C. albicans. The difference in activity obtained from the aqueous extracts of garlic in the different studies can be attributed to the geographical location, age of plant, time of harvesting, freshness of plant materials, and even the time of harvesting of plant materials. Also, the aqueous extract of garlic had activity on A. niger and T. rubrum. The aqueous extract of onion had activity only against T. rubrum which is in line with the study conducted by Bakht et al., 2013 [26] and Irkin and Korukluoglu, 2007 [27]. Aqueous extract of garlic gave an inhibition zone diameter of 20 mm against A. niger which is in line with that obtained by Avasthi et al., 2010 [28] (34 mm), Irkin and Korukluoglu, 2007 [25] (16 mm) and Suleiman and Abdallah, 2014 [27] (14 mm). Again, the difference in the inhibition zone diameter can be attributed to the geographical location, age of plant, time of harvesting, freshness of plant materials, physical factors (temperature, light, and water), contamination by field microbes and even the time of harvesting of plant materials. Thus, the aqueous extract of garlic had activity on the three tested fungi. The propanolic extract of onion had more activity on C. albicans than its methanol extract whereas the methanolic extract of garlic had more activity on C. albicans than its propanolic extract as shown in Table 3. This indicates that the active components inhibiting the growth of C. albicans are more soluble in propanol (onion) and methanol (garlic) in each case. The propanolic extract of garlic had more activity on A. niger than its benzyl alcohol, methanol and aqueous extract as shown in Table 4. Only the propanolic extract of garlic had activity on T. rubrum, the propanolic extract of the other test spices didn't. The benzyl alcohol extract of onion gave an inhibition zone diameter of 22 mm against T. rubrum which is higher than that obtained with its aqueous and methanolic extracts. A. sativum had more activity than A. cepa. Flavonoids and organosulphur compounds present in onion has been found to be responsible for its antifungal properties [13]. Ajoene present in garlic has been noted to have a potent effect on yeasts, molds and dermato-phytes [16]. Some studies revealed that garlic is as effective as fluconazole at inhibiting C. albicans [17]. When crushed, garlic yields allicin, an antifungal compound and other sulphur containing compounds.

The activity of these spice extracts is comparable to the activity of some synthetic antifungal agents' invitro. In a study conducted by Atai et al., 2009 [29] , Nystatin was used as a positive control. Whereas ginger gave an IZD of 11 mm against C. albicans, nystatin produced an IZD of 12 mm. Similarly, fluconazole produced an IZD of

Table 4. Inhibition zone diameter of tested spice extracts on Trichophyton rubrum.

Spices	Aqueous Extracts (mm)	Methanolic Extracts (mm)	Soxhlet Methanolic Extracts (mm)	Propanol Extracts (min)	Benzyl Alcohol Extracts (mm)
C. annuum	10 mm	4 mm	-	-	-
C. annuum seeds	-	-	-	-	-
C. chinense	19 mm	-	-	-	-
C. chinense seeds	18 mm	16 mm	17 mm	-	10 mm
A. melegueta	-	15 mm	22 mm	-	12 mm
A. melegueta seeds		17 mm	18 mm	-	-
A. cepa	12 mm	10 mm	-	-	22 mm
Z. officinale	-	9 mm	17 mm	-	-
A. sativum	21 mm	19 mm	39 mm	27 mm	-

30 mm when used as positive control whereas aqueous garlic extracts produced an IZD of 27 mm against C. albicans in a study conducted by Iwalokun et al., 2004 [17].

Soxhlet extraction is a better extraction method than maceration but its high cost could be the reason while it is not usually employed. Macerated methanol extracts of C. chinense seeds, A. melegueta pods, its seeds, Z. officinale and A. sativum gave inhibition diameters of 16 mm, 15 mm, 17 mm, 9 mm, and 19 mm respectively against T. rubrum whereas their Soxhlet methanol extracts gave IZD of 17 mm, 22 mm, 18 mm, 17 mm, and 39 mm respectively. Also, macerated methanol extracts of C. chinense, its seeds, A. melegueta pods and seeds, A. cepa, Z. officinale and A. sativum gave inhibition zone diameters of 0 mm, 0 mm, 15 mm, 10 mm, 11 mm, 5 mm, and 14 mm respectively against A. niger whereas their Soxhlet methanol extracts gave IZD of 14 mm, 17 mm, 19 mm, 12 mm, 19 mm, 22 mm, and 25 mm respectively as shown in Table 4.

The zones of inhibition obtained from extracts of these spices shows that they have great potential as remedies for diseases caused by the test fungi and can even serve as natural prophylactics.

4. CONCLUSION

The study demonstrated that the six spices used had activity on the yeast (Candida albicans), the mould (Aspergillus niger) and the dermatophyte (Trichophyton rubrum). Allium sativum had the greatest activity on all the test organisms signifying that it had greater promise as an antifungal agent. Active constituents with potent activity against tested fungi were also present in the separated seeds of C. annuum, C. chinense and A. melegueta. This encouraged the use of these spices for controlling fungal infections and as natural prophylactics. They offered the advantage of being used indefinitely in large amounts.

REFERENCES

1. Tapsell, L.C., Hemphill, I., Cobiac, L, Patch, C.S., Sullivan, D.R., Fenech, M., Roodenrys, S., Keogh, J.B., Clifton, P.M., Williams, P.G., Fazio, V.A. and Inge, K.E. (2006) Health Benefits of Herbs and Spices: The Past, the Present, the Future. Medicinal Journal of Australia, 185, S4-S24.
2. Tarun, A., Rachana, S., Amar, D.S. and Imran, W. (2012) Antibacterial Activity of Water, Methanol and Ethanol Extracts of Ocimum gratissimum, Vernonia amygdalina, and Aframomum melegueta. Asian Journal of Plant Science and Research, 1, 36-40.
3. Souza, E.L., Stamford, T.L.M., Lima, E.O., Trajano, V.N. and Filiho, J.B. (2005) Antimicrobial Effectiveness of Spices: An Approach for Use in Food Conservation Systems. Brazilian Archives of Biology and Technology, 48, 92-130.
4. Dias, G.B., Gomes, V.M., Moraes, T.M., Zottich, U.P., Rabelo, G.R., Carvalho, A.O., Moulin, M., Gon&ccedilalves, L.S., Rodrigues, R. and Da Cunha, M. (2013) Characterization of Capsicum Species Using Anatomical and Molecular Data. Genetics Molecular Research, 12, 6488-6501.
5. Wahyuni, Y., Ballester, A.R., Sudarmonowati, E., Bino, R.J. and Bovy, A.G. (2013) Secondary Metabolites of Capsicum Species and Their Importance in the Human Diet. Journal of Natural Products, 76,783-793.
6. Oyagade, J.O., Awotoye, O.O., Adewunmi, J.T. and Thorpe, H.T. (1999) Antimicrobial Activity of Some Nigerian Medicinal Plants, Screening for Antibacterial Activity. Journal of Bioscience research Communication, 11, 193-197.
7. Galal, A.M. (1996) Antimicrobial Activity of 6-Paradol and Related Compounds. International Journal of Pharmacognosy, 31, 37-43.

8. Okwu, D.E. (2001) An Improving Nutritive Value of Cassava Tapioca Meal with Local Spices. Journal of Nutraceutical, Functional and Medical Food, 3, 43-51.
9. Chen, I.N., Chang, C.C., Wang, C.Y., Shyu, T.Y., Ng, C.C. and Chang, T.L. (2008) Antioxidant and Antimicrobial Activity of Zingiberaceous Plants in Taiwan. Plants Foods Hum Nutrition, 63, 15-20.
10. Ali, B.H., Blunden, G., Tanira, M.O. and Nemmar, A. (2008) Some Phytochemical, Pharmacological and Toxicological Properties of Ginger (Zingiber officinale Roscoe): A Review of Recent Research. Food Chemistry and Toxicology, 46, 409-420.
11. Tan, B.K.H. and Vanitha, J. (2004) Immunomodulatory and Antibacterial Effects of Some Traditional Chinese Medicinal Herbs: A Review. Curriculum of Medicinal Chemistry, 11, 1423-1430.
12. Sampath Kumar, K.P., Debjit, B., Chiran, J., Biswa, J. and Pankaj, T. (2010) Allium cepa: A Traditional Medicinal Herb and Its Health Benefits. Journal of Chemical and Pharmaceutical Research, 2, 283-291.
13. Block, E., Gulati, H., Putman, D., Sha, D., You, N. and Zhao, S.-H. (1997) Allium Chemistry: Synthesis of 1-[Alk(en) ylsulfinyl]propyl Alk (en)yl Disulfides (Cepaenes), Antithrombotic Flavorants from Homogenates of Onion (Allium cepa). Journal of Agriculture and Food Chemistry, 45, 4414-4422.
14. Block, E. (2010) Garlic and Other Alliums: The Lore and the Science. Royal Society of Chemistry, Cambridge, 152-166.
15. Reuter, H.D. and Sendl, A. (1994) Allium sativum and Allium ursinum: Chemistry, Pharmacology, and Medicinal Applications. In: Wagner, H. and Farnsworth, N.R., Eds., Economic and Medicinal Plants Research, Academic Press, London, 55-113.
16. Dikasso, D., Lemma, H., Urga, K., Debella, A., Addis, G., Tadele, A. and Yirsaw, K. (2001) Investigation on the Antifungal Properties of Freshly Pressed Garlic Juice on Major Pathogenic Fungi. Journal of Ethiopian Medical Practitioners, 3, 16-20.
17. Iwalokun, B.A., Ogunledun, A., Ogbolu, D.O., Bamiro, S.B. and Jimi-Omojola, J. (2004) In Vitro Antimicrobial Properties of Aqueous Garlic Extract against Multidrug-Resistant Bacteria and Candida Species from Nigeria. Journal of Medicinal Food, 7, 327-333.
18. Ledezma, E. and Apitz-Castro, R. (2006) Ajoene the Main Active Compound of Garlic (Allium sativum): A New Antifungal Agent. Revista Iberoamericana de Micologia, 23, 75-80.
19. Stephen, P.D., Norman, A.H. and Sean, P.G., Eds. (2004) Hugo & Russell's Pharmaceutical Microbiology. 7th Edition, Blackwell Science, Hoboken, 44-58.
20. DeBoer, H.J., Kool, A., Broberg, A., Mziray, W.R., Hedberg, I. and Levenfors, J.J. (2005) Antifungal Activity of Some Herbal Remedies from Tanzania. Journal of Ethnopharmacology, 96, 461-469.
21. Odetunde, S.K., Adekola, I.T., Arungbeto, M.O. and Lawal, A.K. (2015) Antimicrobial Effect and Phytochemical Analysis of Aframomum melegueta on Some Selected Bacteria and Fungi. European Journal of Biotechnology and Bioscience, 3, 15-19.
22. Okwu, D.E. (2004) Phytochemicals and Vitamin Content of Indigenous Spices of South Eastern Nigeria. Journal of Sustain Agricultural Environment, 6, 30-34.
23. Odebiyi, O.O. and Sofowora, E.A. (1978) Phytochemical Screening of Nigerian. Medicinal Plants Part II: Iloydia, 41, 1-25.
24. Supreetha, S., Mannur, S., Simon, S.P., Jain, J., Tikare, S. and Mahuli, A. (2011) Antifungal Activity of Ginger Extract on Candida albicans: An in Vitro Study. Journal of Dental Sciences and Research, 2, 1-5.
25. Suleiman, A.A., Eltayeb, F.M., Suleiman, S.A. and Osman, N.A. (2015) Antimicrobial Activity of Zingiber officinale (Ginger) Oil against Bacteria Isolated from Children's Throat. International Journal of Microbiology, 1, 1-6.
26. Bakht, J., Khan, S. and Shafi, M. (2013) Antimicobial Potentials of Fresh Allium cepa against Gram Positive and Gram Negative Bacteria and Fungi. Pakistan Journal of Botany, 45, 1-6.
27. Irkin, R. and Korukluoglu, M. (2007) Control of Aspergillus niger with Garlic, Onion and Leek Extracts. African Journal of Biotechnology, 6, 384-387.

28. Avasthi, S., Gautam, A.K. and Bhadaura, R. (2010) Antifungal Activity of Plant Products against Aspergillus niger: A Potential Application in the Control of a Spoilage Fungus. Biological Forum—An International Journal, 2, 53-55.
29. Atai, Z., Atapour, M. and Mohseni, M. (2009) Inhibition Effect of Ginger Extract on Candida albicans. American Journal of Applied Sciences, 6, 1067-1069.

CHAPTER 17

A HapMap Leads to a Capsicum Annuum SNP Infinium Array: A new Tool for Pepper Breeding

Amanda M Hulse-Kemp[1,4], Hamid Ashrafi1,[4,5], Joerg Plieske[3], Jana Lemm[3], Kevin Stoffel[1], Theresa Hill[1], Hartmut Luerssen[3], Charit L Pethiyagoda[2], Cindy T Lawley[2], Martin W Ganal[3] and Allen Van Deynze[1]

[1]Department of Plant Sciences, University of California-Davis, Davis, California 95616, USA; [2]Illumina Incorporated, 5200 Illumina Way, San Diego, CA 92122, USA and

[3]TraitGenetics GmbH, Am Schwabeplan 1b, Stadt Seeland OT, Gatersleben, Germany.

[4]These authors contribute equally to this work.

[5]Current Address: Department of Horticultural Sciences, North Carolina State University, 2721 Founders Drive, 234 Kilgore Hall, Raleigh, NC 27518, USA.

ABSTRACT

The *Capsicum* genus (Pepper) is a part of the Solanacae family. It has been important in many cultures worldwide for its key nutritional components and uses as spices, medicines, ornamentals and vegetables. Worldwide population growth is associated with demand for more nutritionally valuable vegetables while contending with decreasing resources and available land. These conditions require increased efficiency in pepper breeding to deal with these imminent challenges. Through resequencing of inbred lines we have completed a valuable haplotype map (HapMap) for the pepper genome based on single-nucleotide polymorphisms (SNP). The identified SNPs were annotated and classified based on their gene annotation in the pepper draft genome sequence and phenotype of the sequenced inbred lines. A selection of one marker per gene model was utilized to create the PepperSNP16K array, which simultaneously genotyped 16 405 SNPs, of which 90.7% were found to be informative. A set of 84 inbred and hybrid lines and a mapping population of 90 interspecific F_2 individuals were utilized to validate the array. Diversity analysis of the inbred lines shows a distinct separation of bell versus chile/hot pepper types and separates them into five distinct germplasm groups. The interspecific population created between Tabasco (*C. frutescens* chile type) and P4 (*C. annuum* blocky type) produced a linkage map with 5546 markers separated into 1361 bins on twelve 12 linkage groups representing 1392.3 cM. This publically available genotyping platform can be used to rapidly assess a large number of markers in a reproducible high-throughput manner for pepper. As a standardized tool for genetic analyses, the PepperSNP16K can be used worldwide to share findings and analyze QTLs for important traits leading to continued improvement of pepper for consumers. Data and information on the array are available through the Solanaceae Genomics Network.

INTRODUCTION

The *Capsicum* genus in the Solanaceae family, commonly known as pepper or paprika, has been very important in many cultures worldwide for spices, medicines, ornamentals and vegetables and are key components of many food dishes. The fruits provide a high nutritional value and are a rich source of Vitamins A, B and C, iron, potassium, magnesium, beta carotene, folic acid and fiber. Cultivated peppers are typically divided into two categories, bell and chile. Bell peppers (or blocky type) refer to *C. annuum* varieties with blocky shaped fruits that are sweet (non-pungent) and can come in many colors from green (not fully ripe) to oranges, reds and purples. Whereas chile peppers (or hot types) refer to a large number of varieties that tend to have an elongated shape and can vary greatly in spice (pungency) from mild to extremely spicy, including ancho, anaheim, cayenne, fresno, habanero, jalapeno, poblano and serrano, just to name a few. The chile category contains types of all five domesticated species of *Capsicum*: *C. annuum, C. frutescens, C. chinense, C. pubescens, and C. baccatum.*

Production of these two pepper classifications, primarily composed of *C. annuum* varieties, totaled 44 800 acres of bell peppers and 19 400 acres of chile peppers in 2015 for the US. The yield per acre was much higher for bell peppers at 376 cwt/acre compared with chile peppers at 223 cwt/acre. Overall US crop value in 2015 for bell peppers was $806M while the value for chile peppers was $135M.[1] The past 20 years have shown a steady trend in the increase in production of peppers worldwide (2.9% per year in area harvested) as reported by the Food and Agriculture Organization of United Nations statistics service (2013, production data available at http://faostat3.fao.org/faostat-gateway/). Peppers are grown in at least 44 countries all over the world and as the worldwide population increases, demand for food and vegetable crops grows, whereas available land and other resources decrease. These conditions require a revolution in breeding technologies for pepper and vegetables in order to deal with these daunting challenges.

Most pepper species are diploid and have 12 pairs of chromosomes ($2n = 2\times = 24$) which makes the *Capsicum* species amenable to traditional breeding methods as they are inter-fertile to varying degrees owing to their similar genome structures. The pepper genome is quite complex and contains a large amount of repetitive DNA sequences, which has caused inflation in the genome size to around 3.25–3.48 Gb[2,3] compared with many other solanaceous species with genomes from 0.85–1.2 Gb.[4] Because of the large genome size of pepper, development of sequence-based resources have been delayed compared with other Solanaceae, which have significantly smaller genome sizes, such as tomato (900 Mb) and potato (844 Mb), which had genome sequences released in 2012 and 2011, respectively.[5,6]

At the advent, the price for sequencing was very expensive, so discovery and analysis of the large pepper genome started with investigation of the gene space. Using 454 sequencing technology the first transcriptomes were developed in pepper using RNA from the fruit of two pepper parental lines CM334, Taean and their resulting hybrid.[7,8] These were quickly followed by another assembly of unigenes that were derived from Sanger sequences of a hot pepper, Bugang.[9,10] With a full transition to next-generation sequencing and the resulting marked reduction in sequencing cost and increase in output, discovery and analysis of the pepper genome moved from gene space-based studies to large scale genomic analyses. The Ashrafi *et al.*[9] publication led this transition with the inclusion of three next-generation-sequencing-based assemblies. Genomics then jump started a new era in pepper breeding with the recent release of multiple *C. annuum* draft genome sequences including the Zunla-1 and varieties of *C. annuum* var. galbriusculum,[11] the hot pepper landrace CM334, and *C. chinense*.[2] These efforts also produced more transcriptome sequences in order to assist in annotation of the genomic sequences.

The transcriptome and genome assemblies make it possible to assess genetic differences and the structure of pepper germplasm for subsequent utilization in improvement of the crop. It is desirable to have methods to systematically assess diversity within *Capsicum* in an efficient and reproducible manner that will provide a large number of markers. Single-nucleotide polymorphism (SNP) markers are the preferred marker type as they have the possibility of occurring randomly throughout the genome, have a known sequence, and have the possibility to be the causative allele for beneficial phenotypic traits. With advancements in microarray technology, it is possible to perform high-density genotyping of SNPs in an extremely high-throughput manner, based on fixed probes on the microarrays. The unigene assembly produced by Ashrafi *et al.*[9] was used to generate the first high-density system for analyzing SNPs in Capsisum in the form of an Affymetrix GeneChip.[10] Hill *et al.*[10] were able to assess the diversity among 40 *C. annuum* lines representing the primary breeding germplasm plus three additional species. However, this tool did not allow for identification of the specific SNP genotype at polymorphic loci and is no longer available publicly. A current publicly available tool for assaying a large number of loci with the ability to obtain exact genotypes would be a benefit to the pepper breeding community. Owing to the large genome size of pepper, whole-genome sequencing strategies that avoid large amounts of missing data would be costly and are usually unnecessary for QTL studies.[12] In order to assist the pepper breeding community, we describe here the development of the PepperSNP16K array with one marker per annotated gene, and validation of the array by genetic mapping and diversity analysis for large-scale genotyping in pepper.

MATERIALS AND METHODS

Resequencing and SNP development

Leaf tissues from 22 lines representing multiple different chile and bell phenotypes of *C. annuum* were obtained and DNA was collected using the Qiagen (Valencia, CA, USA) DNeasy plant kit following manufacturer's protocols, including RNase digestion. All DNA was quantified using Picogreen (Thermo Fisher Scientific, Waltham, MA, USA) and then prepared for sequencing using the Illumina TruSeq library preparation kit (FC-930-1023). Library products were size-selected for insert sizes ranging from 300–500 bp with Agencourt AMPure XP beads (Beckman Coulter, Indianapolis, IN, USA). The final libraries were quantified and analyzed on the Agilent Bioanalyzer (Santa Clara, CA, USA). Libraries were sequenced on the Illumina HiSeq2000 to generate 2×100 bp paired-end reads.

Adapters as well as 13 bases from the start and the last 5 bases of each read were removed from all raw reads due to poor quality. The remaining read sequences were quality trimmed and any reads fewer than 40 bases were removed using CLC Genomics Workbench 7.0 (https://www.qiagenbioinformatics.com/). Reads were aligned to the Pepper genome v1.5[2] using CLC with default parameters. The CLC mapping files were exported as BAM files and were used in SAMtools[13] as follows. In order to generate a file with all SNPs between all samples, SAMtools was used to merge BAM files for all samples to make a master BAM file. The merged file was sorted and then analyzed with SAMtools pileup (-cv) with SAMtools version 0.1.7a to call all variants in the merged file with the entire set of samples. Then variants are called for each individual sorted BAM file also with SAMtools pileup and options -c and -l. These resulting 22 files were then parsed and processed with a pipeline of in-house Perl scripts (Supplementary File 1) in order from 01 to 05 and run sequentially. To compensate for errors in sequencing and base calls, if a base represented >90% of reads, it was considered homozygous. This correction was required due to an issue that we identified with the SAMtools version that we used. In the pileup file of SAMtools (version 0.1.7a) for diploid genomes, a variant position is a non-reference allele from all the reads that are interrogating that putative variant position. We empirically found that at 90%, the call for variant is correct in 78–90% of the times depending on the crop (highest in pepper).[9] After all individual pileup files were parsed and corrected, they were merged one-by-one to the parsed master pileup file to make a final Genotypes Table file that contained all polymorphic positions within the samples. Subsequently, the SNP caller script was run with Genotype Table file as input, requiring at least a set of two genotypes to be homozygous and different from each other with a minimum depth of three reads for each. After identifying all SNPs that met these criteria, the overall data set was filtered to remove SNPs that had nearby adjacent SNPs in the vicinity of 50 bases.

The filtered file with a list of positions for SNPs was used to categorize SNPs into three different sets, representing SNPs that (1) were found within blocky type peppers, (2) were found within hot type peppers and (3) were found to occur within both blocky and hot type peppers. The annotation information for Pepper version 1.5 was used to determine which SNPs were located in genes throughout the reference and used to annotate the determined SNPs.

Array design

The upstream 50 bases and downstream 50 bases from each SNP in the set were obtained. These sequences were submitted to the Illumina Assay Design Tool (ADT) to obtain a design score for the Illumina Infinium technology. Infinium utilizes a bead-based array for genotyping with single base pair extension from a 50 bp oligonucleotide probe to assay the SNP base with fluorescently labeled nucleotides. The technology utilizes two fluorophores, therefore there are marker types that can be assayed with one-bead type (Infinium II) and other markers that require two-bead types

(Infinium I). The markers were filtered to remove those with design score <0.8 and to select a single one-bead type marker (Infinium II) per gene across the Pepper v1.5 genome[2] in order to obtain 19 000 markers with >9% minor allele frequency (2 out of 22) in the sequenced samples. A total of 2909 markers were selected from the blocky type data set; 3622 markers were selected from the hot type data set; and 12 469 markers were selected from the set within both blocky and hot types. Of the 19 000 SNPs targeted for manufacture of the array, 16 405 were represented following manufacture quality control resulting in the PepperSNP16K array.

Genotyping with the array

Genomic DNA was fluorescently quantified with Picogreen or other assays to standardize samples at 50 ng μl^{-1} for each of the 84 lines representing a diverse germplasm background. Of the lines, 78 individuals were hybrid varieties and 6 individuals were inbred lines, which included the parents of known mapping populations. The hybrids were provided by a number of breeding companies in coded form but for most of the material, a description of the type of the respective line/variety was provided. For mapping, the parents of a cross between a hot type (Tabasco) and a blocky type (P4) described in Nagy *et al.*,[14] and 90 individuals of an F_2 mapping population were used. Samples were processed according to Illumina instructions and hybridized to the PepperSNP16K array and the arrays were analyzed with the Illumina iScan to measure fluorescent genotyping data. All image files were uploaded into a single GenomeStudio project containing the 177 individuals. Data from all markers were clustered using GenomeStudio Genotyping Module (v1.9.4, Illumina, Inc., San Diego, CA, USA). All markers were then viewed and manually curated, for construction of an optimized cluster file for pepper. The cluster file is available at http://www.solgenomics.net. Genotypes for each sample were exported using the final cluster file and utilized in further analyses. The quality of markers and their minor allele frequencies were determined.

Genetic linkage analysis

Genotypes from both parents (Tabasco a hot type and P4 a blocky type), F_1 and 90 individuals were obtained. The data were transformed into mapping data format ('ABH') for markers that were found to be segregating in a co-dominant pattern. The mapping data format files were used to map in JoinMap v4.0[15] using default parameters with respect to grouping and ordering. The map orders were exported and haplotype maps were examined to manually remove problematic markers which were elevating double crossover numbers and increasing map size. After removing the problematic markers, the final linkage groups were generated in Map Manager QTX[16] with no prior assumptions on marker position using the following settings: linkage evaluation F_2, search linkage criterion P=0.05, map function Kosambi, and cross-type line cross. The resulting order of markers were validated using CheckMatrix software (www.atgc.org/XLinkage/). Any linkage groups that produced errantly ordered markers were manually reanalyzed for problematic markers, which were subsequently removed before repeating the process to determine the final orders. The final maps were drawn with MapChart version 2.2.[17]

The positions in the linkage map compared with the marker position on the Pepper reference genome (v1.5) were plotted. The positions of all markers on the array on the updated Pepper reference genome V1.55 (http://peppergenome.snu.ac.kr/) were determined using the SNP and flanking sequence by BLAST alignment. The correlation of position on the linkage map to the updated reference version 1.55 was also plotted. Correlation with the interspecific map of a cross between *C. frutescens* accession BG2814-6 x *C. annuum* 'NuMex RNaky' (FA map) produced by Hill *et al.*[18] was analyzed based on markers positioned in shared genes across the two technologies. Linkage map positions in each of the respective maps were plotted to determine the R^2 between the two maps. Correlation with the (FA) map was used to orient the linkage groups. Array markers and linkage map positions are available on the Sol Genomics Network database (https://solgenomics.net).

Analysis of population structure

We used the software package STRUCTURE to reveal the population structure of the investigated pepper lines and hybrids.[19] For the SNP marker set, STRUCTURE was run for K=1–20. For each value of K, five replications were performed with a length of burn-in period of 100 and 1000 MCMC reps after burn-in. To determine the most probable value of K, we applied the *ad hoc* criterion described by Evanno *et al.*[20]

RESULTS

Array development

Coverage for each sample ranged from 19- to 80-fold for each of the resequenced pepper lines. Once these reads were processed, an average depth of 12.9× was mapped back to the Pepper reference genome v1.5

covering an average of 86.8% of the genome. The mapped reads were utilized with SAMtools and an in-house pipeline to identify SNPs. The Illumina Infinium array was designed with 19 000 putative SNP markers, of which 16 405 SNPs (86.3%) passed through the manufacturing process with adequate quality and representation to be included in the manifest. This set is composed of 2531 SNPs, which are unique within blocky types, 3058 SNPs, which are unique within hot types, and 10 816 SNPs, which are found within both hot and blocky types. The position of SNPs along Pepper genome version 1.5 were utilized to annotate those SNPs that fell within predicted genes (Supplementary Table 1). The proportion of individuals that had adequate signal for a marker making it amenable to genotyping, or 'call frequency', was calculated (Table 1). A total of 1528 or 9.3% of markers were deemed as functionally failed due to the inability to call any samples for these markers. This resulted in the remaining 14 877 markers being classified as 'functional'. These functional markers were then classified into two groups, those which represented monomorphic loci in which no difference in genotype was seen among samples assayed, and those which showed differences and were thus polymorphic. A total of 13 800 markers of the 16 405 assays on the array were found to be polymorphic, which is a success rate of 84.1% for the array overall with our samples. The 1077 functional but monomorphic markers could either be falsely identified SNPs or are polymorphisms not represented in our germplasm run on the array. The GenTrain score of the polymorphic markers was assayed to determine the behavior of the clusters in the genotype plots, with 1 representing very well-discriminated and tight clusters. The GenTrain scores for all polymorphic markers are shown in Table 1 and, as expected for the diploid pepper genome, the majority of the markers exhibited very high GenTrain scores. The minor allele frequency observed within the sample set was also calculated for all polymorphic markers (Table 1). This showed that there were 13 760, 12 655, 11 965 and 9921 SNPs represented for minor allele frequencies of greater than 1%, 3%, 5% and 10%, respectively. The 14 877 functional markers comprise the PepperSNP16K array. The resulting cluster file for the array is available at http://www.solgenomics.net.

Genetic map construction

The initial genetic map generated with the array was created with an F_2 population derived from a cross between a hot pepper and a sweet pepper parent (Tabasco x P4). Most linkage groups easily fell into place and generated expected heat map plots across the resulting order; however, three chromosomes (Chr01, Chr07 and Chr08) needed manual adjustment following visualization of the initial orders determined using CheckMaxtrix (www.atgc.org/XLinkage/). After removal of a few problematic markers,

Table 1: Validation statistics for markers on the PepperSNP16K array based on genotyping of 84 inbred and hybrid lines for (a) call frequency for all markers on the array, (b) marker clustering statistics or GenTrain score for all polymorphic markers and (c) minor allele frequencies for all polymorphic SNPs determined using 84 inbred and hybrid lines; mapping samples were not included.

	Marker	
	Count	*Percentage*
(a) Call frequency		
0.000–0.750	1579	9.63
0.750–0.980	1688	10.29
0.980–0.999	2742	16.71
1.000	10 396	63.37
Total	16 405	100.00
(b) GenTrain Score		
< 0.2	71	0.51
< 0.3	684	4.96
< 0.6	2426	17.58
< 0.8	4139	29.99
> 0.8	6480	46.96
Total	13 800	100.00
(c) Minor allele frequency		
0.00–0.05	1835	13.30
0.05–0.10	2044	14.81
0.10–0.15	1990	14.42
0.15–0.20	1491	10.80
0.20–0.25	1285	9.31
0.25–0.30	1192	8.64
0.30–0.35	989	7.17
0.35–0.40	1051	7.62
0.40–0.45	985	7.14
0.45–0.50	938	6.80
Total	13 800	100.00

the software was accurately able to determine the ordering for Chr07 (Figure 1). However, Chr01 and Chr08 were problematic due to a translocation event that is present between the parents in this interspecific cross. As this is the case, each of the pieces as depicted by Hill *et al.*[18] were broken out according to their groupings based on LOD score, and the groups were identified based on overlap with the FA map in that study and overlap with the draft genome assembly. The groups identified as the top of Chr01, the bottom of Chr01, wild Chr08 and a combination of both Chr01 pieces were ordered independently and visualized with CheckMatrix (Figure 2). It was observed that when the top and bottom pieces of Chr01 were attempted to be ordered together, the software generated a non-optimal ordering of the markers (Figure 2a). Owing to this, the order from the independent ordering of each piece was used to establish a final order (Figure 2b), then distances between each marker were calculated for that set order (Figure 2c). Finally, a total of 5546 markers were mapped into 12 linkage groups representing a total of 1392.3 cM (Figure 3). These represent largely markers from the set found within both blocky and hot types (3963). However, as the parents were not included in the set of samples sequenced for marker development and classification of markers into phenotype sets, the population was also able to map a number of markers from the unique sets including 713 from the blocky set and 918 markers that were from the hot set. As the marker classification is dependent on the samples used, in this case, based on the lines sequenced, the classifications that are provided in the 'Type' column in Supplementary Table 1 should be utilized loosely.

This map corresponds to 1361 unique recombination bins with an average of 113 bins per chromosome. On average there are 462 markers on each linkage group. This translates to a marker on average across the genome every 610 Mb or 0.25 cM, based on chromosome sizes in Pepper Genome Version 1.5. Due to the translocation between the parents, Chr01 represents the largest map distance at 199.1 cM, whereas Chr08 represents the smallest map distance at 46.7 cM, which is consistent with the findings of Hill *et al.*18 Overall the FA map produced by Hill *et al.*18 and the map produced here are exceptionally similar with an R^2 of 0.9861 (Figure 4) based on 822 markers in shared genes. Most chromosomes show a high number of overlapping markers, except for the smallest, chromosome 08, that is also genetically the shortest, contains the least number of markers and can be seen in Figure 5 to have only a few overlapping markers. The total sizes were also highly similar at 1380 cM for the FA map and 1392 cM in the map produced with the PepperSNP16K array.

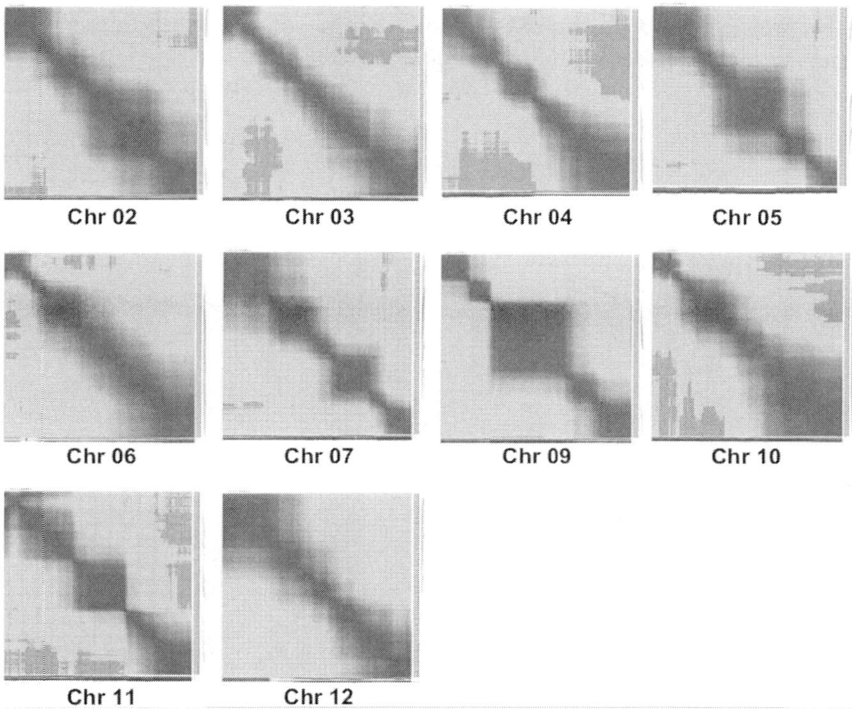

Figure 1 Heat map plots produced with CheckMatrix software for genetic map orders of all chromosomes, except Chr01/08. All orders produced from initial mappings except for Chromosome 07, which is shown after manual removal of problematic markers.

Synteny to reference genome sequences

Most linkage groups from the interspecific genetic map showed a very good correlation with the most recent version (1.55) of the pepper draft reference genome (Figure 5) as well as the original published version (Supplementary Figure 1); however, not all loci were able to be translated to version 1.55 from version 1.5 (Supplementary Table 1). Generally, the order of markers between original version 1.5 and the updated version 1.55 was highly similar with R^2 of 0.9754 (Supplementary Figure 1). It was also observed that chromosomes 01 and 08 that were involved in a translocation between the two parents of the population, show association with the opposite chromosome from the draft reference sequence, which are circled in the figures. Chromosome 07 which initially caused some problems in ordering, again shows abnormalities when mapped back to the reference recurrence. It is possible that there are some changes in genome structure between the parents in this chromosome as well that may be causing these issues with ordering and what is observed back to the CM334-based reference, as CM334 is a landrace in *C. annuum*.

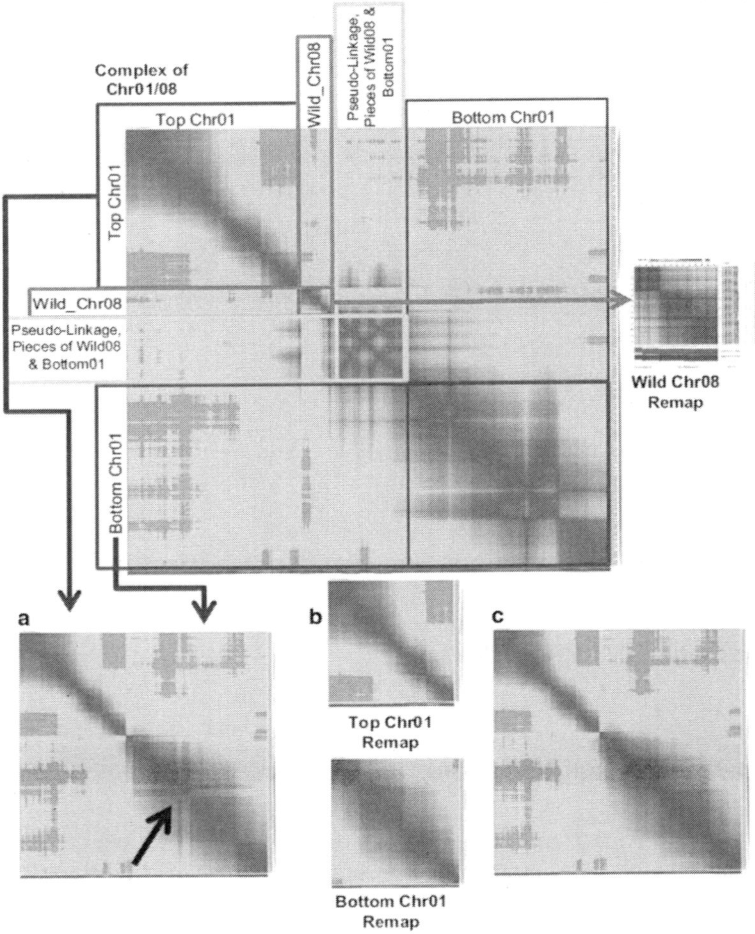

Figure 2 Initial ordering and subsequent ordering of the separate pieces of the Chromosome 01 and 08 complex due to translocation between *C. annuum* and *C. frutescens* parents. Remapping of Chromosome 01 by (**a**) reordering all markers from both top and bottom pieces together and (**b**) mapping top and bottom separately and (**c**) calculating distances between markers with a set order based on the mapping of the top and bottom pieces separately.

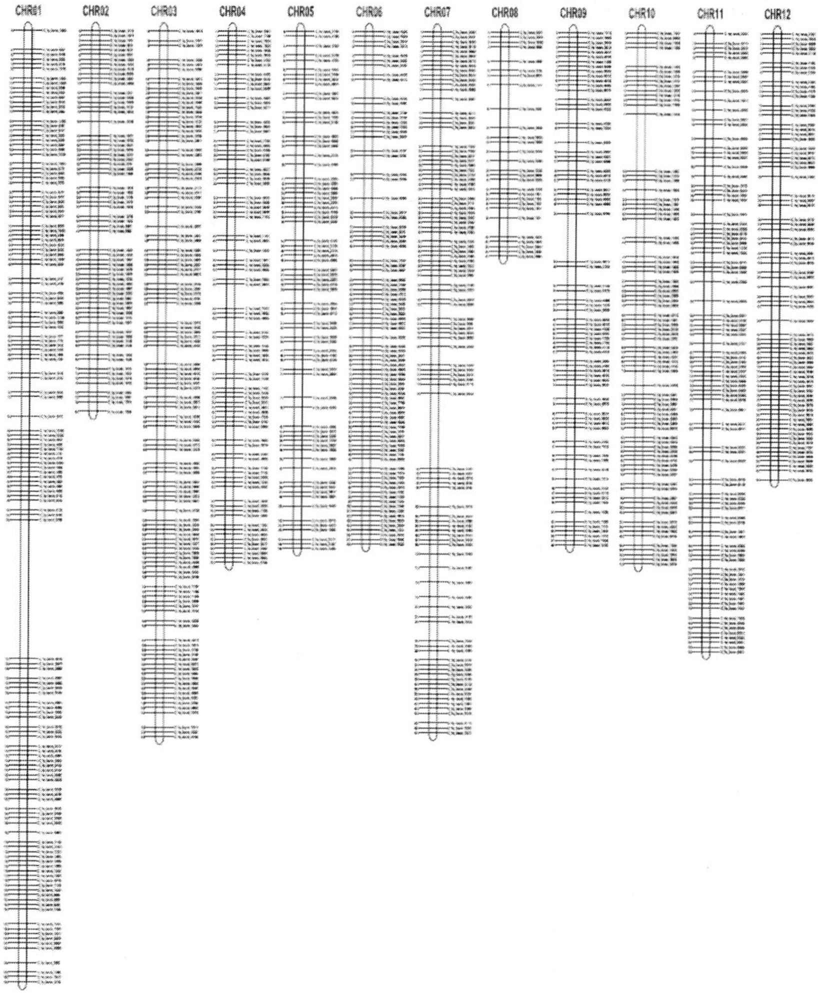

Figure 3 Interspecific linkage map of the 12 pepper chromosomes. Map determined using 90 F_2individuals from a cross between a Tabasco (hot type parent), and P4, a blocky type parent. Only one marker is listed on the right per centiMorgan, even if there were more markers co-segregating.

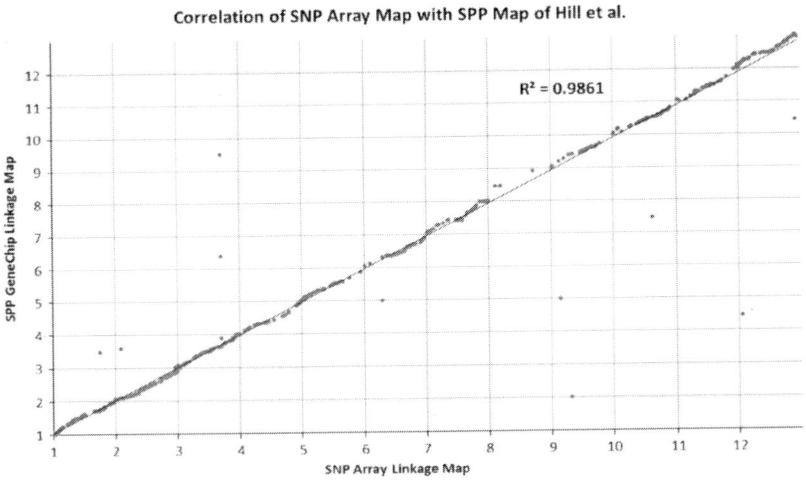

Figure 4 Correlation of interspecific map with FA map produced by Hill *et al.*18 Positions across linkage groups are normalized. Numbers along each axis corresponds to each pepper chromosome.

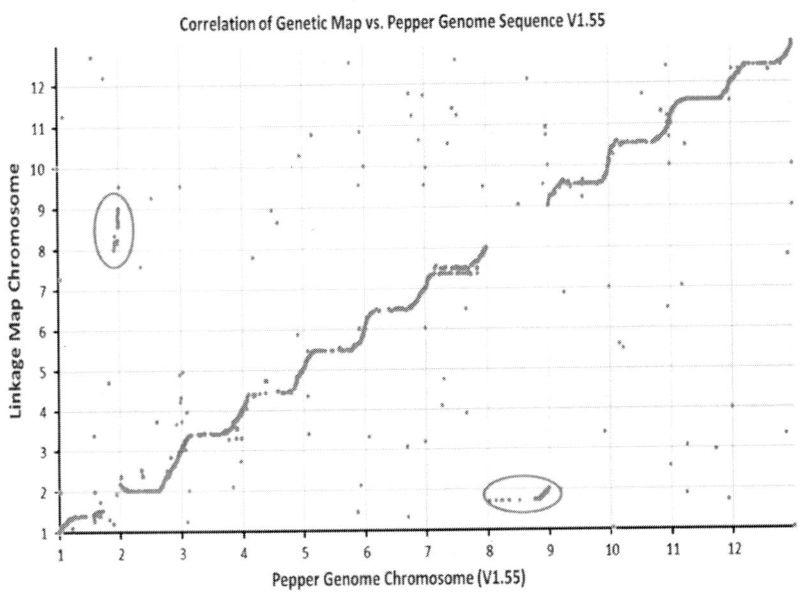

Figure 5 Correlation of interspecific linkage map with pepper draft genome sequence version 1.55.

Diversity analysis

As expected, the distinction between the two pepper types can be clearly visualized by STRUCTURE analysis (Figure 6) utilizing 11 027 polymorphic SNPs for the analysis after filtering. The analysis with respect to the *K* value revealed a maximum value between 4 and 5. Thus, we have used a *K* value of 5 to analyze the population structure and evaluate the results with respect to known information about the type and species classification of the material.

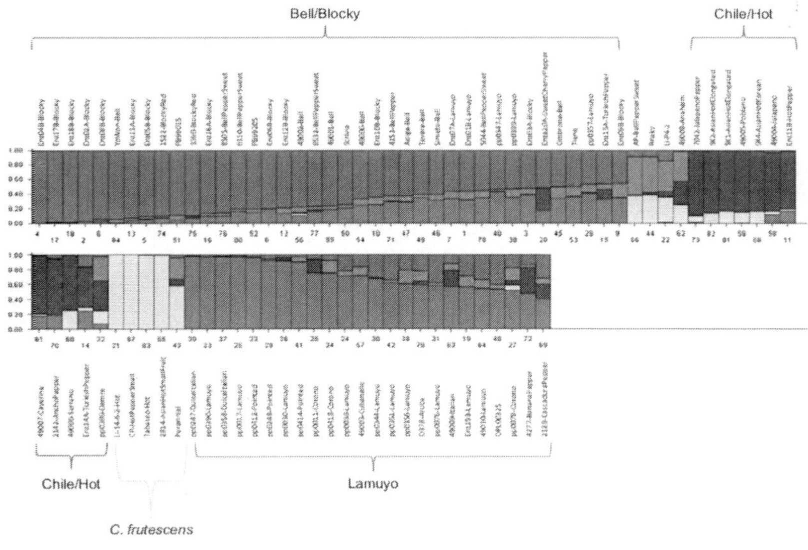

Figure 6 STRUCTURE analysis of inbred pepper lines using the data produced with the PepperSNP16K array analyzed in STRUCTURE for *K*=5.

The subgroup splits are generally observable along distinct fruit types. Assessing the five groups based on the available knowledge concerning the material, the red and pink groups are blocky types while the green and blue groups represent the chile/hot types. The largest group A (red) is represented by mainly bell or blocky types together with a limited set of Lamuyo types. The second largest group B (pink) is mainly represented by Lamuyo, Pointed, Corono and Italian types. Although there is some overlap with group A, this group B is sufficiently distinct from the A group. The third largest group C (blue) is represented by mainly hot types including a considerable number of Asian hot types. The next group D (yellow) is represented by hot pepper such as Tabasco that are different species (*C. frutescens*) or appear to contain a considerable proportion of exotic introgressions. The fifth and smallest group E (green) contains varieties that are *C. annuum* probably with significant introgressions from wild species. It is known that in pepper, specific disease resistance genes have been introgressed from other species. In summary, the classification described in the STRUCTURE analysis is mainly in agreement with what is known about the different pepper types and the data from microsatellite analysis,14 where similar groups could be identified and confirmed through lines that were analyzed in both studies. The results also correlate with the analysis completed with a smaller set of 40 lines analyzed with the Affymetrix GeneChip.10

The genotypes of lines were used to identify the non-polymorphic region surrounding the *PUN1/CSY1* gene (Ca02g19260) by correlating the STRUCTURE analysis according to phenotypic group with the SNP markers around the gene. Although a marker specifically in the pun1 gene has not been included on the Pepper16K array, the physical location surrounding the genome position of the identified pun1 locus in the draft genome sequence revealed 23 SNPs for which the non-pungent group types identified using STRUCTURE are monomorphic. This region was found to correspond to a physical distance according to the genome sequence of 1.016 Megabase pairs and a genetic distance of 1.6 cM, spanning from 50.6 to 52.2 cM on Chromosome 02 (Figure 7).

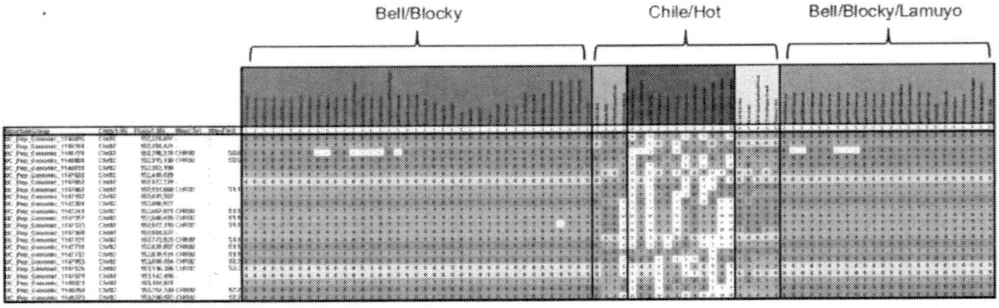

Figure 7 Monomorphic haplotype block in non-pungent inbred types surrounding PUN1 locus. Germplasm groups depicted were determined with the STRUCTURE program.

DISCUSSION

We have developed a publically available high-throughput, high-density SNP genotyping platform for pepper, the PepperSNP16K array. Interrogation of the 14 877 functional markers on the array allowed the production of the first interspecific genetic map and the first small-scale germplasm analysis using the array. This resource will allow for researchers as well as breeders to obtain easy-to-work with, reproducible data with very low rates of missing data that can be directly integrated into their breeding platforms through marker-assisted selection and discovery of markers associated with economically important traits for pepper.

The array will be valuable for investigating the *Capsicum* genus with limited ascertainment bias, but it must be taken into consideration that while many species have the same chromosome number and are interfertile among the genera, significant differences in genome/chromosome structure can be present that will create difficulties when attempting to create a genetic map with a population generated with interspecific parents. This was seen in the interspecific cross used here between *C. frutescens* and *C. annuum* that created varying amounts of difficulties for three chromosomes. As previous researchers have demonstrated, despite the occurrence of a translocation in *C. frutescens* relative to *C. annuum*, it was possible to identify the groupings of markers that corresponded to the two separate pieces of chromosome 01, which correspond to the top of chromosome 01 and the bottom translocated region, which corresponds to cultivated chromosome 08 and the wild chromosome 08 that corresponds to the bottom of cultivated chromosome 01 (as outlined in Hill *et al.*[18,21]—Figure 6). Hill *et al.* reported that there is a pseudolinkage region that occurs between chromosomes 01 and 08 in the interspecific cross, which caused issues with local ordering of markers within that region. In their case, they had about three times the amount of markers included in their genetic map, which allowed for them to generate a consecutive linkage block that corresponded to chromosome 01 and showed pseudolinkage to chromosome 08. Although with a smaller number of markers, we were unable to generate a consecutive linkage group for chromosome 01, so it was necessary to break it into two pieces that were ordered independently, then joined to determine the final distances between the markers over the entire chromosome (Figure 2).

The results and correlation with the Hill *et al.*[18] map were extremely valuable in this study in order to assist with issues with problematic chromosomes involving the translocation; however, there are a number of differences between the arrays used in the two studies. In the previous study a GeneChip array was utilized in order to identify single-position polymorphisms (SPPs); although this approach offers the possibility to discover *de novo* markers in the same way as through sequencing, there are limitations to SPPs relative to SNPs. Calling of SPPs is susceptible to both missing and erroneous calls owing to the algorithm/technology not being able to discriminate heterozygous calls,[10] although the degree of redundancy (13 probes per bp) resulted in highly accurate calls in RIL populations. Therefore, for individuals or mapping populations such as F_2s that have a significant proportion of heterozygous loci, these positions will have to be corrected prior to direct utilization of the data. The correction of these issues can lead to downstream problems as association software or genetic mapping/ordering of markers is highly susceptible to calling errors, particularly in high-density maps. In addition, a discrete genotype is not obtained by calling of SPPs, just the relative hybridization intensity is obtained, not the actual genotype at the locus that can be obtained with SNP genotyping methods. Determination of 'genotypes' from the GeneChip array that delivers just hybridization intensities requires a downstream analysis following actual reading to infer the genotype, whereas in the Infinium analysis, the genotype is obtained directly following reading of the SNP array. These differences make processing of the GeneChip array computationally intensive and time consuming following obtaining results from the laboratory, while the output from the SNP arrays are ready for analysis directly following data acquisition.

Post-processing of genotyping-by-sequencing and other genotyping through sequencing methods also require bioinformaticians with knowledge of the data to process from raw data to genotypes and can take weeks to months. Once genotype data are obtained, the missing data rates are quite high, utilized data for published analyses are typically up to 17–20%.[22,23,24] In order to achieve acceptable missing rates for downstream analyses, imputation algorithms have to be utilized. However, the accuracy of imputation methods can be detrimental to downstream analyses, which depend on good genotypic data.[25] For this reason the low missing rates produced with array technology make utilization of the genotyping data produced straightforward. The genotyping rate achieved with the PepperSNP16K array was 99.2% for functional polymorphic markers, thus producing a missing rate of <1% overall. Similar genotyping rates have been seen with other recently produced arrays for cotton,[26] soybean[27] and rice.[28] Similar challenges with calling heterozygotes are found in GBS data due to low coverage in sequencing each base.[25] The SNP arrays thus allow for high-quality, high-density genotype data that will be ready for analysis from plant material within 72 h.

The quick turnaround from plant material to high-quality data can be a valuable tool for breeders when the applicability of the data to breeding programs is understood. *De novo* clustering of markers into the haploid number of chromosomes in pepper maps has previously been rare.[18] The array has been shown to accurately separate polymorphic markers into 12 linkage groups corresponding to the haploid number of chromosomes in pepper. The approach to use a selection of gene-associated SNPs on the array should provide a fairly even distribution of markers along the genome as genes have been shown to have a distribution that is similar to the distribution of genetic recombination and across the reference sequence.[2] Analysis of the 84 hybrid and inbred lines genotyped on the PepperSNP16K showed that the available germplasm in *Capsicum* is largely divided based on phenotypic characteristics of the fruit. In this study, the germplasm was shown to split into five distinct groups. The available diversity within each phenotypic group appears to be fairly variable, which will allow for improvement of groups through interbreeding within a phenotypic class.

Although diversity within groups is still available, diversity within specific regions responsible for particular traits may be fixed within the group, which can create difficulties for breeding for traits controlled by loci in such areas within specific populations. One such genomic area that accounts for one of the most important traits for breeding in the *Capsicum* genus is at the *PUN1/CSY1* gene (Ca02g19260). This locus is responsible for the presence of capsaicin in pungent varieties (chile/hot) or lack of capsaicin in non-pungent (bell) varieties.[29,30] The gene is located on Chromosome 02 at 152.61 Mb on version 1.55 of the pepper draft genome sequence.[2] Surrounding this locus Hill *et al.*[10] was able to identify 42 markers that were monomorphic among non-pungent lines, which extended to over 8.74 cM on Chromosome 02. Here we were able to reduce the region down to a ~1 Mb region corresponding to just over 1.6 cM of genetic distance. It is likely that the inclusion of additional samples in the diversity analysis shown here compared with the previous investigation was able to reduce the size of the region likely affected by the selective sweep surrounding the *PUN1* gene and the phenotypic split between pungent and non-pungent pepper types. In breeding populations, it will be important to utilize population scale data to identify regions with low diversity, if there are important genes which may attribute beneficial characteristics in those areas. If so, genetic diversity will need to be taken in from outside sources to allow for inclusion of diversity within these areas.

Datasets such as the one that was developed here will allow for breeders to have additional population-scale information for addressing breeding at the genetic scale for such key low diversity areas. The genetic map produced will also allow for localization of 4724 additional genes on the genetic map that were unable to be localized in previous high-density mapping efforts, which will allow for progression of fine mapping efforts for genes responsible for agronomically important traits.

CONCLUSIONS

We describe a publicly available standardized high-throughput genotyping array for pepper, the PepperSNP16K. This array and the accompanying resources developed here will be a significant addition to the pepper breeder's toolbox which will allow them to focus on improving pepper in a time of growing demand and decreasing resources. Due to the standardized nature of the array, research using the array will occur on a worldwide common platform that will permit assimilation of results on a common reference tool. This will allow the deployment of findings for marker-assisted breeding programs worldwide.

REFERENCES

1. USDA National Agriculture Statistics Service. 1994–2013; https://www.nass.usda.gov/.
2. Kim S, Park M, Yeom SI, Kim YM, Lee JM, Lee HA *et al. Genome sequence of the hot pepper provides insights into the evolution of pungency in Capsicum species. Nat Genet* 2014; **46**: 270–278.
3. Moscone EA, Baranyi M, Ebert I, Greilhuber J, Ehrendorfer F, Hunziker AT. *Analysis of nuclear DNA content in Capsicum (Solanaceae) by flow cytometry and Feulgen densitometry. Ann Bot* 2003; **92**: 21–29.
4. Zonneveld BJM, Leitch IJ, Bennett MD. *First nuclear DNA amounts in more than 300 angiosperms. Annals of Botany* 2005; **96**: 229–244.
5. Sato S, Tabata S, Hirakawa H, Asamizu E, Shirasawa K, Isobe S *et al. The tomato genome sequence provides insights into fleshy fruit evolution. Nature* 2012; **485**: 635–641.
6. Xu X, Pan SK, Cheng SF, Zhang B, Mu DS, Ni PX *et al. Genome sequence and analysis of the tuber crop potato. Nature* 2011; **475**: 189–U194.
7. Lu FH, Cho MC, Park YJ. *Transcriptome profiling and molecular marker discovery in red pepper, Capsicum annuum L. TF68. Mol Biol Rep* 2012; **39**: 3327–3335.
8. Lu F-H, Yoon M-Y, Cho Y-I, Chung J-W, Kim K-T, Cho M-C *et al.Transcriptome analysis and SNP/SSR marker information of red pepper variety YCM334 and Taean. Scientia Horticulturae* 2011; **129**: 38–45.
9. Ashrafi H, Hill T, Stoffel K, Kozik A, Yao J, Chin-Wo SR *et al. De novo assembly of the pepper transcriptome (Capsicum annuum): a benchmark for in silico discovery of SNPs, SSRs and candidate genes. BMC Genomics* 2012; **13**: 571.
10. Hill TA, Ashrafi H, Reyes-Chin-Wo S, Yao J, Stoffel K, Truco MJ *et al. Characterization of Capsicum annuum genetic diversity and population structure based on parallel polymorphism discovery with a 30K unigene Pepper GeneChip. PLoS One* 2013; **8**: e56200.
11. Qin C, Yu C, Shen Y, Fang X, Chen L, Min J *et al. Whole-genome sequencing of cultivated and wild peppers provides insights into Capsicum domestication and specialization. Proc Natl Acad Sci USA* 2014; **111**: 5135–5140.
12. Xu X, Chao J, Cheng X, Wang R, Sun B, Wang H *et al. Mapping of a novel race specific resistance gene to phytophthora root rot of pepper (Capsicum annuum) using bulked segregant analysis combined with specific length amplified fragment sequencing strategy. PLoS One* 2016; **11**: e0151401.
13. Li H, Handsaker B, Wysoker A, Fennell T, Ruan J, Homer N *et al.The Sequence Alignment/Map format and SAMtools. Bioinformatics* 2009; **25**: 2078–2079.
14. Nagy I, Stagel A, Sasvari Z, Roder M, Ganal M. *Development, characterization, and transferability to other Solanaceae of microsatellite markers in pepper (Capsicum annuum L.). Genome* 2007; **50**: 668–688.
15. Van Ooijen J JoinMap 4. Software for the calculation of genetic linkage maps in experimental populations Kyazma BV, Wageningen, Netherlands 2006.

16. Manly KF, Cudmore RH Jr, Meer JM. *Map Manager QTX, cross-platform software for genetic mapping. Mamm Genome* 2001; **12**: 930–932.
17. Voorrips RE. *MapChart: software for the graphical presentation of linkage maps and QTLs. J Hered* 2002; **93**: 77–78.
18. Hill T, Ashrafi H, Chin-Wo SR, Stoffel K, Truco MJ, Kozik A *et al.Ultra-high density, transcript-based genetic maps of pepper define recombination in the genome and synteny among related species. G3 (Bethesda)* 2015; **5**: 2341–2355.
19. Pritchard JK, Stephens M, Donnelly P. *Inference of population structure using multilocus genotype data. Genetics* 2000; **155**: 945–959.
20. Evanno G, Regnaut S, Goudet J. *Detecting the number of clusters of individuals using the software STRUCTURE: a simulation study. Mol Ecol* 2005; **14**: 2611–2620.
21. Wu F, Eannetta NT, Xu Y, Durrett R, Mazourek M, Jahn MM *et al.A COSII genetic map of the pepper genome provides a detailed picture of synteny with tomato and new insights into recent chromosome evolution in the genus Capsicum. Theor Appl Genet* 2009; **118**: 1279–1293.
22. Poland JA, Brown PJ, Sorrells ME, Jannink JL. *Development of high-density genetic maps for barley and wheat using a novel two-enzyme genotyping-by-sequencing approach. PLoS One* 2012; **7**: e32253.
23. Mascher M, Muehlbauer GJ, Rokhsar DS, Chapman J, Schmutz J, Barry K *et al. Anchoring and ordering NGS contig assemblies by population sequencing (POPSEQ). Plant J* 2013; **76**: 718–727.
24. Boutet G, Alves Carvalho S, Falque M, Peterlongo P, Lhuillier E, Bouchez O *et al. SNP discovery and genetic mapping using genotyping by sequencing of whole genome genomic DNA from a pea RIL population. BMC Genomics* 2016; **17**: 121.
25. Xavier A, Muir WM, Rainey KM. *Impact of imputation methods on the amount of genetic variation captured by a single-nucleotide polymorphism panel in soybeans. BMC Bioinformatics* 2016; **17**: 55.
26. Hulse-Kemp AM, Lemm J, Plieske J, Ashrafi H, Buyyarapu R, Fang DD *et al. Development of a 63K SNP array for cotton and high-density mapping of intraspecific and interspecific populations of Gossypium spp. G3 (Bethesda)* 2015; **5**: 1187–1209.
27. Lee YG, Jeong N, Kim JH, Lee K, Kim KH, Pirani A *et al.Development, validation and genetic analysis of a large soybean SNP genotyping array. Plant J* 2015; **81**: 625–636.
28. Singh N, Jayaswal PK, Panda K, Mandal P, Kumar V, Singh B *et al.Single-copy gene based 50K SNP chip for genetic studies and molecular breeding in rice. Sci Rep* 2015; **5**: 11600.
29. Wahyuni Y, Ballester AR, Sudarmonowati E, Bino RJ, Bovy AG. *Metabolite biodiversity in pepper (Capsicum) fruits of thirty-two diverse accessions: variation in health-related compounds and implications for breeding. Phytochemistry* 2011; **72**: 1358–1370.
30. Stewart C Jr, Kang BC, Liu K, Mazourek M, Moore SL, Yoo EY *et al.The Pun1 gene for pungency in pepper encodes a putative acyltransferase. Plant J* 2005; **42**: 675–688.

CHAPTER 18

Cinnamon: A Multifaceted Medicinal Plant

Pasupuleti Visweswara Rao[1,2] and Siew Hua Gan[2]

[1]*Faculty of Agro Based Industry, Universiti Malaysia Kelantan, Jeli Campus, Locked Bag No. 100, 17600 Jeli, Kelantan, Malaysia*

[2]*Human Genome Centre, School of Medical Sciences, Universiti Sains Malaysia, 16150 Kubang Kerian, Kelantan, Malaysia*

ABSTRACT

Cinnamon (Cinnamomum zeylanicum, and Cinnamon cassia), the eternal tree of tropical medicine, belongs to the Lauraceae family. Cinnamon is one of the most important spices used daily by people all over the world. Cinnamon primarily contains vital oils and other derivatives, such as cinnamaldehyde, cinnamic acid, and cinnamate. In addition to being an antioxidant, anti-inflammatory, antidiabetic, antimicrobial, anticancer, lipid-lowering, and cardiovascular-disease-lowering compound, cinnamon has also been reported to have activities against neurological disorders, such as Parkinson's and Alzheimer's diseases. This review illustrates the pharmacological prospective of cinnamon and its use in daily life.

1. INTRODUCTION

The bark of various cinnamon species is one of the most important and popular spices used worldwide not only for cooking but also in traditional and modern medicines. Overall, approximately 250 species have been identified among the cinnamon genus, with trees being scattered all over the world [1, 2].

Cinnamon is mainly used in the aroma and essence industries due to its fragrance, which can be incorporated into different varieties of foodstuffs, perfumes, and medicinal products [3]. The most important constituents of cinnamon are cinnamaldehyde and trans-cinnamaldehyde (Cin), which are present in the essential oil, thus contributing to the fragrance and to the various biological activities observed with cinnamon [4]. A study onCinnamomum osmophloeum (C. osmophloeum) indicated that the essential oil from cinnamon leaves contains a high level of Cin. Consequently, C. osmophloeum is also used as an alternative spice for C. cassia [5]. One of the major constituents of essential oil extracted from C. zeylanicum named (E)-cinnamaldehyde has an antityrosinase activity [6], while cinnamaldehyde is the principal compound responsible for this activity [7].

Cinnamon bark contains procyanidins and catechins [8]. The components of procyanidins include both procyanidin A-type and B-type linkages [9–11]. These procyanidins extracted from cinnamon and berries also possess antioxidant activities [10, 12].

2. METHODOLOGY

The current review was conducted using a complete and organized search of the available literature on the medicinal plant cinnamon from 1982 to 2013. The searches were performed using various databases, including PubMed (http://www.ncbi.nlm.nih.gov/pubmed), Science Direct (http://www. sciencedirect.com/), (http://www..com/), Scirus (http://www. scirus.com/), and (http://www.scholar.google.com/).

3. TRADITIONAL USES

In addition to being used as a spice and flavoring agent, cinnamon is also added to flavor chewing gums due to its mouth refreshing effects and ability to remove bad breath [13]. Cinnamon can also improve the health of the colon, thereby reducing the risk of colon cancer [14].

Cinnamon is a coagulant and prevents bleeding [15]. Cinnamon also increases the blood circulation in the uterus and advances tissue regeneration [16]. This plant plays a vital role as a spice, but its essential oils and other constituents also have important activities, including antimicrobial [17–20], antifungal [21], antioxidant [22–26], and antidiabetic [27–33].

Cinnamon has been used as anti-inflammatory [34–36], antitermitic [36], nematicidal [37, 38], mosquito larvicidal [39], insecticidal [40], antimycotic, [40–43] and anticancer agent [44–47]. Cinnamon has also been traditionally used as tooth powder and to treat toothaches, dental problems, oral microbiota, and bad breath [48, 49].

4. CHEMICAL CONSTITUENTS

Cinnamon consists of a variety of resinous compounds, including cinnamaldehyde, cinnamate, cinnamic acid, and numerous essential oils [50] (Table 1). Singh et al. [51] reported that the spicy taste and fragrance are due to the presence of cinnamaldehyde and occur due to the absorption of oxygen. As cinnamon ages, it darkens in color, improving the resinous compounds [51]. Sangal reported various physiochemical properties of cinnamon (Table 2). The presence of a wide range of essential oils, such as trans-cinnamaldehyde, cinnamyl acetate, eugenol, L-borneol, caryophyllene oxide, b-caryophyllene, L-bornyl acetate, E-nerolidol, α-cubebene, α-terpineol, terpinolene, and α-thujene, has been reported [35, 36].

Table 1: Chemical constituents of different parts of cinnamon [136] (Vangalapati et al., 2012 [2]).

Part of the plant	Compound
Leaves	Cinnamaldehyde: 1.00 to 5.00% Eugenol: 70.00 to 95.00%
Bark	Cinnamaldehyde: 65.00 to 80.00% Eugenol: 5.00 to 10.00%
Root bark	Camphor: 60.00%
Fruit	*trans*-Cinnamyl acetate (42.00 to 54.00%) and caryophyllene (9.00 to 14.00%)
C. zeylanicum buds	Terpene hydrocarbons: 78.00% *alpha*-Bergamotene: 27.38% *alpha*-Copaene: 23.05% Oxygenated terpenoids: 9.00%
C. zeylanicum flowers	(E)-Cinnamyl acetate: 41.98% *trans-alpha*-Bergamotene: 7.97% Caryophyllene oxide: 7.20%

Table 2: Physicochemical properties of cinnamon(Sangal., 2011 [1]).

Parameter	Leaf oil	Bark oil
Specific gravity (20°C)	1.030–1.050	1.010–1.030
Optical rotation (°) (20°C)	1°96′–0°40′	Slightly laevorotatory
Refractive index (20°C)	1.529–1.537	1.573–1.591
Aldehyde content	4%	65–76%
Eugenol content	77.3–90.5%	4–10%
Solubility characteristics	Soluble in 1.5 volumes of 70% alcohol	Soluble in 2.0–3.0 volumes of 70% alcohol

The chemical structures of some important constituents of cinnamon are shown in Figures 1, 2, 3, 4, and 5.

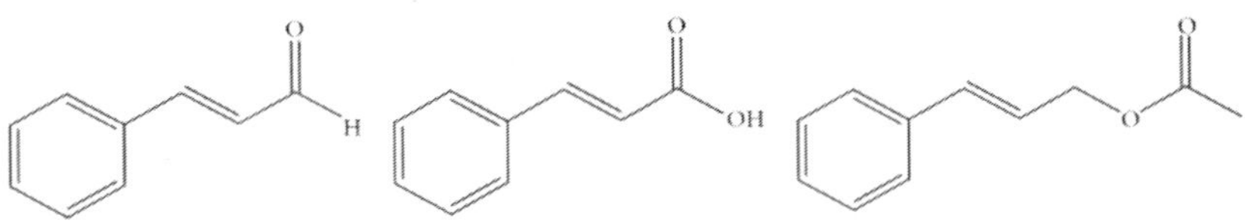

(a) Cinnamaldehyde (b) Cinnamic acid (c) Cinnamyl acetate

Figure 1: Cinnamyl group-containing compounds.

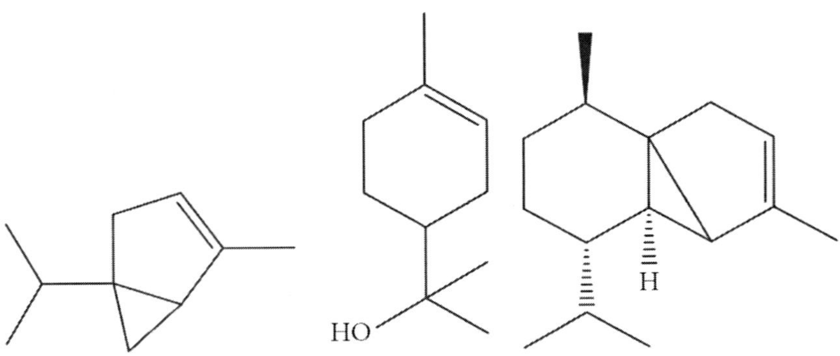

(a) α-Thujene (b) α-Terpineol (c) α-Cubebene

Figure 2: Endocyclic double bond-containing compounds.

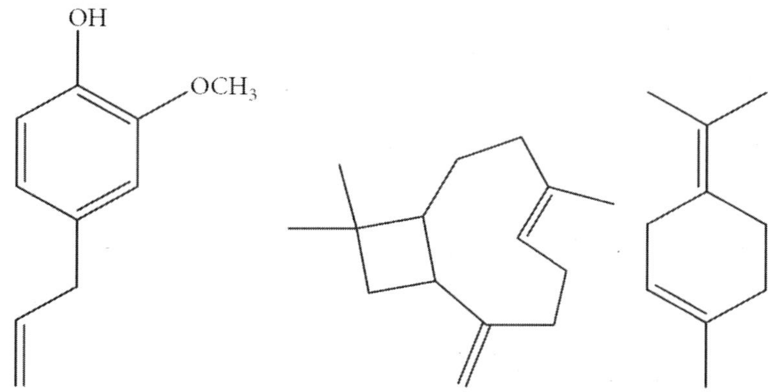

(a) Eugenol (b) β-Caryophyllene (c) Terpinolene

Figure 3: Unconjugated exocyclic double bond-containing compounds.

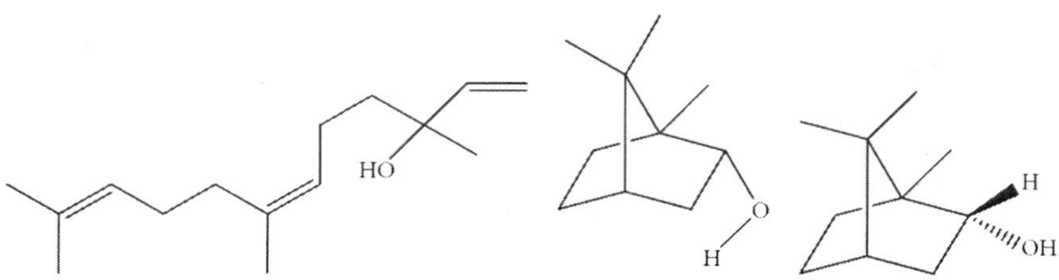

(a) E-Nerolidol (b) L-Borneol (c) Borneol

Figure 4: Hydroxy-substituted aliphatic compounds.

5. ANTIOXIDANT ACTIVITY

Antioxidant compounds present in foodstuffs play a vital role in human life, acting as health-protecting agents. In addition to this role, antioxidants are one of the key additives used in fats and oils. Even in the food processing industry, antioxidants have been used to delay or prevent food spoilage. Spices and medicinal plants have received rapid consideration as sources of beneficial antioxidants against various diseases [52]. Antioxidants have been considered the most important drivers in the progress and existence of humans, as they respond to free radicals and damage in metabolic diseases and age-related syndromes of humans and other animals [53, 54].

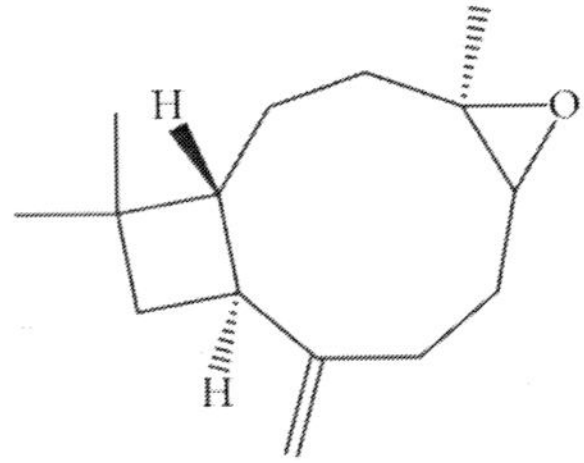

(a) Caryophyllene oxide

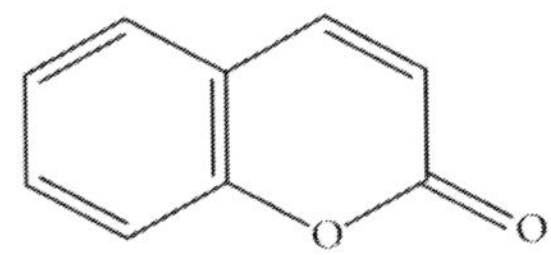

(b) Coumarin

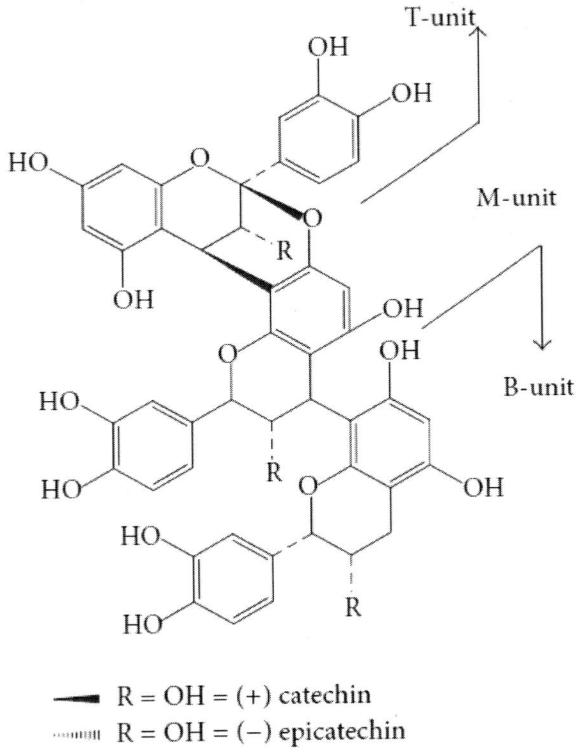

(c) Procyanidin type-A polymer

Figure 5: Other miscellaneous compounds containing oxirane, 2-pyranone, and pyran groups.

Mancini-Filho et al. reported various extracts of cinnamon, such as ether, aqueous, and methanolic extracts that have shown considerable antioxidant activities [22]. A study on rats reported that the administration of the bark powderof C. verum (10%) for 90 days produced antioxidant activities as indicated by cardiac and hepatic antioxidant enzymes, lipid conjugate dienes, and glutathione (GSH) [55]. A research group reported that cinnamon oil potentially exhibits superoxide-dismutase- (SOD-) like activity as indicated by the inhibition of the inhibiting capacity of pyrogallol autoxidation [56].

The aqueous and alcoholic extract (1 : 1) of cinnamon potentially significantly inhibits fatty acid oxidation and lipid peroxidation in vitro [23]. Different flavonoids isolated from cinnamon have free-radical-scavenging activities and antioxidant properties [57]. A study of the inhibitory effects of cinnamaldehyde and other compounds of cinnamon on nitric oxide production revealed that cinnamaldehyde possesses potential activity against the production of nitric oxide as well as the expression of inducible nitric oxide. The highest inhibitory activities were reported as 81.5%, 71.7%, and 41.2% at 1.0, 0.5, and 0.1 $\mu g/\mu L$, respectively [58]. Lin et al. reported the in vivo antioxidant activity of two different extracts, the ethanolic and hot water extracts of the dry bark of C. cassia. The ethanolic extract of C. cassia exhibited significant inhibition (96.3%) compared to the natural antioxidant α-tocopherol (93.74%) [59]. Overall, cinnamon exhibited higher antioxidant activities compared to that of other dessert spices [60].

The essential oils and some of the major compounds present in cinnamon, including (E)-cinnamaldehyde, eugenol, and linalool, were investigated in reference to peroxynitrite-induced nitration and lipid peroxidation. Eugenol and the essential oils were more effective than the other two compounds [61]. In a comparative study among 26 spices, cinnamon showed the highest antioxidant activity, indicating that it can be applied as an antioxidant used in foods [62]. Another study investigated the effectiveness of a mixture of spices on oxidative stress markers as well as the antioxidant activity in high fructose-fed insulin-resistant rats. The mixture, which consisted of 1 g/100 g cinnamon bark, showed a significant antioxidant activity compared to the fructose alone group [63]. Volatile oils from C. zeylanicum showed significant biological activities [64].

Forty-one different volatile compounds in the bark oil of cinnamon have been identified and were found to vary significantly in percentage composition depending on the growth stages and segments of the C. cassia tree [65]. To extract essential oil for industrial use, the yields and compositions of bark oil during Cinnamomum cassia growth (1–3 years old for the branch bark and 5–12 years old for the stem bark) were determined. These researchers also found that the branch bark fraction tended to yield more essential oil compared to the entire branch, indicating that selecting the bark based on the tree growth stages as well as separating the stem barks into top, center, and lower sections within a tree can significantly improve the extraction efficiency of essential oils.

A preliminary study on C. malabatrum leaves was conducted in various types of extracts (n-hexane, alcoholic, and aqueous extracts) to determine the presence of phenolic compounds, which indicate antioxidant activity. All of the extracts had moderate amounts of phenolic compounds and showed potential activity against hydrogen peroxide, nitric oxide, and lipid peroxide free radicals [66]. A recent study investigated the antioxidant properties of several parts (i.e., the leaves, barks, and buds) of C. cassia. The ethanolic extract of all of the plant parts had significant antioxidant properties compared with the extraction using the supercritical fluid [67]. The supercritical extracts showed decreased activity compared to the ethanol extracts, indicating that the active components are constituents with high polarity.

C. tamala has potential antioxidant activities in diabetic rats [68], while C. osmophloeum, a species from Taiwan, has significant in vitro and in vivo antioxidant activities under oxidative stress [69]. The antioxidant activity of C. zeylanicum has been investigated using various methods. In addition to the antioxidant activity, cinnamon can be used as a preservative in cakes and other food products [70]. A recent study reported that pectin film coated with cinnamon leaf extract yielded high antioxidant and antibacterial activities [71]. Dong et al. reported that cinnamaldehyde (E) extracted from C. cassia is the main compound and is present in levels as high as 72.7% compared to other volatile components [72]. Cinnamaldehyde (E) is well known for its antityrosinase activity [6, 73]. Currently, much attention is given to tyrosinase inhibitors due to their actions in suppressing hyperpigmentation as well as the unsightly browning effects observed in mushrooms, fruits, and vegetables when they are exposed to sunlight or air. Hence, antityrosinase agents are associated with a wide range of applications, such as cosmetics, medicine, and food [74, 75].

6. ANTI-INFLAMMATORY ACTIVITIES

Several studies on medicinal plants and their components have indicated the anti-inflammatory activities of cinnamon [76–81]. Various studies reported the anti-inflammatory activity of cinnamon and its essential oils [34–36]. To date, there are several flavonoid compounds (e.g., gossypin, gnaphalin, hesperidin, hibifolin, hypolaetin, oroxindin, and quercetin) that have been isolated and have anti-inflammatory activities [82–86].

A recent study reported that 2′-hydroxycinnamaldehyde isolated from C. cassia bark exhibited an inhibitory effect on the production of nitric oxide by inhibiting the activation of the nuclear factor kappa-light-chain-enhancer of activated B cells (NF-κB), indicating that this substance can potentially be used as an anti-inflammatory agent [87]. The ethanolic extract of C. cassia showed significant anti-inflammatory effects by reducing the activation of Src/spleen-tyrosine-kinase- (Src/Syk-) mediated NF-κB [88, 89]. Various compounds contained in C. ramulus showed anti-inflammatory effects by suppressing the expression of inducible nitric oxide synthesis (iNOS), cyclooxygenase-2 (COX-2), and nitric oxide (NO) production in the central nervous system (CNS). By this mechanism, C. ramulus could be a potential source for the therapeutic treatment or prevention of inflammation-mediated neurodegenerative diseases [90]. Furthermore, the aqueous extract of cinnamon decreases the lipopolysaccharide-induced tumor necrosis factor-α levels in the serum [91].

7. NEUROLOGICAL DISORDERS

Cinnamophilin is a novel thromboxane A2 receptor antagonist isolated from C. philippinensis [92]. A study reported that cinnamophilin confers protection against ischemic damage in rat brains when administered at 80 mg/kg at different time intervals (2, 4, and 6 h) after insult. The effects were found to have a considerable effect (by 34–43%) on abridged brain infarction [93] and further enhance neurobehavioral outcomes. Cinnamophilin also dramatically condenses the oxygen glucose deprivation-induced neuronal damage in organotypic hippocampal slices in experimental rats. A substance called procyanidin type-A trimer (trimer 1) isolated from cinnamon's water-soluble extract showed that trimer 1 may reduce cell swelling by controlling the movement of intracellular calcium $[Ca^{2+}]_i$ [94]. Trimer 1 also considerably alleviates the oxygen glucose deprivation-induced diminishing effects on glutamate uptake. The protective effects of trimer 1 in attenuating the diminution in glutamate uptake are possibly arbitrated via their effects on the mitochondria [94].

Parkinson's disease (PD) is the second major widespread neurodegenerative disorder after Alzheimer's disease, with a prevalence of 2% in people 65 years and older [95]. PD protein 7 (PARK7) is an autosomal recessive form of early-onset parkinsonism caused by alterations in the *DJ-1* gene [96]. Khasnavis and Pahan reported that sodium benzoate, a cinnamon metabolite, upregulates *DJ-1* by modulating mevalonate metabolites [97, 98]. Cinnamon and its metabolite sodium benzoate also upregulate the neurotropic factors BDNF (brain-derived neurotropic factors) as well as neurotrophin-3 (NT-3) in the mouse central nervous system [99]. PARK7 is one of the main neuroprotective proteins that protects cells from damage and from the further detrimental effects of oxidative stress; therefore, this protein may be an effective molecule that can be incorporated into the therapeutic intervention of Parkinson's disease [98].

A natural compound isolated from cinnamon extract (CEppt) significantly reduces the formation of toxic β-amyloid polypeptide ($A\beta$) oligomers and prevents its toxicity on neuronal pheochromocytoma (PC12) cells [100]. The study indicated that CEppt resolved the reduced permanence, fully improved deficiencies in locomotion, and totally eradicated the tetrameric species of $A\beta$ in the brain of the fly model of Alzheimer's disease, leading to a noticeable reduction in the 56 kDa $A\beta$ oligomers, reducing plaques and improving the cognitive performance of transgenic mice models [100].

Another study reported that the aqueous extract of C. zeylanicum can reduce tau aggregation and filament formation, two of the main features of Alzheimer's disease. The extract can also encourage the complete fragmentation of recombinant tau filaments and cause the considerable modification of the morphology of paired helical filaments from Alzheimer's disease brain [101], indicating the potential of cinnamon in the treatment of Alzheimer's disease.

8. ANTIDIABETIC ACTIVITY

A substance from cinnamon has been isolated and coined as "insulin-potentiating factor" (IPF) [102], while the antidiabetic effects of cinnamon bark have been shown in streptozotocin-induced diabetic rats [33]. Several studies have also revealed that cinnamon extracts lower not only blood glucose but also cholesterol levels [103–107].

A study comparing the insulin-potentiating effects of many spices revealed that the aqueous extract of cinnamon was 20-fold higher than the other spices [108]. Methylhydroxychalcone polymer (MHCP) is the purified polymer of hydroxychalcone with the ability to stimulate glucose oxidation [30, 109]. Anderson et al. isolated and characterized the polyphenol type-A polymers from cinnamon and found that these substances act as insulin-like molecules [9]. Following this characterization, a new compound from hydroxycinnamic acid derivatives named naphthalenemethyl ester, which has blood glucose-lowering effects, has been identified [27], further confirming cinnamon's antidiabetic effects.

Several polyphenols have been isolated from cinnamon. These polyphenols include rutin (90.0672%), catechin (1.9%), quercetin (0.172%), kaempferol (0.016%), and isorhamnetin (0.103%) [67, 110]. Cao et al. (2007) demonstrated that the aqueous extract of cinnamon containing polyphenols purified by high performance liquid chromatography (HPLC) showed insulin-like activity [111]. The aqueous extract of cinnamon markedly decreased the absorption of alanine in the rat intestine. Alanine plays a vital role in gluconeogenesis, is altered back to pyruvate in the liver, and is utilized as a substrate for gluconeogenesis [112]. However, another study conducted on diabetic postmenopausal women supplemented with cinnamon showed poor glycemic control [113], even though cinnamon is generally believed to be useful for diabetes. However, it is plausible that differences in the dose of cinnamon used, as well as baseline glucose and lipid levels, have led to these variations.

In a recent study [114], suitable doses of cinnamon (5, 10, and 20 mg/kg) of the linalool chemotype were found to help with glycemic control in diabetics due to enhanced insulin secretion. It is plausible that the amelioration of oxidative stress and the proinflammatory environment in the pancreas may confer protection to pancreatic βcells [114], which should be further investigated.

9. ANTIMICROBIAL ACTIVITY

To date, several antimicrobial activities of cinnamon and its oils have been reported in various studies [20, 28, 115]. For example, Matan et al. reported the effects of cinnamon oils on different bacterial (Pediococcus halophilus and Staphylococcus aureus), fungal (Aspergillus flavus, Mucor plumbeus, Penicillium roqueforti, andEurotium sp.), and yeast species (Candida lipolytica, Pichia membranaefaciens, Debaryomyces hansenii, andZygosaccharomyces rouxii) [19], indicating that cinnamon is a natural antimicrobial agent.

Goñi et al. described the antibacterial activity of a combination of cinnamon and clove oils against Gram-positive organisms (Listeria monocytogenes, Enterococcus faecalis, Staphylococcus aureus, and Bacillus cereus), as well as against Gram-negative bacteria (Salmonella choleraesuis, Escherichia coli, Pseudomonas aeruginosa, andYersinia enterocolitica) [116]. A study from Hili et al. indicated that cinnamon oils have potential action against various bacteria (Pseudomonas aeruginosa, Staphylococcus aureus, and Escherichia coli) and yeast (Torulopsis utilis, Schizosaccharomyces pombe, Candida albicans, and Saccharomyces cerevisiae) [18]. A recent study reported the activity of the aqueous extract of cinnamon and other plants against oral microflora. Overall, the essential oil from cinnamon is more potent than other tested plant extracts, such as Azadirachta indica andSyzygium aromaticum [117].

10. ANTICANCER ACTIVITY

The aqueous extract and the fraction of cinnamon (procyanidins) from HPLC inhibit vascular endothelial growth factor subtype 2 (VEGFR2) kinase activity, thereby inhibiting the angiogenesis involved in cancer. The results of the study revealed that cinnamon could potentially be used in cancer prevention [44]. Cinnamaldehydes have been synthesized and tested as inhibitors against angiogenesis [118]. Jeong et al. reported that CB403, a chemical that can be synthesized from 2′-hydroxycinnamaldehyde derived from cinnamaldehyde, can inhibit tumor growth. Overall, the antitumor and growth-inhibitory properties of CB403 in animal-based studies as well as in cell culture-based studies indicate the potential of cinnamon to be used as an anticancer agent [119].

Cabello et al. reported that cinnamic aldehyde inhibits the activity of NF-κB and the production of tumor necrosis factor alpha (TNFα-) induced interleukin-8 (IL-8) in A375 cells [120]. This inhibition provides additional support to the existing unrecognized role of cinnamic acid as a potential anticancer agent [120]. Fang and others reported the anticancer effect of trans-cinnamaldehyde from C. osmophloeum, finding thattrans-cinnamaldehyde showed potential effects in restraining tumor cell growth and in enhancing tumor cell apoptosis [121].

A preliminary study on cinnamon and cardamom against azoxymethane-(AOM-) induced colon cancer in Swiss albino mice has been conducted [122]. Treatments with the aqueous extracts of cinnamon and cardamom augment the activities of the detoxifying and antioxidant enzyme glutathione-s-transferase (GST) with a concomitant reduction in lipid peroxidation levels in animals with colon cancer compared to controls [122]. The essential oils extracted from C. cassia inhibit alpha melanocyte-stimulating hormone's induced melanin production, thereby suppressing oxidative stress in murine B16 melanoma cells [7].

11. CARDIOVASCULAR DISEASES

One of the active components isolated from C. cassia named 2-methoxycinnamaldehyde (2-MCA) decreases the expression of vascular cell adhesion molecule-1 (VCAM-1) in TNFα-activated endothelial cells, suggesting that ischemia/reperfusion (I/R) injury is ameliorated due to the induction of hemeoxygenase- (HO-) 1 [123]. A recent study reported the potential effects of two compounds, cinnamic aldehyde and cinnamic acid, isolated from C. cassia against myocardial ischemia [124], indicating that cinnamon also has the potential to be used to treat cardiovascular diseases.

Several studies have reported the protective effects of cinnamaldehyde on the cardiovascular system. Cinnamophilin is one of the important lignans isolated from C. philippinensis and has been confirmed to have thromboxane A2 (TXA_2) receptor blocking activity in rats as well as in guinea pigs [125]. Cinnamophilin acts as a potential thromboxane synthase inhibitor and TXA_2 receptor antagonist and may be helpful when incorporated in the treatment of diseases involving TXA_2 disorders [125], such as platelet aggregation [126] and cancers [127]. Cinnamophilin mainly inhibits thromboxane receptor-mediated vascular smooth muscle cell proliferation and may have the potential for use in the prevention of vascular diseases and atherosclerosis [128].

Cinnamaldehyde produces hypotensive effects, which are possibly mainly due to peripheral vasodilatation in anesthetized dogs and guinea pigs [129]. The vasodilatation induced by cinnamaldehyde in dogs lasted and remained over the recovery period of the fall in blood pressure to the baseline [130]. A recent study showed that cinnamaldehyde expands rat vascular smooth muscle in an endothelium-independent manner. The ability of cinnamaldehyde in vasodilatory function may be because it impedes both Ca^{2+} influx and Ca^{2+} release [131]. Cinnamaldehyde averts the progress of hypertension in types 1 and 2 diabetes by abridging vascular contractility, in addition to its insulinotropic effect in insulin deficiency [132].

12. CHOLESTEROL- AND LIPID-LOWERING EFFECTS

The administration of cinnamon to mice positively affected the lipid profile, whereby the high density lipoprotein (HDL) cholesterol levels decreased, and plasma triglycerides were reduced [27]. Another study by [133] found a reduction in the total cholesterol, triglycerides, and low-density lipoproteins in rats administeredCinnamomum cassia powder (15%) for 35 days. Additionally, cinnamon oils reduced the cholesterol levels in broiler chickens [134]. A study by Khan et al. reported that the administration of cinnamon at 1, 3, and 6 g doses per day caused a reduction in serum glucose, triglyceride, total cholesterol, and LDL cholesterol levels in humans [104].

13. ADVANCED GLYCATION END PRODUCTS (AGES)

Different types of phenolic and flavonoid compounds have been isolated from cinnamon. Epicatechin, catechin, and procyanidin B2, which are the phenolic compounds isolated from cinnamon, showed noteworthy and potentially inhibitory activities on the formation of AGEs. These antiglycation activities of the phenolic compounds not only are attributed to their antioxidant activities but also are associated with the entrapping capabilities of reactive carbonyl species, such as methylglyoxal (MGO), an intermediate reactive carbonyl of AGE formation [10, 135]. The inhibition of AGE formation by trapping the reactive carbonyl species could be a logical therapeutic approach to treat diabetes and its complications [10].

14. CONCLUSIONS

Cinnamon has been used as a spice in daily life without any side effects. Several reports have dealt with the numerous properties of cinnamon in the forms of bark, essential oils, bark powder, phenolic compounds, flavonoids, and isolated components. Each of these properties plays a key role in the advancement of human health. The antioxidant and antimicrobial activities may occur through the direct action on oxidants or microbes, whereas the anti-inflammatory, anticancer, and antidiabetic activities occur indirectly via receptor-mediated mechanisms. The significant health benefits of numerous types of cinnamon have been explored. Further investigations are necessary to provide additional clinical evidence for the traditional uses of this spice against cancer and inflammatory, cardioprotective, and neurological disorders.

ABBREVIATIONS

GSH:	Glutathione
SOD:	Superoxide dismutase
Src/Syk:	Src/spleen tyrosine kinase
iNOS:	Inducible nitric oxide synthesis
COX-2:	Cyclooxygenase-2

NO:	Nitric oxide
CNS:	Central nervous system
PARK7:	PD protein 7
NT-3:	Neurotrophin-3
Aβ:	β-Amyloid polypeptide
PC 12:	Pheochromocytoma 12
IPF:	Insulin-potentiating factor
MHCP:	Methylhydroxychalcone polymer
VEGFR2:	Vascular endothelial growth factor subtype 2
TNFα:	Tumor necrosis factor-alpha interleukin-8 (IL-8)
AOM:	Azoxymethane
GST:	Glutathione-s-transferase
2-MCA:	2-Methoxycinnamaldehyde
VCAM-1:	Vascular cell adhesion molecule-1
HO:	Heme oxygenase (HO)
TXA_2:	Thromboxane A2
HDL cholesterol:	High-density lipoprotein cholesterol
MGO:	Methylglyoxal.

ACKNOWLEDGMENT

The corresponding author Dr. Pasupuleti Visweswara Rao expresses thanks to Mr. Balam Satheesh Krishna, Department of Chemistry, S. V. University, for the discussion and help in chemical structure related issues.

REFERENCES

1. Sangal, "Role of cinnamon as beneficial antidiabetic food adjunct: a review," Advances in Applied Science Research, vol. 2, no. 4, pp. 440–450, 2011.

2. M. Vangalapati, N. Sree Satya, D. Surya Prakash, and S. Avanigadda, "A review on pharmacological activities and clinical effects of cinnamon species," Research Journal of Pharmaceutical, Biological and Chemical Sciences, vol. 3, no. 1, pp. 653–663, 2012.
3. T.-C. Huang, H.-Y. Fu, C.-T. Ho, D. Tan, Y.-T. Huang, and M.-H. Pan, "Induction of apoptosis by cinnamaldehyde from indigenous cinnamon Cinnamomum osmophloeum Kaneh through reactive oxygen species production, glutathione depletion, and caspase activation in human leukemia K562 cells," Food Chemistry, vol. 103, no. 2, pp. 434–443, 2007.
4. H.-F. Yeh, C.-Y. Luo, C.-Y. Lin, S.-S. Cheng, Y.-R. Hsu, and S.-T. Chang, "Methods for thermal stability enhancement of leaf essential oils and their main Constituents from Indigenous Cinnamon (Cinnamomum osmophloeum)," Journal of Agricultural and Food Chemistry, vol. 61, no. 26, pp. 6293–6298, 2013.
5. C.-W. Chang, W.-L. Chang, S.-T. Chang, and S.-S. Cheng, "Antibacterial activities of plant essential oils against Legionella pneumophila," Water Research, vol. 42, no. 1-2, pp. 278–286, 2008.
6. Marongiu, A. Piras, S. Porcedda et al., "Supercritical CO_2 extract of Cinnamomum zeylanicum: chemical characterization and antityrosinase activity," Journal of Agricultural and Food Chemistry, vol. 55, no. 24, pp. 10022–10027, 2007.
7. S.-T. Chou, W.-L. Chang, C.-T. Chang, S.-L. Hsu, Y.-C. Lin, and Y. Shih, "Cinnamomum cassia Essential Oil inhibits α-MSH-induced melanin production and oxidative stress in murine B16 melanoma cells," International Journal of Molecular Sciences, vol. 14, no. 9, pp. 19186–19201, 2013.
8. G.-I. Nonaka, S. Morimoto, and I. Nishioka, "Tannins and related compounds. Part 13. Isolation and structures of trimeric, tetrameric, and pentameric proanthicyanidins from cinnamon," Journal of the Chemical Society, Perkin Transactions 1, pp. 2139–2145, 1983.
9. R. A. Anderson, C. L. Broadhurst, M. M. Polansky et al., "Isolation and characterization of polyphenol type-A polymers from cinnamon with insulin-like biological activity," Journal of Agricultural and Food Chemistry, vol. 52, no. 1, pp. 65–70, 2004.
10. X. Peng, K.-W. Cheng, J. Ma et al., "Cinnamon bark proanthocyanidins as reactive carbonyl scavengers to prevent the formation of advanced glycation endproducts," Journal of Agricultural and Food Chemistry, vol. 56, no. 6, pp. 1907–1911, 2008.
11. T. Tanaka, Y. Matsuo, Y. Yamada, and I. Kouno, "Structure of polymeric polyphenols of cinnamon bark deduced from condensation products of cinnamaldehyde with catechin and procyanidins," Journal of Agricultural and Food Chemistry, vol. 56, no. 14, pp. 5864–5870, 2008.
12. K. R. Määttä-Riihinen, M. P. Kähkönen, A. R. Törrönen, and I. M. Heinonen, "Catechins and procyanidins in berries of vaccinium species and their antioxidant activity," Journal of Agricultural and Food Chemistry, vol. 53, no. 22, pp. 8485–8491, 2005.
13. V. Jakhetia, R. Patel, P. Khatri et al., "Cinnamon: a pharmacological review," Journal of Advanced Scientific Research, vol. 1, no. 2, pp. 19–12, 2010.
14. G. T. Wondrak, N. F. Villeneuve, S. D. Lamore, A. S. Bause, T. Jiang, and D. D. Zhang, "The cinnamon-derived dietary factor cinnamic aldehyde activates the Nrf2-dependent antioxidant response in human epithelial colon cells," Molecules, vol. 15, no. 5, pp. 3338–3355, 2010.
15. N. Hossein, Z. Zahra, M. Abolfazl, S. Mahdi, and K. Ali, "Effect of Cinnamon zeylanicum essence and distillate on the clotting time," Journal of Medicinal Plants Research, vol. 7, no. 19, pp. 1339–1343, 2013.
16. St. Minich and L. Msom, Chinese Herbal Medicine in Women's Health, Women's Health, 2008.

17. S.-T. Chang, P.-F. Chen, and S.-C. Chang, "Antibacterial activity of leaf essential oils and their constituents from Cinnamomum osmophloeum," Journal of Ethnopharmacology, vol. 77, no. 1, pp. 123–127, 2001.
18. P. Hili, C. S. Evans, and R. G. Veness, "Antimicrobial action of essential oils: the effect of dimethylsulphoxide on the activity of cinnamon oil," Letters in Applied Microbiology, vol. 24, no. 4, pp. 269–275, 1997.
19. N. Matan, H. Rimkeeree, A. J. Mawson, P. Chompreeda, V. Haruthaithanasan, and M. Parker, "Antimicrobial activity of cinnamon and clove oils under modified atmosphere conditions," International Journal of Food Microbiology, vol. 107, no. 2, pp. 180–185, 2006.
20. L. B. Gende, I. Floris, R. Fritz, and M. J. Eguaras, "Antimicrobial activity of cinnamon (Cinnamomum zeylanicum) essential oil and its main components against paenibacillus larvae from argentine," Bulletin of Insectology, vol. 61, no. 1, pp. 1–4, 2008.
21. S.-Y. Wang, P.-F. Chen, and S.-T. Chang, "Antifungal activities of essential oils and their constituents from indigenous cinnamon (Cinnamomum osmophloeum) leaves against wood decay fungi," Bioresource Technology, vol. 96, no. 7, pp. 813–818, 2005.
22. J. Mancini-Filho, A. van-Koiij, D. A. P. Mancini, F. F. Cozzolino, and R. P. Torres, "Antioxidant activity of cinnamon (Cinnamomum zeylanicum, breyne) extracts," Bollettino Chimico Farmaceutico, vol. 137, no. 11, pp. 443–447, 1998.
23. S. Shobana and K. Akhilender Naidu, "Antioxidant activity of selected Indian spices," Prostaglandins Leukotrienes and Essential Fatty Acids, vol. 62, no. 2, pp. 107–110, 2000.
24. S. Mathew and T. E. Abraham, "Studies on the antioxidant activities of cinnamon (Cinnamomum verum) bark extracts, through various in vitro models," Food Chemistry, vol. 94, no. 4, pp. 520–528, 2006.
25. S. Mathew and T. E. Abraham, "In vitro antioxidant activity and scavenging effects of Cinnamomum verum leaf extract assayed by different methodologies," Food and Chemical Toxicology, vol. 44, no. 2, pp. 198–206, 2006.
26. N. Kim, H. Sung, and W. Kim, "Effect of solvents and some extraction conditions on antioxidant activity in cinnamon extracts," Korean Journal of Food Science and Technology, vol. 25, no. 3, pp. 204–209, 1993.
27. S. H. Kim, S. H. Hyun, and S. Y. Choung, "Anti-diabetic effect of cinnamon extract on blood glucose in db/db mice," Journal of Ethnopharmacology, vol. 104, no. 1-2, pp. 119–123, 2006.
28. S. Prabuseenivasan, M. Jayakumar, and S. Ignacimuthu, "In vitro antibacterial activity of some plant essential oils," BMC Complementary and Alternative Medicine, vol. 6, article 39, 2006.
29. Q. Jia, X. Liu, X. Wu et al., "Hypoglycemic activity of a polyphenolic oligomer-rich extract of Cinnamomum parthenoxylon bark in normal and streptozotocin-induced diabetic rats," Phytomedicine, vol. 16, no. 8, pp. 744–750, 2009.
30. K. J. Jarvill-Taylor, R. A. Anderson, and D. J. Graves, "A hydroxychalcone derived from cinnamon functions as a mimetic for insulin in 3T3-L1 adipocytes," Journal of the American College of Nutrition, vol. 20, no. 4, pp. 327–336, 2001.
31. Z. Lu, Q. Jia, R. Wang et al., "Hypoglycemic activities of A- and B-type procyanidin oligomer-rich extracts from different Cinnamon barks," Phytomedicine, vol. 18, no. 4, pp. 298–302, 2011.
32. P. Subash Babu, S. Prabuseenivasan, and S. Ignacimuthu, "Cinnamaldehyde—a potential antidiabetic agent," Phytomedicine, vol. 14, no. 1, pp. 15–22, 2007.
33. S. Onderoglu, S. Sozer, K. M. Erbil, R. Ortac, and F. Lermioglu, "The evaluation of long-term effects of cinnamon bark and olive leaf on toxicity induced by streptozotocin administration to rats," Journal of Pharmacy and Pharmacology, vol. 51, no. 11, pp. 1305–1312, 1999.
34. L. K. Chao, K.-F. Hua, H.-Y. Hsu, S.-S. Cheng, J.-Y. Liu, and S.-T. Chang, "Study on the Antiinflammatory activity of essential oil from leaves of Cinnamomum

osmophloeum," Journal of Agricultural and Food Chemistry, vol. 53, no. 18, pp. 7274–7278, 2005.

35. Y.-T. Tung, M.-T. Chua, S.-Y. Wang, and S.-T. Chang, "Anti-inflammation activities of essential oil and its constituents from indigenous cinnamon (Cinnamomum osmophloeum) twigs," Bioresource Technology, vol. 99, no. 9, pp. 3908–3913, 2008.

36. Y.-T. Tung, P.-L. Yen, C.-Y. Lin, and S.-T. Chang, "Anti-inflammatory activities of essential oils and their constituents from different provenances of indigenous cinnamon (Cinnamomum osmophloeum) leaves," Pharmaceutical Biology, vol. 48, no. 10, pp. 1130–1136, 2010.

37. I.-K. Park, J.-Y. Park, K.-H. Kim et al., "Nematicidal activity of plant essential oils and components from garlic (Allium sativum) and cinnamon (Cinnamomum verum) oils against the pine wood nematode (Bursaphelenchus xylophilus)," Nematology, vol. 7, no. 5, pp. 767–774, 2005.

38. J.-O. Kong, S.-M. Lee, Y.-S. Moon, S.-G. Lee, and Y.-J. Ahn, "Nematicidal activity of cassia and cinnamon oil compounds and related compounds toward Bursaphelenchus xylophilus (Nematoda: Parasitaphelenchidae)," Journal of Nematology, vol. 39, no. 1, pp. 31–36, 2007.

39. S.-S. Cheng, J.-Y. Liu, K.-H. Tsai, W.-J. Chen, and S.-T. Chang, "Chemical composition and mosquito larvicidal activity of essential oils from leaves of different Cinnamomum osmophloeum provenances," Journal of Agricultural and Food Chemistry, vol. 52, no. 14, pp. 4395–4400, 2004.

40. S.-S. Cheng, J.-Y. Liu, C.-G. Huang, Y.-R. Hsui, W.-J. Chen, and S.-T. Chang, "Insecticidal activities of leaf essential oils from Cinnamomum osmophloeum against three mosquito species," Bioresource Technology, vol. 100, no. 1, pp. 457–464, 2009.

41. V. Dhulasavant, S. Shinde, M. Pawar, and N. S. Naikwade, "Antihyperlipidemic activity of Cinnamomum tamala Nees. on high cholesterol diet induced Hyperlipidemia," International Journal of PharmTech Research, vol. 2, no. 4, pp. 2517–2521, 2010.

42. K. A. Amin and T. M. A. El-Twab, "Oxidative markers, nitric oxide and homocysteine alteration in hypercholesterolimic rats: role of atorvastatine and cinnamon," International Journal of Clinical and Experimental Medicine, vol. 2, no. 3, pp. 254–265, 2009.

43. T. Bandara, I. Uluwaduge, and E. R. Jansz, "Bioactivity of cinnamon with special emphasis on diabetes mellitus: a review," International Journal of Food Sciences and Nutrition, vol. 63, no. 3, pp. 380–386, 2012.

44. J. Lu, K. Zhang, S. Nam, R. A. Anderson, R. Jove, and W. Wen, "Novel angiogenesis inhibitory activity in cinnamon extract blocks VEGFR2 kinase and downstream signaling," Carcinogenesis, vol. 31, no. 3, pp. 481–488, 2010.

45. H.-K. Kwon, W. K. Jeon, J.-S. Hwang et al., "Cinnamon extract suppresses tumor progression by modulating angiogenesis and the effector function of $CD8^+$ T cells," Cancer Letters, vol. 278, no. 2, pp. 174–182, 2009.

46. H.-K. Kwon, J.-S. Hwang, J.-S. So et al., "Cinnamon extract induces tumor cell death through inhibition of NFκB and AP1," BMC Cancer, vol. 10, no. 1, article 392, 2010.

47. S. J. Koppikar, A. S. Choudhari, S. A. Suryavanshi, S. Kumari, S. Chattopadhyay, and R. Kaul-Ghanekar, "Aqueous Cinnamon Extract (ACE-c) from the bark of Cinnamomum cassia causes apoptosis in human cervical cancer cell line (SiHa) through loss of mitochondrial membrane potential," BMC Cancer, vol. 10, no. 1, article 210, 2010.

48. K. Aneja, R. Joshi, and C. Sharma, "Antimicrobial activity of dalchini (Cinnamomum zeylanicum bark) extracts on some dental caries pathogens," Journal of Pharmacy Research, vol. 2, no. 9, pp. 1387–1390, 2009.

49. Gupta, A. Kumari, A. P. Garg, R. Catanzaro, and F. Marotta, "Comparative study of cinnamon oil and clove oil on some oral microbiota," Acta Bio-Medica: Atenei Parmensis, vol. 82, no. 3, article 197, 2011.

50. U. M. Senanayake, T. H. Lee, and R. B. H. Wills, "Volatile constituents of cinnamon (Cinnamomum zeylanicum) oils," Journal of Agricultural and Food Chemistry, vol. 26, no. 4, pp. 822–824, 1978.

51. G. Singh, S. Maurya, M. P. deLampasona, and C. A. N. Catalan, "A comparison of chemical, antioxidant and antimicrobial studies of cinnamon leaf and bark volatile oils, oleoresins and their constituents," Food and Chemical Toxicology, vol. 45, no. 9, pp. 1650–1661, 2007.
52. M. Suhaj, "Spice antioxidants isolation and their antiradical activity: a review," Journal of Food Composition and Analysis, vol. 19, no. 6-7, pp. 531–537, 2006.
53. B. Halliwell, "Free radicals and antioxidants—quo vadis?" Trends in Pharmacological Sciences, vol. 32, no. 3, pp. 125–130, 2011.
54. B. Halliwell, "Reactive species and antioxidants. Redox biology is a fundamental theme of aerobic life," Plant Physiology, vol. 141, no. 2, pp. 312–322, 2006.
55. J. N. Dhuley, "Anti-oxidant effects of cinnamon (Cinnamomum verum) bark and greater cardamon (Amomum subulatum) seeds in rats fed high fat diet," Indian Journal of Experimental Biology, vol. 37, no. 3, pp. 238–242, 1999.
56. S. J. Kim, D. Han, K. D. Moon, and J. S. Rhee, "Measurement of superoxide dismutase-like activity of natural antioxidants," Bioscience, Biotechnology, and Biochemistry, vol. 59, no. 5, pp. 822–826, 1995.
57. M. Okawa, J. Kinjo, T. Nohara, and M. Ono, "DPPH (1,1-diphenyl-2-Picrylhydrazyl) radical scavenging activity of flavonoids obtained from some medicinal plants," Biological and Pharmaceutical Bulletin, vol. 24, no. 10, pp. 1202–1205, 2001.
58. H.-S. Lee, B.-S. Kim, and M.-K. Kim, "Suppression effect of Cinnamomum cassia bark-derived component on nitric oxide synthase," Journal of Agricultural and Food Chemistry, vol. 50, no. 26, pp. 7700–7703, 2002.
59. C.-C. Lin, S.-J. Wu, C.-H. Chang, and L.-T. Ng, "Antioxidant activity of Cinnamomum cassia," Phytotherapy Research, vol. 17, no. 7, pp. 726–730, 2003.
60. M. A. Murcia, I. Egea, F. Romojaro, P. Parras, A. M. Jiménez, and M. Martínez-Tomé, "Antioxidant evaluation in dessert spices compared with common food additives. Influence of irradiation procedure," Journal of Agricultural and Food Chemistry, vol. 52, no. 7, pp. 1872–1881, 2004.
61. S. Chericoni, J. M. Prieto, P. Iacopini, P. Cioni, and I. Morelli, "In vitro activity of the essential oil of Cinnamomum zeylanicum and eugenol in peroxynitrite-induced oxidative processes," Journal of Agricultural and Food Chemistry, vol. 53, no. 12, pp. 4762–4765, 2005.
62. B. Shan, Y. Z. Cai, M. Sun, and H. Corke, "Antioxidant capacity of 26 spice extracts and characterization of their phenolic constituents," Journal of Agricultural and Food Chemistry, vol. 53, no. 20, pp. 7749–7759, 2005.
63. R. Suganthi, S. Rajamani, M. K. Ravichandran, and C. V. Anuradha, "Effect of food seasoning spices mixture on biomarkers of oxidative stress in tissues of fructose-fed insulin-resistant rats," Journal of Medicinal Food, vol. 10, no. 1, pp. 149–153, 2007.
64. G. K. Jayaprakasha and L. J. M. Rao, "Chemistry, biogenesis, and biological activities of Cinnamomum zeylanicum," Critical Reviews in Food Science and Nutrition, vol. 51, no. 6, pp. 547–562, 2011.
65. S. Geng, Z. Cui, X. Huang, Y. Chen, D. Xu, and P. Xiong, "Variations in essential oil yield and composition during Cinnamomum cassia bark growth," Industrial Crops and Products, vol. 33, no. 1, pp. 248–252, 2011.
66. R. Aravind, T. Aneesh, A. Bindu, and K. Bindu, "Estimation of phenolics and evaluation of antioxidant activity of Cinnamomum malabatrum (Burm. F). Blume," Asian Journal of Research in Chemistry, vol. 5, no. 5, pp. 628–632, 2012.
67. C.-H. Yang, R.-X. Li, and L.-Y. Chuang, "Antioxidant activity of various parts of Cinnamomum cassiaextracted with different extraction methods," Molecules, vol. 17, no. 6, pp. 7294–7304, 2012.
68. S. Kumar, N. Vasudeva, and S. Sharma, "GC-MS analysis and screening of antidiabetic, antioxidant and hypolipidemic potential of Cinnamomum tamala oil in streptozotocin induced diabetes mellitus in rats," Cardiovascular Diabetology, vol. 11, no. 1, pp. 1–11, 2012.

69. F.-L. Hsu, W.-H. Li, C.-W. Yu et al., "In vivo antioxidant activities of essential oils and their constituents from leaves of the Taiwanese Cinnamomum osmophloeum," Journal of Agricultural and Food Chemistry, vol. 60, no. 12, pp. 3092–3097, 2012.
70. H. Kordsardouei, M. Barzegar, and M. A. Sahari, "Application of Zataria multiflora Boiss. and Cinnamon zeylanicum essential oils as two natural preservatives in cake," Avicenna Journal of Phytomedicine, vol. 3, no. 3, pp. 238–247, 2013.
71. J. F. Ayala-Zavala, B. Silva-Espinoza, M. Cruz-Valenzuela et al., "Pectin-cinnamon leaf oil coatings add antioxidant and antibacterial properties to fresh-cut peach," Flavour and Fragrance Journal, vol. 28, no. 1, pp. 39–45, 2013.
72. Y. Dong, N. Lu, and R. B. Cole, "Analysis of the volatile organic compounds in Cinnamomum cassia bark by direct sample introduction thermal desorption gas chromatography-mass spectrometry," Journal of Essential Oil Research, vol. 25, no. 6, pp. 458–463, 2013.
73. Y. Shi, Q.-X. Chen, Q. Wang, K.-K. Song, and L. Qiu, "Inhibitory effects of cinnamic acid and its derivatives on the diphenolase activity of mushroom (Agaricus bisporus) tyrosinase," Food Chemistry, vol. 92, no. 4, pp. 707–712, 2005.
74. L. Georgiev, M. Chochkova, I. Totseva et al., "Anti-tyrosinase, antioxidant and antimicrobial activities of hydroxycinnamoylamides," Medicinal Chemistry Research, vol. 22, no. 9, pp. 4173–4182, 2013.
75. S. Parvez, M. Kang, H.-S. Chung, and H. Bae, "Naturally occurring tyrosinase inhibitors: mechanism and applications in skin health, cosmetics and agriculture industries," Phytotherapy Research, vol. 21, no. 9, pp. 805–816, 2007.
76. J. Lin, A. R. Opoku, M. Geheeb-Keller et al., "Preliminary screening of some traditional zulu medicinal plants for anti-inflammatory and anti-microbial activities," Journal of Ethnopharmacology, vol. 68, no. 1–3, pp. 267–274, 1999.
77. N. Mascolo, F. Capasso, A. Menghini, and M. P. Fasulo, "Biological screening of Italian medicinal plants for anti-inflammatory activity," Phytotherapy Research, vol. 1, no. 1, pp. 28–31, 1987.
78. H. Tunón, C. Olavsdotter, and L. Bohlin, "Evaluation of anti-inflammatory activity of some Swedish medicinal plants. Inhibition of prostaglandin biosynthesis and PAF-induced exocytosis," Journal of Ethnopharmacology, vol. 48, no. 2, pp. 61–76, 1995.
79. R. W. Li, G. David Lin, S. P. Myers, and D. N. Leach, "Anti-inflammatory activity of Chinese medicinal vine plants," Journal of Ethnopharmacology, vol. 85, no. 1, pp. 61–67, 2003.
80. S. Sosa, M. J. Balick, R. Arvigo et al., "Screening of the topical anti-inflammatory activity of some Central American plants," Journal of Ethnopharmacology, vol. 81, no. 2, pp. 211–215, 2002.
81. N. Matu and J. van Staden, "Antibacterial and anti-inflammatory activities of some plants used for medicinal purposes in Kenya," Journal of Ethnopharmacology, vol. 87, no. 1, pp. 35–41, 2003.
82. García-Lafuente, E. Guillamón, A. Villares, M. A. Rostagno, and J. A. Martínez, "Flavonoids as anti-inflammatory agents: implications in cancer and cardiovascular disease," Inflammation Research, vol. 58, no. 9, pp. 537–552, 2009.
83. H. P. Kim, K. H. Son, H. W. Chang, and S. S. Kang, "Anti-inflammatory plant flavonoids and cellular action mechanisms," Journal of Pharmacological Sciences, vol. 96, no. 3, pp. 229–245, 2004.
84. T. Guardia, A. E. Rotelli, A. O. Juarez, and L. E. Pelzer, "Anti-inflammatory properties of plant flavonoids. Effects of rutin, quercetin and hesperidin on adjuvant arthritis in rat," Farmaco, vol. 56, no. 9, pp. 683–687, 2001.
85. G. Stoner and L.-S. Wang, "Natural products as anti-inflammatory agents," in Obesity, Inflammation and Cancer, pp. 341–361, Springer, 2013.
86. N. Cho, K. Y. Lee, J. Huh et al., "Cognitive-enhancing effects of Rhus verniciflua bark extract and its active flavonoids with neuroprotective and anti-inflammatory activities," Food and Chemical Toxicology, vol. 58, pp. 355–361, 2013.

87. S. H. Lee, S. Y. Lee, D. J. Son et al., "Inhibitory effect of 2′-hydroxycinnamaldehyde on nitric oxide production through inhibition of NF-κB activation in RAW 264.7 cells," Biochemical Pharmacology, vol. 69, no. 5, pp. 791–799, 2005.
88. T. Yu, S. Lee, W. S. Yang et al., "The ability of an ethanol extract of Cinnamomum cassia to inhibit Src and spleen tyrosine kinase activity contributes to its anti-inflammatory action," Journal of Ethnopharmacology, vol. 139, no. 2, pp. 566–573, 2012.
89. H. S. Youn, J. K. Lee, Y. J. Choi et al., "Cinnamaldehyde suppresses toll-like receptor 4 activation mediated through the inhibition of receptor oligomerization," Biochemical Pharmacology, vol. 75, no. 2, pp. 494–502, 2008.
90. S.-H. Hwang, Y. G. Choi, M.-Y. Jeong, Y.-M. Hong, J.-H. Lee, and S. Lim, "Microarray analysis of gene expression profile by treatment of Cinnamomi ramulus in lipopolysaccharide-stimulated BV-2 cells," Gene, vol. 443, no. 1-2, pp. 83–90, 2009.
91. J.-W. Hong, G.-E. Yang, Y. B. Kim, S. H. Eom, J.-H. Lew, and H. Kang, "Anti-inflammatory activity of cinnamon water extract in vivo and in vitro LPS-induced models," BMC Complementary and Alternative Medicine, vol. 12, no. 1, article 237, 2012.
92. S.-M. Yu, F.-N. Ko, T.-S. Wu, J.-Y. Lee, and C.-M. Teng, "Cinnamophilin, a novel thromboxane A_2receptor antagonist, isolated from Cinnamomum philippinense," European Journal of Pharmacology, vol. 256, no. 1, pp. 85–91, 1994.
93. E.-J. Lee, H.-Y. Chen, Y.-C. Hung et al., "Therapeutic window for cinnamophilin following oxygen-glucose deprivation and transient focal cerebral ischemia," Experimental Neurology, vol. 217, no. 1, pp. 74–83, 2009.
94. K. S. Panickar, M. M. Polansky, D. J. Graves, J. F. Urban, and R. A. Anderson, "A procyanidin type A trimer from cinnamon extract attenuates glial cell swelling and the reduction in glutamate uptake following ischemia-like injury in vitro," Neuroscience, vol. 202, pp. 87–98, 2012.
95. M. C. de Rijk, L. J. Launer, K. Berger et al., "Prevalence of Parkinson's disease in Europe: a collaborative study of population-based cohorts," Neurology, vol. 54, no. 11, supplement 5, pp. S21–S23, 2000.
96. V. Bonifati, B. A. Oostra, and P. Heutink, "Linking DJ-1 to neurodegeneration offers novel insights for understanding the pathogenesis of Parkinson's disease," Journal of Molecular Medicine, vol. 82, no. 3, pp. 163–174, 2004.
97. S. Brahmachari, A. Jana, and K. Pahan, "Sodium benzoate, a metabolite of cinnamon and a food additive, reduces microglial and astroglial inflammatory responses," The Journal of Immunology, vol. 183, no. 9, pp. 5917–5927, 2009.
98. S. Khasnavis and K. Pahan, "Sodium benzoate, a metabolite of cinnamon and a food additive, upregulates neuroprotective parkinson disease protein DJ-1 in astrocytes and neurons," Journal of Neuroimmune Pharmacology, vol. 7, no. 2, pp. 424–435, 2012.
99. Jana, K. K. Modi, A. Roy, J. A. Anderson, R. B. van Breemen, and K. Pahan, "Up-regulation of neurotrophic factors by cinnamon and its metabolite sodium benzoate: therapeutic implications for neurodegenerative disorders," Journal of Neuroimmune Pharmacology, vol. 8, no. 3, pp. 739–755, 2013.
100. Frydman-Marom, A. Levin, D. Farfara et al., "Orally administrated cinnamon extract reduces β-amyloid oligomerization and corrects cognitive impairment in Alzheimer's disease animal models," PLoS ONE, vol. 6, no. 1, Article ID e16564, 2011.
101. W. Peterson, R. C. George, F. Scaramozzino et al., "Cinnamon extract inhibits tau aggregation associated with alzheimer's disease in vitro," Journal of Alzheimer's Disease, vol. 17, no. 3, pp. 585–597, 2009.
102. Khan, N. A. Bryden, M. M. Polansky, and R. A. Anderson, "Insulin potentiating factor and chromium content of selected foods and spices," Biological Trace Element Research, vol. 24, no. 3, pp. 183–188, 1990.
103. S. M. Blevins, M. J. Leyva, J. Brown, J. Wright, R. H. Scofield, and C. E. Aston, "Effect of cinnamon on glucose and lipid levels in non-insulin-dependent type 2 diabetes," Diabetes Care, vol. 30, no. 9, pp. 2236–2237, 2007.

104. Khan, M. Safdar, M. M. A. Khan, K. N. Khattak, and R. A. Anderson, "Cinnamon improves glucose and lipids of people with type 2 diabetes," Diabetes Care, vol. 26, no. 12, pp. 3215–3218, 2003.
105. Mang, M. Wolters, B. Schmitt et al., "Effects of a cinnamon extract on plasma glucose, HbA1c, and serum lipids in diabetes mellitus type 2," European Journal of Clinical Investigation, vol. 36, no. 5, pp. 340–344, 2006.
106. P. Crawford, "Effectiveness of cinnamon for lowering hemoglobin A1C in patients with type 2 diabetes: a randomized, controlled trial," The Journal of the American Board of Family Medicine, vol. 22, no. 5, pp. 507–512, 2009.
107. M. Safdar, A. Khan, M. M. A. K. Khattak, and M. Siddique, "Effect of various doses of cinnamon on blood glucose in diabetic individuals," Pakistan Journal of Nutrition, vol. 3, no. 5, pp. 268–272, 2004.
108. L. Broadhurst, M. M. Polansky, and R. A. Anderson, "Insulin-like biological activity of culinary and medicinal plant aqueous extracts in vitro," Journal of Agricultural and Food Chemistry, vol. 48, no. 3, pp. 849–852, 2000.
109. R. A. Anderson, C. L. Broadhurst, and M. M. Polansky, "Isolation and characterization of chalcone polymers from cinnamon with insulin like biological activities," American Journal of Clinical Nutrition, vol. 84, no. 3, pp. 1432–1436, 2006.
110. H.-B. Li, C.-C. Wong, K.-W. Cheng, and F. Chen, "Antioxidant properties in vitro and total phenolic contents in methanol extracts from medicinal plants," LWT-Food Science and Technology, vol. 41, no. 3, pp. 385–390, 2008.
111. H. Cao, M. M. Polansky, and R. A. Anderson, "Cinnamon extract and polyphenols affect the expression of tristetraprolin, insulin receptor, and glucose transporter 4 in mouse 3T3-L1 adipocytes," Archives of Biochemistry and Biophysics, vol. 459, no. 2, pp. 214–222, 2007.
112. S. I. Kreydiyyeh, J. Usta, and R. Copti, "Effect of cinnamon, clove and some of their constituents on the Na^+-K^+-ATPase activity and alanine absorption in the rat jejunum," Food and Chemical Toxicology, vol. 38, no. 9, pp. 755–762, 2000.
113. K. Vanschoonbeek, B. J. W. Thomassen, J. M. Senden, W. K. W. H. Wodzig, and L. J. C. van Loon, "Cinnamon supplementation does not improve glycemic control in postmenopausal type 2 diabetes patients," The Journal of Nutrition, vol. 136, no. 4, pp. 977–980, 2006.
114. S.-C. Lee, W.-X. Xu, L.-Y. Lin, J.J.. Yang, and C.-T. Liu, "Chemical composition and hypoglycemic and pancreas-protective effect of leaf essential oil from indigenous cinnamon (Cinnamomum osmophloeumKanehira)," Journal of Agricultural and Food Chemistry, vol. 61, no. 20, pp. 4905–4913, 2013.
115. R. Becerril, R. Gómez-Lus, P. Goñi, P. López, and C. Nerín, "Combination of analytical and microbiological techniques to study the antimicrobial activity of a new active food packaging containing cinnamon or oregano against E. coli and S. aureus," Analytical and Bioanalytical Chemistry, vol. 388, no. 5-6, pp. 1003–1011, 2007.
116. P. Goñi, P. López, C. Sánchez, R. Gómez-Lus, R. Becerril, and C. Nerín, "Antimicrobial activity in the vapour phase of a combination of cinnamon and clove essential oils," Food Chemistry, vol. 116, no. 4, pp. 982–989, 2009.
117. H. Parthasarathy and S. Thombare, "Evaluation of antimicrobial activity of Azadirachta indica, Syzygium aromaticum and Cinnamomum zeyalnicumagainst oral microflora," Asian Journal of Experimental Sciences, vol. 27, no. 2, pp. 13–16, 2013.
118. B.-M. Kwon, S.-H. Lee, Y.-K. Cho et al., "Synthesis and biological activity of cinnamaldehydes as angiogenesis inhibitors," Bioorganic and Medicinal Chemistry Letters, vol. 7, no. 19, pp. 2473–2476, 1997.
119. H.-W. Jeong, D. C. Han, K.-H. Son et al., "Antitumor effect of the cinnamaldehyde derivative CB403 through the arrest of cell cycle progression in the G_2/M phase," Biochemical Pharmacology, vol. 65, no. 8, pp. 1343–1350, 2003.

120. M. Cabello, W. B. Bair III, S. D. Lamore et al., "The cinnamon-derived Michael acceptor cinnamic aldehyde impairs melanoma cell proliferation, invasiveness, and tumor growth," Free Radical Biology and Medicine, vol. 46, no. 2, pp. 220–231, 2009.
121. S.-H. Fang, Y. K. Rao, and Y.-M. Tzeng, "Cytotoxic effect of trans-cinnamaldehyde from cinnamomum osmophloeum leaves on Human cancer cell lines," International Journal of Applied Science and Engineering, vol. 2, no. 2, pp. 136–147, 2004.
122. S. Bhattacharjee, T. Rana, and A. Sengupta, "Inhibition of lipid peroxidation and enhancement of GST activity by cardamom and cinnamon during chemically induced colon carcinogenesis in Swiss albino mice," Asian Pacific Journal of Cancer Prevention, vol. 8, no. 4, pp. 578–582, 2007.
123. J. S. Hwa, Y. C. Jin, Y. S. Lee et al., "2-Methoxycinnamaldehyde from Cinnamomum cassia reduces rat myocardial ischemia and reperfusion injury in vivo due to HO-1 induction," Journal of Ethnopharmacology, vol. 139, no. 2, pp. 605–615, 2012.
124. Song, H. Li, J. Sun, and S. Wang, "Protective effects of cinnamic acid and cinnamic aldehyde on isoproterenol-induced acute myocardial ischemia in rats," Journal of Ethnopharmacology, vol. 150, no. 1, pp. 125–130, 2013.
125. S.-M. Yu, T.-S. Wu, and C.-M. Teng, "Pharmacological characterization of cinnamophilin, a novel dual inhibitor of thromboxane synthase and thromboxane A2 receptor," British Journal of Pharmacology, vol. 111, no. 3, pp. 906–912, 1994.
126. P. Jurasz, D. Alonso-Escolano, and M. W. Radomski, "Platelet-cancer interactions: mechanisms and pharmacology of tumour cell-induced platelet aggregation," British Journal of Pharmacology, vol. 143, no. 7, pp. 819–826, 2004.
127. Nie, M. Che, A. Zacharek et al., "Differential expression of thromboxane synthase in prostate carcinoma: role in tumor cell motility," The American Journal of Pathology, vol. 164, no. 2, pp. 429–439, 2004.
128. F.-N. Ko, S.-M. Yu, Y.-F. Kang, and C.-M. Teng, "Characterization of the thromboxane (TP-) receptor subtype involved in proliferation in cultured vascular smooth muscle cells of rat," British Journal of Pharmacology, vol. 116, pp. 1801–1808, 1995.
129. M. Harada and S. Yano, "Pharmacological studies on Chinese cinnamon. II. Effects of cinnamaldehyde on the cardiovascular and digestive systems," Chemical and Pharmaceutical Bulletin, vol. 23, no. 5, pp. 941–947, 1975.
130. M. Harada, Y. Hirayama, and R. Yamazaki, "Pharmacological studies on Chinese cinnamon. V. Catecholamine releasing effect of cinnamaldehyde in dogs," Journal of Pharmacobio-Dynamics, vol. 5, no. 8, pp. 539–546, 1982.
131. Y.-L. Xue, H.-X. Shi, F. Murad, and K. Bian, "Vasodilatory effects of cinnamaldehyde and its mechanism of action in the rat aorta," Vascular health and risk management, vol. 7, pp. 273–280, 2011.
132. M. El-Bassossy, A. Fahmy, and D. Badawy, "Cinnamaldehyde protects from the hypertension associated with diabetes," Food and Chemical Toxicology, vol. 49, no. 11, pp. 3007–3012, 2011.
133. S. Rahman, H. Begum, Z. Rahman, F. Ara, M. J. Iqbal, and A. K. M. Yousuf, "Effect of cinnamon (Cinnamomum cassia) as a lipid lowering agent on hypercholesterolemic rats," Journal of Enam Medical College, vol. 3, no. 2, pp. 94–98, 2013.
134. M. Ciftci, U. G. Simsek, A. Yuce, O. Yilmaz, and B. Dalkilic, "Effects of dietary antibiotic and cinnamon oil supplementation on antioxidant enzyme activities, cholesterol levels and fatty acid compositions of serum and meat in broiler chickens," Acta Veterinaria Brno, vol. 79, no. 1, pp. 33–40, 2010.
135. X. Peng, J. Ma, J. Chao et al., "Beneficial effects of cinnamon proanthocyanidins on the formation of specific advanced glycation endproducts and methylglyoxal-induced impairment on glucose consumption," Journal of Agricultural and Food Chemistry, vol. 58, no. 11, pp. 6692–6696, 2010.
136. K. Jayaprakasha, L. J. Rao, and K. K. Sakariah, "Chemical composition of volatile oil from Cinnamomum zeylanicum buds," Zeitschrift fur Naturforschung C Journal of Biosciences, vol. 57, no. 11-12, pp. 990–993, 2002.

CHAPTER 19

Morphological Characterisation of Some Important Indian Garcinia Species

Utpala Parthasarathy and O. P. Nandakishore

Indian Institute of Spices Research, Calicut, Kerala 673012, India

ABSTRACT

The genus Garcinia has over 200 species distributed in the tropics of the world. About 35 species occur in India, many of which are endemic and economically important with immense medicinal properties. However, lack of awareness, coupled with habitat destruction, leads to genetic erosion of this forest resource and many species are threatened. The Indian Institute of Spices Research (IISR), Calicut, has Garcinia genetic resources' collection of 15 species of Western Ghats and Eastern Himalaya species. The morphological characterisation of the species of these two different eco systems indicates that there are variations within the species of the same ecosystem while there are similarities in the species of two different ecosystems.

1. INTRODUCTION

The genus Garcinia (Family: Clusiaceae) consists of over 200 species distributed in the tropics of the world chiefly in Asia, Africa, and Polynesia. They are evergreen polygamous trees, shrubs, and herbs. About 35 species are reported to exist in India, many of which are endemic and economically important with immense medicinal properties [1].

In India, species of Garcinia grow extensively in semiwild state, in the Konkan region of Maharashtra, Goa, coastal areas of Karnataka and Kerala, and evergreen forests of Assam, Khasi, Jantia hills, Nagaland, West Bengal, and Gujarat. In Malabar and Konkan regions of Southern India, they are used in garnishing curries and also as a replacement for tamarind. In North Eastern India, the sundried slices of the fruits are used for culinary purposes and as folk medicine. Some species like Garcinia cambogia, G. indica, and G. cowa are cultivated in certain parts of India. G. pedunculata, G. kydia, G. cowa, and G. lanceaefolia are the most important species in north eastern parts of India. Many species of Garcinia have fruit with edible arils and are eaten locally. The best-known species is the mangosteen (G. mangostana), which is now cultivated throughout Southeast Asia and other tropical countries. The seeds of G. indica fruits yield valuable edible fat known as kokum butter. The fruits of Garcinia are a food source for several animals [2]. Most species in Garcinia are known for their gum resin which is used as purgative or cathartic. Fruits of some Garcinia species are also one of the richest sources of red pigments in the plant kingdom. Fruit and syrup of G. indica are very popular in Konkan region and are antioxidant and antibacterial [3].

Garcinia is the source for a natural diet ingredient (-) hydroxycitric acid. HCA (1,2dihydroxypropane-1,2,3-tricarboxylic acid) which is an antiobesity compound is present in the fruit rind and leaves of Garcinia and is known to inhibit lipid and fatty acid synthesis in living systems [4]. HCA is also a hypocholesterolemic agent [5–7]. On a dry weight basis, HCA constitutes about 20–30% of the fruit. Lack of knowledge, coupled with habitat destruction, leads to genetic erosion of this forest resource and many species are threatened [8]. They need to be studied and conserved. In this study, we tried to characterize the morphological features of the species collected from Western Ghats and Himalayan Foot hills. Nine species studied here are the most common species of the two ecosystems, but the awareness about the crop or its medicinal value is very less. Most Garcinia species occur only as natural populations and are known only locally. Many Garcinia species have edible arils and are eaten locally. Some species fruits are highly esteemed in one region but are unknown just a few hundred kilometers away. Perhaps, trees are cut due to lack of awareness and popularization of importance of Garcinia will help in conserving the local populations of this genus. Here it is worth to mention that the climatic parameters of both ecosystems are almost the same specially the altitude and rainfall pattern. The altitude varies from 100 to 600 MLS, while the annual rainfall varies from 1500 to 4500 mm.

2. METHODOLOGY

The morphological features of the Indian Garcinia were observed for mature and bearing trees.

A total of 9 species were studied here from two distinct geographical locations, namely, Western Ghats and Northeast Himalayan foothills.

Species selected for this study were G. indica, G. gummi-gutta, G. cowa, G. subelliptica and G. mangostana from Western Ghats; G. pedunculata, G. lanceaefolia, and G. kydia from NE Himalayas; and G. xanthochymus from both the ecosystems.

G. gummi-gutta is also known as G. cambogia and G. xanthochymus as G. tinctoria. Garcinia cowa and G. kydiaare considered as synonymous though they show some variations in their morphological features [9].

The plant height was calculated by fixed angle of elevation method, in which the distance from the tree is measured, at which the top of the tree coincides with the tip of the right triangle making an angle of 45° with horizontal. The tree height is calculated by adding the distance between tree and observer to the height of the observer to the position of stretched hands [10].

Dimensions of the plant leaves were measured by placing them on a standard linear graph sheet with each divisions of 1 cm and subdivisions of 1 mm. Fruit and seed sizes were measured using a Vernier caliper. The morphological features of plant leaf and flowers were observed and compared with available literatures.

3. DATASET DESCRIPTION

The dataset associated with this Dataset Paper consists of 26 items which are described as follows.

Dataset Item 1 (Table). Plant morphology of 9 Indian Garcinia species. The average plant height was calculated for the bearing plants. Garcinia shows a variation of height from 8 m to 25 m. The important feature of Garciniaplants is the copious exudates from wounded leaf and barks. The exudates vary from deep yellow to cream-yellow colour. Though red coloured blazes are a common feature in Garcinia, species like G. pedunculata, G. mangostana, and G. kydia do not produce blaze and G. subelliptica is found to have yellowish tender leaves.

Column 1: Species

Column 2: Height (m)

Column 3: Trunk Colour

Column 4: Trunk Texture

Column 5: Branches

Dataset Item 2 (Table). Leaf morphology of 9 Indian Garcinia species. Mature leaves were collected from corresponding trees and observed for morphological features such as dimensions, colour, texture, shape, and venations. Garcinia species are evergreen cauline plants, having opposite decussate foliar arrangement. Garciniaspecies have simple petiolate leaves. The petioles are so small in plants like G. indica, G. subelliptica and G. lanceaefolia that they almost appear to be sessile type leaves. Venation of the leaves is pinnate in all the cases, though the appearance of veins differs. In case of G. cowa, midvein raised abaxially; secondary veins are 12–18 pairs, near margin joining together; tertiary veins are conspicuous on both surfaces. In case of G. kydia,midvein slightly raised abaxially and secondary veins are visible.

Column 1: Species

Column 2: Crown

Column 3: Petiole

Column 4: Venation

Dataset Item 3 (Table). Leaf features of 9 Indian Garcinia species. Leaf margins of Garcinia leaves are entire except G. kydia, G. xanthochymus, and G. subelliptica. These three species have involutes leaf margins. Leaves show variation in shape and lamina size. Leaf shape varies a great from lanceolate to obovate and elliptic.

Column 1: Species

Column 2: Lamina Size (cm)

Column 3: Shape

Column 4: Apex

Column 5: Base

Column 6: Colour and Texture

Dataset Item 4 (Table). Flower morphology of 9 Indian Garcinia species. It is interesting to note that in north eastern region the flowering season is summer (April-June) while in Western Ghats it is spring (December–March). Garcinia shows two types of flowers, tetramerous (4-petaled) and pentamerous (5-petaled).

Column 1: Species

Column 2: Flower Nature

Column 3: Season

Column 4: Location

Dataset Item 5 (Table). Flower parts of 9 Indian Garcinia species. Garcinia flowers are mostly sessile (without flower stalk) except G. kydia which has pedicels of 4–7 mm length. Garcinia flowers except G. subelliptica are actinomorphic. Flower buds are round in shape and vary from 0.7 cm to 1 cm in diameter. All the flowers did not contain nectar. Garcinia also shows variation in flower colour.

Column 1: Species

Column 2: Colour

Column 3: Sepals

Column 4: Petals

Dataset Item 6 (Table). Flower parts of 9 Indian Garcinia species. The stamens of Garcinia flowers are epipetalous. Anthers are mostly sessile, 2 celled, and longitudinally dehiscent. Garcinia flowers are hypogenous.

Column 1: Species

Column 2: Stamens

Column 3: Stigma

Column 4: Ovary

Dataset Item 7 (Image). Flower of the Indian species Garcinia indica.

Dataset Item 8 (Image). Flower of the Indian species Garcinia gummi-gutta.

Dataset Item 9 (Image). Flower of the Indian species Garcinia cowa.

Dataset Item 10 (Image). Flower of the Indian species Garcinia kydia.

Dataset Item 11 (Image). Flower of the Indian species Garcinia lanceaefolia.

Dataset Item 12 (Image). Flower of the Indian species Garcinia pedunculata.

Dataset Item 13 (Image). Flower of the Indian species Garcinia subelliptica.

Dataset Item 14 (Image). Flower of the Indian species Garcinia xanthochymus.

Dataset Item 15 (Image). Flower of the Indian species Garcinia mangostana.

Dataset Item 16 (Table). Fruit morphology of 9 Indian Garcinia species. Garcinia fruits are fleshy berry with smooth fleshy pericarp with high diversity in shape and size. The size varies from 2 to 10 cm in diameter. Arils are acidic and juicy, and the acid content varies from species to species. Garcinia fruits possess an appetising acidic flavour. Fruit size was measured using a Vernier Caliper with horizontal length in diameter.

Column 1: Species

Column 2: Harvest

Column 3: Shape

Column 4: Size (cm)

Column 5: Colour

Column 6: Aril Colour

Dataset Item 17 (Table). Seed morphology of 9 Indian Garcinia species. Seed size is the average length of seeds measured using a Vernier calliper. The number of seeds varies from 1 to 8 in different species.

Column 1: Species

Column 2: Number of Seeds

Column 3: Seed Size (cm)

Column 4: Other Fruit Features

Dataset Item 18 (Image). Fruit of the Indian species Garcinia indica.

Dataset Item 19 (Image). Fruit of the Indian species Garcinia gummi-gutta.

Dataset Item 20 (Image). Fruit of the Indian species Garcinia cowa.

Dataset Item 21 (Image). Fruit of the Indian species Garcinia kydia.

Dataset Item 22 (Image). Fruit of the Indian species Garcinia lanceaefolia.

Dataset Item 23 (Image). Fruit of the Indian species Garcinia pedunculata.

Dataset Item 24 (Image). Fruit of the Indian species Garcinia subelliptica.

Dataset Item 25 (Image). Fruit of the Indian species Garcinia xanthochymus.

Dataset Item 26 (Image). Fruit of the Indian species Garcinia mangostana.

4. CONCLUDING REMARKS

The study indicates that NE Region of India is a good source of diversity for Garcinia as it is the case in Western Ghats. Morphological characterizations of the species of these two different eco systems indicate variation within the species of the same ecosystem and similarities in the species of two different ecosystems. However, the collection surveys indicated that though many species exist, the population density is dangerously low and is reduced to one or two trees in a given location making survival of these species very difficult. This is further aggravated by the fact that seeds fail to produce seedlings due to various physiological and environmental factors making multiplication and maintenance of these species impossible. Though the fruit product ofGarcinia is available in local markets, the awareness of Garcinia as a tree is very rare. In this study, summering of the fruit, leaf, and tree structure has been done which may give awareness about the crop to the people.

ACKNOWLEDGMENTS

The authors acknowledge the Director of IISR, Calicut, Kerala, India, for providing facilities and CSIR for financial support to the SRF.

REFERENCES

1. E. Roberts, Vegetable Materia Medica of India and Ceylon, Bishen Singh Mahendra Pal Singh, Dehra Dun, India, 1984.
2. The Wealth of India, Raw Materials, vol. 4, CSIR, New Delhi, India, 1956.
3. P. S. Negi, G. K. Jayaprakasha, and B. S. Jena, "Antibacterial activity of the extracts from the fruit rinds of Garcinia cowa and Garcinia pedunculata against food borne pathogens and spoilage bacteria," Food Science and Technology, vol. 41, no. 10, pp. 1857–1861, 2008.
4. Y. S. Lewis and S. Neelakantan, "(-)-Hydroxycitric acid-the principal acid in the fruits of Garcinia cambogia desr," Phytochemistry, vol. 4, no. 4, pp. 619–625, 1965.
5. J. M. Lowenstein, "Effect of (-)-hydroxycitrate on fatty acid synthesis by rat liver in vivo," Journal of Biological Chemistry, vol. 246, no. 3, pp. 629–632, 1971.
6. C. Sullivan, J. G. Hamilton, O. N. Miller, and V. R. Wheatley, "Inhibition of lipogenesis in rat liver by (-)-hydroxycitrate," Archives of Biochemistry and Biophysics, vol. 150, no. 1, pp. 183–190, 1972.
7. C. Sullivan and J. Triscari, "Metabolic regulation as a control for lipid disorders. I. Influence of (-) hydroxycitrate on experimentally induced obesity in the rodent," The American Journal of Clinical Nutrition, vol. 30, no. 5, pp. 767–776, 1977.
8. M. Cheek, "Garcinia kola," in IUCN Red List of Threatened Species, IUCN Red List of Threatened Species, 2004, http://www.iucnredlist.org/.

9. J. D. Hooker, Garcinia, vol. 1, Flora of British India, Periodical experts, Delhi, India, 1875.
10. M. Beals, L. Gross, and S. Harrell, "Estimation of tree height: right triangle trigonometry," 2000, http://www.tiem.utk.edu/~gross/bioed/bealsmodules/triangle.html.

CHAPTER 20

Analysis of Plant Growth-Promoting Effects of Fluorescent Pseudomonas Strains Isolated from *Mentha piperita* Rhizosphere and Effects of Their Volatile Organic Compounds on Essential Oil Composition

Maricel V. Santoro, Pablo C. Bogino, Natalia Nocelli, Lorena del Rosario Cappellari, Walter F. Giordano and Erika Banchio*

Department of Biología Molecular, Facultad de Ciencias Exactas, Químicas y Naturales, Universidad Nacional de Río Cuarto, Río Cuarto, Argentina

ABSTRACT

Many species or strains of the genus Pseudomonas have been characterized as plant growth promoting rhizobacteria (PGPR). We used a combination of phenotypic and genotypic techniques to analyze the community of fluorescent Pseudomonas strains in the rhizosphere of commercially grown *Mentha piperita* (peppermint). Biochemical techniques, Amplified rDNA Restriction Analysis (ARDRA), and 16S rRNA gene sequence analysis revealed that the majority of the isolated native fluorescent strains were *P. putida*. Use of two Repetitive Sequence-based PCR (rep-PCR) techniques,

BOX-PCR and ERIC-PCR, allowed us to evaluate diversity among the native strains and to more effectively distinguish among them. PGPR activity was tested for the native strains and reference strain *P. fluorescens* WCS417r. Micropropagated *M. piperita* plantlets were exposed to microbial volatile organic compounds (mVOCs) emitted by the bacterial strains, and plant biomass parameters and production of essential oils (EOs) were measured. mVOCs from 11 of the native strains caused an increase in shoot fresh weight. mVOCs from three native strains (SJ04, SJ25, SJ48) induced changes in *M. pierita* EO composition. The mVOCs caused a reduction of metabolites in the monoterpene pathway, for example menthofuran, and an increase in menthol production. Menthol production is the primary indicator of EO quality. The mVOCs produced by native strains SJ04, SJ25, SJ48, and strain WCS417r were analyzed. The obtained mVOC chromatographic profiles were unique for each of the three native strains analyzed, containing varying hydrocarbon, aromatic, and alogenic compounds. The differential effects of the strains were most likely due to the specific mixtures of mVOCs emitted by each strain, suggesting a synergistic effect occurs among the compounds present.

Keywords: *Mentha*, fluorescent *Pseudomonas*, genotyping, ARDRA, essential oils, microbial volatile organic compounds

INTRODUCTION

Peppermint (*Mentha piperita* L.; family *Labiatae*) is a popular, commonly-used flavoring agent worldwide. Fresh or dried leaves of *Mentha* species are used primarily as condiments. Essential oils (EOs) of these plants, which are produced and stored in glandular hairs, are used as food and beverage flavorings, as fragrances, and as fungicides or insecticides in a variety of pharmaceutical and industrial products (Harrewijn et al., 2001; Ram et al., 2006). EOs are secondary metabolites whose production is associated with primary metabolism and with availability of soil nutrients (Shulka et al., 1992).

Among the many known PGPR genera, *Pseudomonas* has received the most research attention because it is widely distributed in various environments and is easy to culture under laboratory conditions (Palleroni, 2005). Members of the "fluorescent *Pseudomonas* group" produce a water-soluble fluorescent green-yellow pigment termed pseudobactin or pyoverdine (Stanier et al., 1966). This characteristic peptide displays iron-binding activity and is associated with pathogenic or growth-promoting properties (Fuchs et al., 2001; Mehri et al., 2011). Fluorescent *Pseudomonas* strains have been isolated from rhizospheric soils of numerouscropplants, including cotton (Hu et al., 2005), rice (Loaces et al., 2011), banana (Naik et al., 2008), rape (Patten and Glick, 2002), sugar cane (Rameshkumar et al., 2012), wheat, and barley (Mavrodi et al., 2001; Parejko et al., 2012). The two major fluorescent *Pseudomonas* species that display PGPR activity are *P. fluorescens* and *P. putida* (Kloepper and Schroth, 1978). Few studies to date have focused on PGPR isolated from rhizospheres of aromatic plant species because of the presumption that EOs may be released in root exudates and display antimicrobial activity (Chen et al., 2004). We used phenotyping and genotyping techniques to isolate, identify, and characterize fluorescent *Pseudomonas* strains from *M. piperita* rhizosphere of *M. piperita*. We then evaluated PGPR activity of these strains on micropropagated *M. piperita* plants, including mVOC-mediated effects on plant biomass and EO production.

MATERIALS AND METHODS

Sample Collection and Strain Isolation

A total of 20 *M. piperita* plants were collected randomly from a commercial crop (San José) grown in the San Javier department, Córdoba province, Argentina (30°30′00.0″S, 64°30′00.0″W). Samples were placed individually in plastic bags, stored at 4°C, transported and processed within 24 h. Five roots were washed in sterile distilled water to remove loosely adherent soil, and then rotary shaken 15 min in 100 ml sterile phosphate buffer, pH 7.5, to obtain a rhizospheric soil suspension. Serial dilutions were prepared from the suspension, and 0.1 ml of each dilution was spread on Nutrient Agar medium (Laboratorios Britania, Buenos Aires, Argentina) and on King B medium (King et al., 1954) and incubated at 28°C for 48 h. After complete evaporation of buffer from soil, total bacteria and fluorescent bacteria under UV light were counted as colony forming units (CFU) per g rhizospheric soil (Mehnaz et al., 2009). Bacterial strains were purified on King B medium, identified, and characterized. Fifty fluorescent strains were isolated (Table 1), and cryopreserved at −80°C in 20% glycerol (v/v). A single colony from each strain was grown aerobically on LB medium (Luria and Burrous, 1955) and incubated on a rotary shaker (200 rpm) for 24 h at 28°C prior to testing. This isolation from roots was performed four times.

Table 1. Phenotypic characterization of fluorescent *Pseudomonas* strains, based on morphological, physiological, and biochemical evaluation of native strains isolated from the *M. piperita* rhizosphere.

	Colony morphology	Gram	Catalase	Oxidase	UV Fluorescence	Pigmentation		Growth				Hydrolysis				
						King a	King b	4°C	42°C	Nacl 5%	Nacl 6.5%	Tween 80	Starch	Gelatin	Casein	Lecithinase
SJ1	Smooth and gleaming	–	+	+	Green	–	Green	–	±	+	–	–	–	–	–	–
SJ4	Smooth and gleaming	–	+	–	+	–	–	+	±	–	+	–	nd	nd	–	–
SJ7b	Smooth and gleaming	–	+	+	–	Orange	Orange	+	–	+	–	+	–	–	+	–
SJ8	Smooth and gleaming	–	+	+	–	nd	Orange	±	nd	+	–	–	–	–	–	–
SJ9	Smooth and gleaming	–	+	+	Light blue	–	Green	–	±	+	+	–	–	–	–	–
SJ10	Smooth and gleaming	–	+	+	Green	–	Green	–	±	+	+	–	–	–	–	–
SJ11	Smooth and gleaming	–	+	+	Green	–	Green	–	±	+	–	–	–	–	–	–
SJ12	Smooth and gleaming	–	+	+	Light blue	–	–	–	±	+	–	–	–	–	–	–
SJ13	Smooth and gleaming	–	+	+	Light blue	–	Green	–	±	+	+	–	–	–	–	–
SJ16	Smooth and gleaming	–	+	+	+	–	–	–	–	+	+	–	–	–	–	–
SJ17	Smooth and gleaming	–	+	+	Green	–	Green	–	+	+	–	–	–	–	–	–
SJ18	Smooth and gleaming	–	+	+	Green	–	Green	–	+	+	–	–	–	–	–	–
SJ20	Smooth and gleaming	–	+	+	+	–	–	–	–	+	+	–	–	–	–	–
SJ21	Smooth and gleaming	–	+	+	Green	–	Green	–	–	+	–	–	–	–	–	–
SJ22	Smooth and gleaming	–	+	+	Green	–	Green	–	–	+	+	–	–	–	–	–
SJ24	Smooth and gleaming	–	+	+	+	–	–	–	–	+	+	–	–	–	–	–
SJ25	Smooth and gleaming	–	+	+	Green	–	Green	–	–	+	+	–	–	–	–	–
SJ27	Smooth and gleaming	–	+	+	Green	–	Green	–	–	+	–	–	–	–	–	–
SJ28	Smooth and gleaming	–	+	+	+	–	–	–	–	+	+	–	–	–	–	–
SJ29	Smooth and gleaming	–	+	+	Green	–	Green	–	–	+	–	–	–	–	–	–
SJ30	Smooth and gleaming	–	+	+	Light blue	–	Green	–	–	+	+	–	–	–	–	–
SJ31	Smooth and gleaming	–	+	+	Green	–	Green	–	–	+	–	–	–	–	–	–
SJ32	Smooth and gleaming	–	+	+	Green	–	Green	–	–	+	–	–	–	–	–	–
SJ33	Smooth and gleaming	–	+	+	Green	nd	–	–	nd	+	–	–	–	–	–	–
SJ34	Smooth and gleaming	–	+	–	Green	–	–	+	–	+	–	+	–	+	+	+
SJ35	Smooth and gleaming	–	+	+	Green	–	Green	–	–	+	–	–	–	–	–	–
SJ36	Smooth and gleaming	–	+	+	Light blue	–	–	–	–	+	–	–	–	–	–	–
SJ37	Smooth and gleaming	–	+	+	Green	–	Green	–	–	+	–	–	–	–	–	–
SJ38	Smooth and gleaming	–	+	+	Green	–	Green	–	–	+	–	–	–	–	–	–
SJ39	Smooth and gleaming	–	+	+	Green	–	Green	–	–	+	–	–	–	–	–	–
SJ40	Smooth and gleaming	–	+	+	Green	–	Green	–	–	+	–	–	–	–	–	–
SJ41	Smooth and gleaming	–	+	+		–	–	–	–	+	+	–	–	–	–	–
SJ42	Smooth and gleaming	–	+	+	Green	–	Green	–	–	+	–	–	–	–	–	–
SJ44	Smooth and gleaming	–	+	+	Green	–	Green	–	–	+	+	–	–	–	–	–
SJ45	Smooth and gleaming	–	+	+	Green	–	–	±	–	+	–	–	–	–	–	–
SJ46	Smooth and gleaming	–	+	–	–	nd	–	+	nd	+	–	–	–	–	–	–
SJ47	Smooth and gleaming	–	+	+	Green	–	Green	–	–	+	+	–	–	–	–	–
SJ48	Smooth and gleaming	–	+	+	Green	–	Green	–	±	+	–	–	–	–	–	–
PAO1	Smooth and gleaming	–	+	+	Green	Blue	Green	–	++	+	+	+	–	+	+	+
KT2440	Smooth and gleaming	–	+	+	Green	–	Green	+	–	+	±	–	–	–	–	–
WCS417r	Smooth and gleaming	–	+	+	Green	–	Green	+	–	+	–	+	–	–	+	+

Analysis of EO Components in Root Exudates

Micropropagated *M. piperita* plantlets were grown on MS liquid medium (Santoro et al., 2011) for 30 days. The medium was then filtered (pore diameter 0.46 μm), and the volatile fraction was extracted with dichloromethane. EOs were analyzed and detected by gas chromatography/mass spectrometry (GC-MS; Figure 1; Table 2). Flame ionization detector (FID) response factors for each compound generated essentially equal areas (differences<5%). Chemical analyses were performed by GC (model Clarus 600, PerkinElmer; Waltham, MA, USA) with a CBP-1 capillary column (30 m × 0.25 mm, film thickness 0.25 mm) and mass-selective detector. Analytical conditions: injector temperature 250°C, detector temperature 270°C; oven temperature programmed from 60°C (3 min) to 240°C at 4°C min^{-1}; carrier gas ¼ helium at constant flow 0.9 ml min^{-1}; source 70 eV. EO components were identified based on mass spectra and retention times, in comparison with standards.

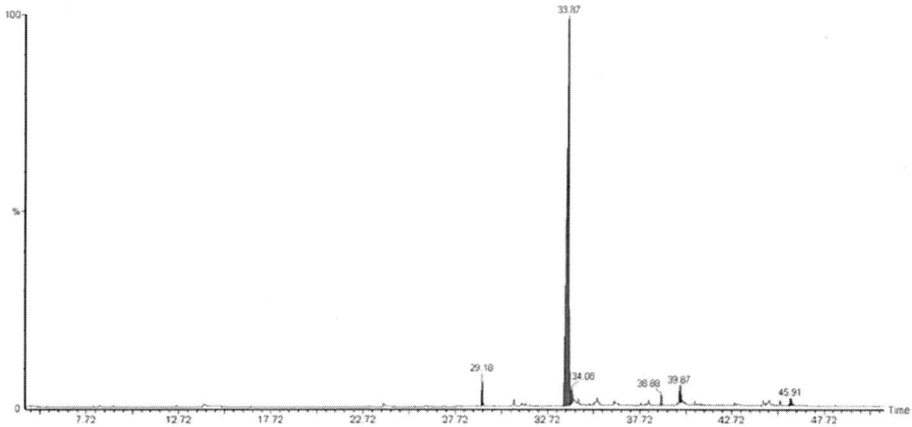

Figure 1. GC profile of *M. piperita* root exudates.

Table 2. Identification and relative percentages of major chemical signals detected in *M. piperita* root exudates, based on GC-MS analysis and comparison with the NIST database, with acceptance of SI ≥ 800.

Retention time	Relative percentage	Identification
29.183	2.31	3,4-dihydro-2H-1,5-(3"-t-butyl)benzodioxepine
33.870	90.82	4-octylbutan-4-olide
34.075	1.90	2(3H)-furanone, 5-heptyldihydro-
38.877	0.79	7,9-di-tert-butyl-1-oxaspiro(4,5)deca-6,9-diene-2,8-dione
39.873	1.35	1,2-benzenedicarboxylic acid, dibutyl ester
39.948	1.63	n-hexadecanoic acid
45.910	1.21	Z-2-octadecen-1-ol

Phenotypic Characterization

Fluorescent bacteria were phenotypically characterized based on morphological, physiological, and biochemical criteria (Lysenko, 1961; Stanier et al., 1966; Palleroni, 2005; Table 1). The tests and parameters included colony morphology, Gram determination, catalase determination (Urzí et al., 2001), oxidase determination (kit fromLaboratorios Britania), fluorescence color under UV illumination, pigment production (King et al., 1954), temperature tolerance, growth at 5 and 6.5% salt concentration, gelatin hydrolysis (Medina and Baresi, 2007), Tween 80 hydrolysis (Sierra, 1957), starch hydrolysis (Gordon and Mihm, 1956), casein hydrolysis (Salisbury and Lykos, 1972), and lecithinase production (Sneath, 1956). Each test was performed twice.

DNA Extraction

A single colony was inoculated on 3 ml LB medium and incubated at 28°C for 24 h. A 2-ml culture sample taken at late exponential growth phase was subjected to DNA extraction using a Genomic DNA Purification Kit (Fermentas, Thermo Fisher Scientific; Pittsburgh, PA, USA) according to the manufacturer's instructions.

Genotypic Testing

Amplified Ribosomal DNA Restriction Analysis (ARDRA)

16S rRNA gene amplification was performed, primers and methodology are specified in Table 3. A single specific band (~1400 bp) was observed by UV transillumination, and endonuclease digestion analysis of each amplicon was performed separately with *Msp*I, *Hinf* I, *Hae*III, and *Alu*I (Achouak et al., 2000; Wu et al., 2009; Mehri et al., 2011; Nievas et al., 2012) according to the manufacturer's (Fermentas') instructions. Reaction mixtures were incubated overnight at 37°C. Total volume of each restriction-digested product was evaluated by 3% (w/v) agarose-ethidium bromide stained gel electrophoresis and photographed for subsequent analysis.

Table 3. Primers sequences used for genotypic analysis of native fluorescent *Pseudomonas* and reference strains.

16S rRNA Gene Nucleotide Sequence Analysis

Genotipic test	Primers	References	Methodology
ARDRA	fD1 (5'-AGAGTTTGATCCTGGCTCAG-3') rD1 (5'-AAGGAGGTGATCCAGCC-3')	Weisburg et al., 1991	Nievas et al., 2012
16S rRNA sequencing	27F (5'-AGAGTTTGATCCTGGCTCAG-3') 1492R (5'-TACGGTTACCTTGTTACGACTT-3')	Willems et al., 2001	Willems et al., 2001
ERIC-PCR	E1 (5'-ATGTAAGCTCCTGGGGATTCAC-3') E2 (5'-AAGTAAGTGACTGGGGTGACGG-3')	Versalovic et al., 1991.	Nievas et al., 2012
BOX-PCR	BOX1AR (5'- CTACGGCAAGGCGACGCTGACG- 3')	Versalovic et al., 1994	Wolska et al., 2011

Nucleotide sequence analysis of the *16S rRNA* gene was performed for 17 native strains (Table 4). Direct PCR was performed with 1 μl DNA template in a 20-μl PCR reaction mixture using primer and methodology specified in Table 3. Purified PCR products (~1400 bp) were sequenced by Macrogen,

Table 4. RFLP patterns of *16SrRNA* genes of native and reference strains of fluorescent *Pseudomonas*.

Strain	*16S rRNA* genotype[a]	RFLP pattern[b]			
		HinfI	*MspI*	*HaeIII*	*AluI*
SJ01	2	A	B	B	B
SJ04	6	B	A	A	C
SJ7b	6	B	A	A	C
SJ08	7	B	B	B	D
SJ09	3	A	A	A	C
SJ10	3	A	A	A	C
SJ11	10	A	B	D	B
SJ12	15	D	B	C	E
SJ13	1	A	A	A	A
SJ16	2	A	B	B	B
SJ17	4	A	A	B	A
SJ18	2	A	B	B	B
SJ20	2	A	B	B	B
SJ21	13	C	B	B	A
SJ22	1	A	A	A	A
SJ24	1	A	A	A	A
SJ25	9	A	B	A	A
SJ27	1	A	A	A	A
SJ28	4	A	A	B	A
SJ29	1	A	A	A	A
SJ30	1	A	A	A	A
SJ31	1	A	A	A	A
SJ32	1	A	A	A	A
SJ33	1	A	A	A	A
SJ34	11	B	B	B	B
SJ35	1	A	A	A	A
SJ36	1	A	A	A	A
SJ37	1	A	A	A	A
SJ38	14	C	A	A	A
SJ39	8	C	B	B	B
SJ40	8	C	B	B	B
SJ41	5	A	B	B	A
SJ42	1	A	A	A	A
SJ44	5	A	B	B	A
SJ45	1	A	A	A	A
SJ46	7	B	B	B	D
SJ47	1	A	A	A	A
SJ48	1	A	A	A	A
WCS417r	12	B	A	B	F
KT2440	3	A	A	A	C

[a]The 16S rRNA genotype represents a combination of RFLP patterns obtained from four restriction enzymes.
[b]Strain with the same letter have the same RFLP pattern obtained from one endonuclease.

Table 5. BLAST analysis of *16SrRNA* gene sequences of native strains isolated from *M. piperita* rhizosphere.

Strain	ARDRA analysis		*16SrRNA* analysis		
	Genotype	Group	Accession number	Significant Alignment	Identity (%)
SJ01	2	B	KF312466.1	*P. putida* C-G-PDA3 – HM755498.1	99
SJ04	6	A	KF312467.1	*P. putida* GM6 – DQ133606.1	99
SJ7b	6	A	KF312468.1	*P. fluorescens* MazP22 – JX865768.1	99
SJ08	7	B	KF312469.1	*Pseudomonas* sp. P97.38 – DQ453823.1	99
SJ13	1	A	KF312470.1	*P. putida* NBRC102090 – AB681701.1	99
SJ16	2	B	KF312471.1	*P. putida* S18 – AY741157.1	99
SJ17	4	A	KF312472.1	*Pseudomonas* sp. MKP213 – KC013979.1	99
SJ25	9	B	KF312473.1	*Pseudomonas* sp. MKP213 – KC013979.1	99
SJ28	4	A	KF312474.1	*P. putida* 1599 – JN679860.1	99
SJ31	1	A	KF312475.1	*Pseudomonas* sp. MKP213 – KC013979.1	99
SJ32	1	A	KF312476.1	*Pseudomonas* sp. ONBA-17 – DQ079062.1	99
SJ45	1	A	KF312477.1	Pseudomonas sp. m4(2012) – JX675235.1	99
SJ46	7	B	KF312478.1	Ps. fluorescens Pf29A – DQ473439.1	99
SJ48	1	A	KF312479.1	*P. putida* NBRC102090 – AB681701.1	99

Inc. (Seoul, Korea) using an automated DNA sequencing system (model 3730XL, Applied Biosystems; Foster City, CA, USA). Sequence identities were determined by BLAST search program (National Center for Biotechnology Information [NCBI]; Bethesda, MD, USA; Altschul et al., 1997). Nucleotide sequences of the *16S rRNA* gene for each analyzed strain were deposited in GenBank with respective accession numbers (Table 5).

To generate a phylogenetic tree reflecting evolutionary relationships among fluorescent *Pseudomonas* strains, phylogenetic analysis was performed using the MEGA 4.0.2 software program (Tamura et al., 2007). Multiple alignments were made with the Clustal W software program (Higgins et al., 1994). Aligned sequences were used to construct a phylogenetic tree based on the neighbor-joining method (Saitou and Nei, 1987), using the Kimura two-parameter model (Kimura, 1980; Figure 2).

ERIC-PCR Fingerprinting

Enterobacterial Repetitive Intergenic Consensus (ERIC) fragments were analyzed for 40 native fluorescent strains. Amplification was performed using primers and methodology specified in Table 3. PCR products were separated by 1.2% (w/v) agarose gel electrophoresis in TAE 1X buffer, stained with ethidium bromide, and photographed for subsequent analysis.

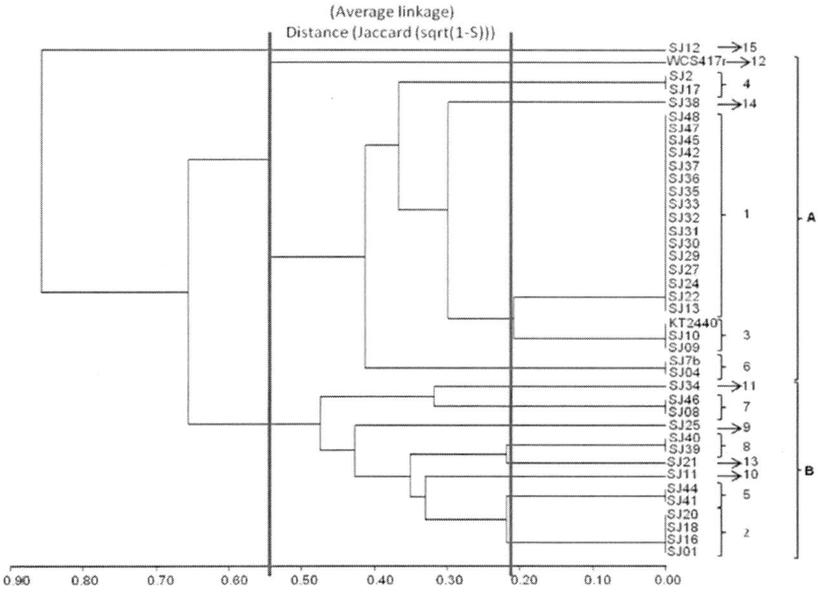

Figure 2. Dendrogram generated from RFLP patterns of *16S rRNA* genes of native and reference strains of fluorescent *Pseudomonas*, using the UPGMA algorithm and Jaccard coefficient.

BOX-PCR Fingerprinting

Repetitive extragenic palindromic sequences (BOX-PCR) were evaluated for the 40 strains. Amplification was performed with primer and methodology specified in Table 3. Amplified products were loaded on 2% (w/v) agarose gel electrophoresis, stained with ethidium bromide, and run at 60 V for 30 min and90 V for 120 min. The obtained gel was photographed for latter analysis.

Analysis of Genotypic Data

Fingerprints generated by ARDRA and by Repetitive Sequence-based PCR (rep-PCR) techniques (ERIC-PCR, BOX-PCR) as above were analyzed manually. Presence or absence of each band in the PCR profile was coded as a binary matrix. Genetic similarity and clustering analysis for each generated matrix were performed using the Infogen software program, v. 2011 (Universidad Nacional de Córdoba, Argentina). Genetic distances among native fluorescent strains were evaluated using a Jaccard coefficient. Clustering was visualized by construction of dendrograms using the

Unweighted Pair-Group Method with Arithmetic Average (UPGMA) algorithm (Di Renzo et al., 2001; Balzarini, 2003).

Evaluation of Growth-Promoting Effects

Bacterial Suspension

Strains were grown aerobically on LB liquid medium for 24 h at 30°C with rotary shaking. Saturated cultures were diluted in sterile distilled water to final concentration 10^9 CFU ml^{-1}.

Exposure to mVOCs

Plastic Petri dishes (90 × 15 mm) with a center partition (I-plate, Fisher Scientific) were used. One side of the dish contained MS solid medium, and the other contained Hoagland medium (Hoagland and Arnon, 1938) added with 3% sucrose (w/v) and 1.5% agar (w/v). Bacterial suspension as above (one drop; 20 µl) was placed on the Hoagland side. Distilled water (one drop) was used as negative control. A single node from a young shoot of aseptically cultured *M. piperita* (Santoro et al., 2011) was placed on the opposite side of the partition. In this way, plants were exposed to mVOCs without physical contact. Dishes were parafilm-sealed, arranged in a completely randomized design, and placed in a growth chamber under controlled conditions of light (16/8-h light/dark cycle), temperature (22 ± 2°C), and relative humidity (~70%). Five plants were used for each treatment, and experiments were replicated four times.

Plant Biomass and Morphology

Plants exposed to mVOCs were harvested after 30 days and evaluated for shoot length, root length, shoot fresh weight, root dry weight, node number, leaf number, and shoot ramification number (Santoro et al., 2011).

Extraction and Analysis of EOs

Each shoot sample was weighed, subjected to hydrodistillation in a micro Clevenger-like apparatus for 14 min, and the volatile fraction was extracted with dichloromethane. An internal standard (1 µl γ-dodecalactone in 100 µl ethanol) was added during the distillation procedure. Oil components were initially identified based on mass spectral and retention time data and confirmed by direct comparison with commercial standards (Sigma-

Aldrich). FID response factors for each compound generated essentially equal areas (differences <5%). For comparison of results under different treatments, response factors for individual compounds were assumed to be equal (Shukui et al., 2005). Chemical analyses of EOs were performed as described in Section Analysis of EO Components in Root Exudates. For each plant, areas corresponding to analyzed compounds were added, and total weight of EOs from each sample was expressed relative to area of the internal standard. Shoots were weighed, and yield was expressed as μg g^{-1} fresh weight.

Statistical Analysis

Data were subjected to analysis of variance (ANOVA) followed by comparison of multiple treatment levels with control, using post hoc Fisher's Least Significant Difference (LSD) test. Infostat software program, v. 2.0 (Group Infostat, Universidad Nacional de Córdoba, Argentina) was used for all statistical analyses.

Chemical Analysis of mVOCs

mVOC Collection

The three bacterial strains that had the greatest effect on plant biomass and EO production (SJ04, SJ25, SJ48) were cultured for mVOC analysis, and WCS417r was included as reference strain. Bacteria were cultured in sealed 100-ml Kitasato flasksin Hoagland medium plus 0.4% TSB for 48 h at 30°C. The flasks were then connected to a vacuum pump to guarantee air flow, and the flaskinlet was covered with a cap that had been pre-drilled and modified with a cotton-filled glass tube to control air flux to the collection trap. Air exited the chamber through a reusable glass collection trap packed with 30 mg Super-Q absorbent (80–100 mesh; Alltech Assoc., Deerfield, IL, USA) to capture mVOCs. Prior to mVOC collection, eachfilter was rinsed with 2 ml dichloromethane and 2 ml hexaneto remove impurities. Flasks were then placed on a heating block at 50°C with magnetic stirring for 30 min. The filters were immediately eluted from absorbent traps with 200 μl dichloromethane, and collected mVOCs were analyzed by GC-MS as described below.

GC-MS Analysis and Conditions

Filter samples were concentrated with nitrogen gas, resuspended in ethanol, and injected into a Perkin-Elmer Clarus600 gas chromatograph equipped with DB5 column (60 m, i.d. 0.25 mm, film thickness 0.25 μm;

J&W Scientific, Folsom, CA, USA) and mass-selective detector. GC-MS conditions: 25-minruns; injection port operated in splitless mode with constant He flow 1.0 ml/min; initial oven temperature 33°C, held for 3 min, ramped at 10°C min^{-1} to 180°C, ramped at40 min^{-1} to 220°C, and held for 5 min; HP quadrupole mass spectrometer operated in electron ionization mode at 70 eV; source temperature 200°C; quadrupole temperature 150°C; continuous scan from m/z 40 to 500. Volatile components were identified by comparison of retention times and recorded mass spectra against the NIST database, with acceptance of similarity index (SI) ≥ 800.

RESULTS AND DISCUSSION

Sample and Strain Isolation

Pseudomonas is the most extensively studied of the known PGPR genera because the species are widely distributed in many different environments and are easy to culture under laboratory conditions. For isolation of fluorescent *Pseudomonas* strains, rhizospheric soil from a commercial *M. piperita* crop as described in Section Sample Collection and Strain Isolation was collected, with total bacteria count 3.05×10^7 CFU g^{-1}. Similar results were obtained in previous studies of rhizospheric soil from other crops (Torsvik et al., 1996; Houlden et al., 2008; Mehnaz et al., 2009). In total 4.7 $\times 10^5$ CFU g^{-1} fluorescent bacteria were counted, corresponding to 1.54% of total bacteria. Fifty colony isolates were collected based on observation of fluorescence under UV light (Table 1).

Various types of soil used for crop production support highly diverse microbial communities, and introduction of crop plants affects those communities by promoting or suppressing growth of indigenous bacteria (Houlden et al., 2008; Agaras et al., 2012). Whether the primary determinant of microbial community structure is soil properties or introduced plants remains controversial. Plants clearly have a substantial effect on microbial communities based on the composition of their root exudates, which is closely associated with nutritional status and growth stage (Houlden et al., 2008). The proportion of fluorescent bacteria relative to total count of culturable bacteria in rhizospheric soil or bulk soil has not been addressed in previous studies. Agaras et al. (2012) reported that fluorescent strains accounted for 10–94% of total *Pseudomonas* in various crop soils and bulk soils. Results of the present study suggest that selection by *M. piperita* root exudates may explain the much lower count ($\sim 10^7$ CFU g^{-1}) of culturable bacteria for this species than for other crop plants ($\sim 10^{10}$ CFU g^{-1}; Agaras et al., 2012; Bogino et al., 2013).

Analysis of Root Exudates

Native bacterial strains have not been isolated from the rhizosphere of aromatic crop plants in previous studies, because EOs present in root exudates are generally presumed to have a toxic effect on microorganisms (Chen et al., 2004). We analyzed root exudates from *in vitro* cultured *M. piperita* by GC-MS (Figure 1), and evaluated fragmentation patterns by comparison with the NIST database, with acceptance of SI ≥ 800 (Table 2). No monoterpenes were detected in the root exudates, indicating that bacterial communities can be established in the *M. piperita* rhizosphere. The absence of EOs in root exudates suggests that the low total bacterial count did not reflect selection based on plant antimicrobial properties, e.g., inhibition or suppression of bacterial growth. Rather, the low count may result from competition or inhibition of established communities through direct growth inhibition by microbial compounds, or from adaptation by bacteria to the soil environment, edaphic conditions, and/or altitude.

Phenotypic Testing

All 50 isolated strains were subjected to phenotypic testing. Three strains representing the main groups of fluorescent *Pseudomonas* were used as references: *P. fluorescens* WCS417r, *P. putida* KT2440, and *P. aeruginosa* PAO I (Lysenko, 1961; Stanier et al., 1966; Palleroni, 2005). Biochemical characteristics of the isolated strains are summarized in Table 1. In terms of biochemical responses, 33 strains (66% of total) were most similar to *P. putida* KT2440, one (SJ34) was similar to *P. fluorescens* WCS417R, and none were similar to *P. aeruginosa* PAO I. The 50 strains were subjected to genotypic testing (Sec. 3.4) for identification of *Pseudomonas* species, and screened for PGPR activity mediated by production of mVOCs (Section Plant Growth-Promoting Effects of mVOCs). Strains SJ04, SJ7b, and SJ46 were included in the genotypic testing because they showed the strongest promoting effects on biomass and EO production of micropropagated *M. piperita* plants.

Genotypic Testing

Three molecular techniques (ARDRA, ERIC-PCR fingerprinting, BOX-PCR fingerprinting) were applied to evaluate similarities through clustering analysis of the isolated fluorescent *Pseudomonas* strains. Forty native strains and two reference strains (*P. fluorescens* WCS417r, *P. putida* KT2440) were evaluated. *P. aeruginosa* PAO I was excluded because none of the native strains showed biochemical similarity to it.

ARDRA

The ARDRA technique (analysis of16S rRNA gene fragments produced by digestion with restriction enzymes) has been widely applied in genetic diversity studies of Pseudomonas and other rhizobacteria. It has been used successfully for analysis of fluorescent Pseudomonas communities isolated from rhizospheres of various crop plants under a variety of environmental and climatic conditions (Prosser, 2002; Mehri et al., 2011; Nievas et al., 2012). It has also been used to study variation of Pseudomonas communities as a function of crop management strategies (Achouak et al., 2000) and application of pesticides or other chemical compounds (Viti and Giovannetti, 2005; Braun et al., 2006; Wang et al., 2008; Wu et al., 2009). The present study is the first to apply ARDRA to PGPR isolates from the *M. piperita* rhizosphere.

We used restriction endonucleases *Alu* I, *Msp* I, *Hinf* I, and *Hae* III because of their ability to discriminate among *Pseudomonas* strains (Achouak et al., 2000; Wu et al., 2009; Mehri et al., 2011). For each analyzed strain, a single amplicon (~1400 bp) was obtained using the primers rD1 and fD1. Only seven of the 40 markers gave monomorphic bands. *Alu*I had the highest polymorphism, generating six different Restriction Fragment Length Polymorphism (RFLP) patterns. *Hinf* I and *Hae* III each generated four RFLP patterns, and *Msp*I generated two RFLP patterns, resulting in less informative restriction (Table 4). *16S rRNA* RFLP patterns obtained using each of the endonucleases were combined to obtain a general genotype for each strain. Genotype 1, the most frequent, was present in 16 native fluorescent strains. The genotype of *P. putida* KT2440 was also found in two other strains, while the genotype of *P. fluorescens*WCS417r was unique (Table 4).

A dendrogram was constructed as described in Section Analysis of Genotypic Data to determine genetic distances among native *Pseudomonas* strains (Figure 2). At 35% similarity, the dendrogram showed a division into two major clusters (A and B). At 45% similarity, cluster A grouped 25 strains having genotypes 1, 3 (including *P. putida* KT2440), 4, 6, 12 (including *P. fluorescens* WCS417r), and 14. Within cluster A, genotypes 1 and 3 were highly related (80%similarity). Cluster B grouped the remaining strains at 35% similarity. These strains had genotypes 2, 5, 7, 8, 9, 10, and 13. Within cluster B, genotypes 8 and 13, and 5 and 2 were highly related (80% similarity).

The genotypic analysis revealed high correlations between *P. putida* KT2440 and ≥ 50% of total fluorescent strains isolated from the *M. piperita* rhizosphere, consistent with the results of biochemical analysis. Strains with genotype 7 differed from *P. putida* KT2440 in both genotypic and phenotypic analyses. Strains with genotype 10, 2, 13, 9, 8, and 5 displayed genotypic divergence from *P. putida* KT2440 in spite of phenotypic similarities. Similar divergence was observed between strains with genotype 11 and *P. fluorescens* WCS417r (Table 4; Figure 2).

In a comparison of the phenotypic and genotypic results, clustering discrepancies may result from the differing characteristics evaluated by the methodologies. Phenotypic testing detects metabolic activities that may be shared among various species within a particular genus, whereas genotypic testing detects similarities or distances among genetic markers generated from a defined gene. Regardless of these considerations, 60% of native fluorescent rhizobacterial strains displayed high biochemical and genetic similarity with *P. putida* KT2440, and 50% of strains showed 80% genotypic similarity with KT2440. *P. putida* is clearly the major fluorescent strain present in the *M. piperita* rhizosphere.

Results have shown that the*16SrRNA* gene is highly conserved, causing difficulty in differentiating among closely related species. It is therefore preferable to apply two or more molecular techniques. Inclusion of techniques that identify sequences repeated along the whole genome permits analysis of more data, and of differences in sequences of more than one gene (Grimont, 2002; Rangel-Castro et al., 2002; Gomila et al., 2007; Naik et al., 2008; Mehri et al., 2011).

16S rRNA Gene Nucleotide Sequence Analysis

The analyzed strains were selected based on their abundance of genotypes and distribution in clusters A and B (Table 4). PGPR activity of strains on micropropagated *M. piperita* plantlets mediated by mVOC production (Section Plant Growth-Promoting Effects of mVOCs) was also taken into consideration.

The BLAST software program was used to analyze the obtained *16S rRNA* gene sequences and to establish their alignment and identity with reference strains (Altschul et al., 1997). All of the analyzed strains had identities within the genus *Pseudomonas*. The native strains had high percentage identity with *P. putida* and *P. fluorescens*. Strains SJ13 and SJ48, and SJ17 and SJ25,had the same identity (Table 5). They should be considered different strains of the same species, in view of their very low similarity in ERIC-PCR and BOX-PCR analyses (Sections ERIC-PCR, BOX-PCR), which deeply probe strain diversity.

A phylogenetic tree was constructed to determine relationships and evolutionary distances among the fluorescent *Pseudomonas* strains isolated in this study, and the major, well-known species of this genus (Figure 3). We utilized the GenBank database because of its size and the ability to download sequences closely related to the fluorescent group (Altschul et al.,

1997). Two major clusters appear in the phylogenetic tree: the first contains *P. putida, P. fluorescens, P. protegens, P. syringae, P. chlororaphis,* and *P. corrugate*; the second contains *P. fulva, P. mendocina, P. nitroreducens, P. stutzeri, P. aeruginosa*, and *P. alcaligenes*. The first cluster includes all the native fluorescent *Pseudomonas* strains isolated from the *M. piperita* rhizosphere, and can be divided into two subclusters centered on *P. putida* and *P. fluorescens*.

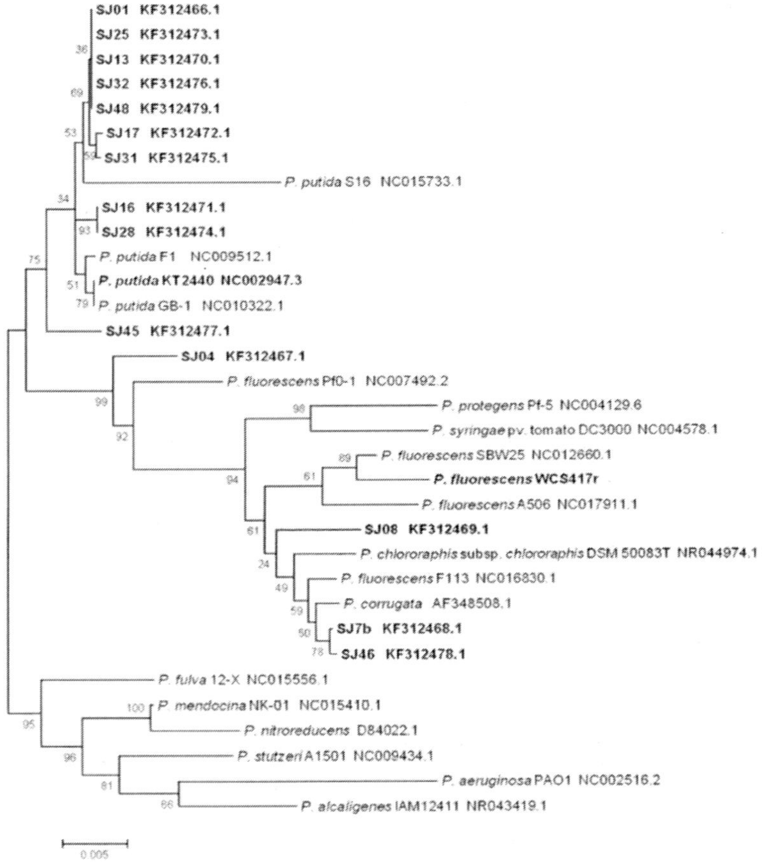

Figure 3. Phylogenetic tree of native and reference strains of fluorescent *Pseudomonas* based on aligned *16S rRNA* sequences. Native strains isolated from the *M. piperita* rhizosphere are indicated by boldface. The tree was constructed based on multiple alignments and the neighbor-joining method, using the MEGA 4.0.2 software program. Bootstrap values on branches are shown as percentages from 1000 replications. Scale bar at bottom represents 5 nt substitution per 1000 nt sequences. GenBank accession numbers are shown.

Most of the native strains analyzed had high similarity with *P. putida*. The strains most strongly related to *P. putida* were SJ16 and SJ28, followed by SJ01, SJ13, SJ17, SJ25, SJ31, SJ32, and SJ48. Strain SJ45 was distant from this group. These findings are consistent with genotypic and phenotypic analyses, and confirm that *P. putida* is the major fluorescent species present in the *M. piperita* rhizosphere.

The remaining four strains were part of the *P. fluorescens* subcluster and were notably distant from the *P. putida* subcluster, again consistent with results of genotypic and phenotypic analyses. Strain SJ04was intermediate in phylogenetic position between the two subclusters, and showed a phylogenetic relationship with *P. fluorescens* in spite of its identity with *P. putida* (Table 4). Strains SJ7b and SJ46 had a stronger phylogenetic relationship with *P. corrugate* than with *P. fluorescens*, consistent with results of BLAST analysis. Strain SJ08 had a phylogenetic relationship with *P. fluorescens* and *P. chlororaphis*.

ERIC-PCR

Of the 40 native fluorescent strains analyzed, only 38 generated an appropriate profile; SJ32 and SJ35 did not show amplification under the conditions applied. The primers used generated 67 markers with sizes ranging from 100 to 1700 bp, represented only by polymorphic bands. The pairs SJ21/SJ22, SJ17/SJ18, and SJ38/SJ39 showed identical fingerprinting (100% similarity); i.e., each of these pairs was likely the same strain.

A dendrogram was constructed to determine genetic distance or proximity among the isolated strains (Figure 4). Clusters were generally observed at greater genetic distances than in ARDRA analysis, most likely because of the higher number of markers generated and their distribution along the whole genome. At 50% similarity, four groups (I-IV) were formed. The main group (II) included 19 strains and was divided into three subgroups (A-C). Within these subgroups, the most closely related strains were SJ16 and SJ20 (70% similarity), SJ29 and SJ47 (70% similarity), and strains SJ45 and SJ46 (65% similarity). Group I included reference strains *P. fluorescens*WCS417r and *P. Putida* KT2440 with 55% similarity. Groups III and IV were smaller than groups I and II.

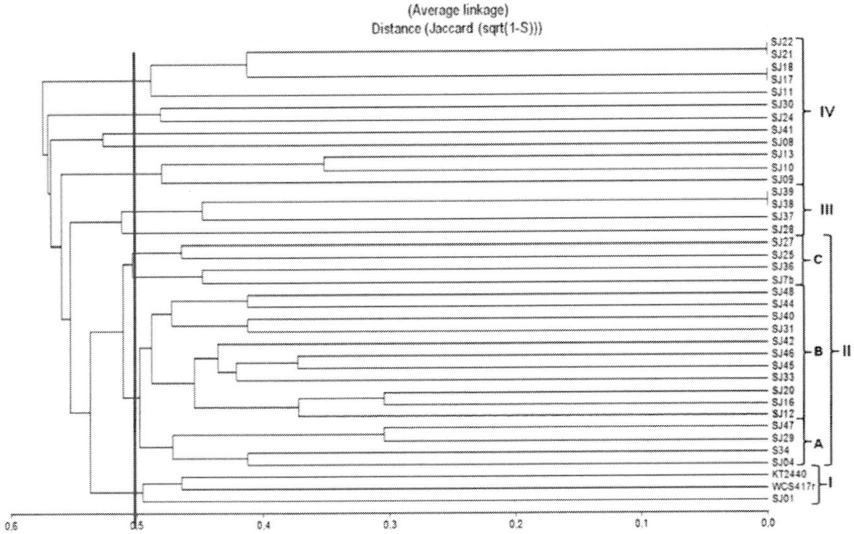

Figure 4. Dendrogram based on ERIC-PCR profiles, generated from native, and reference *Pseudomonas* strains using the UPGMA algorithm and Jaccard coefficient.

ERIC-PCR analysis did not allow us to distinguish between the two fluorescent reference strains because they presented 50% similarity, the same as observed for various native strains (Figure 4). In previous studies, this technique was used to evaluate diversity/similarity within groups of native *Pseudomonas* strains, and gave comparable degrees of similarity (Achouak et al., 2000; Wolska and Szweda, 2008; Selvakumar et al., 2009; Vyas et al., 2009). In contrast, it did not give high degrees of similarity for native fluorescent *Pseudomonas* isolated from the *M. piperita* rhizosphere, suggesting that these populations have considerable genetic diversity in spite of their phylogenetic affiliation with *P. putida*.

BOX-PCR

Each of the 40 analyzed strains generated an appropriate profile using the primer BOXA1R. This technique allowed a high degree of discrimination among native fluorescent strains isolated from the *M. piperita* rhizosphere. It is frequently applied for grouping of *Pseudomonas* strains isolated from soil samples (Naik et al., 2008; Vyas et al., 2009; Mehri et al., 2011; Parejko et al., 2012). Of the 40 strains, 39 generated an appropriate profile, whereas strain SJ25 did not undergo amplification under the conditions applied. In comparison with ERIC-PCR, BOX-PCR analysis generated a larger number of markers. A total of 104 discrete polymorphic bands were scored, ranging in size from 250 to 2000 bp, with 8 to 20 bands for each strain. Six strains presented the same profile, and were presumably identical.

A dendrogram was constructed to determine genetic distance or proximity among the isolated native strains (Figure 5). The BOX-PCR dendrogram was highly complex, and showed formation of a larger number of clusters at higher similarities. All strains grouped at 48% similarity, except for SJ01 which separated from the other strains at lower similarity. A large cluster (1) was formed at 50% similarity. It included the majority of strains (29), and was divided into five subclusters (A–E). The main subcluster (A), formed at ~55% similarity, included 15 of the native strains and reference strain *P. fluorescens* WCS417r. Within subcluster A, 100% similarity was obtained for three groups of strains: SJ09/SJ10/SJ13, SJ38/SJ39, and SJ31/SJ32. A smaller cluster (2) included high-similarity groups; e.g., SJ21/SJ22/SJ27 showed 100% similarity among themselves and 85% similarity with SJ29.SJ17/SJ18 grouped at 85% similarity. Reference strain *P. putida* KT2440 was not part of either cluster 1 or 2, and showed only low similarity (~55%) with cluster 1.

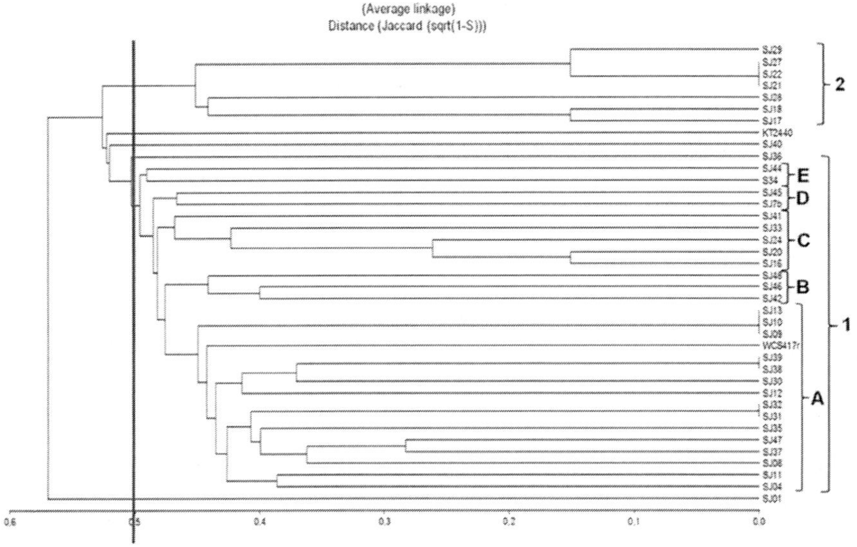

Figure 5. Dendrogram based on BOX-PCR profiles, generated from native, and reference *Pseudomonas* strains using the UPGMA algorithm and Jaccard coefficient.

In contrast to ERIC-PCR, BOX-PCR analysis did not give high similarity levels among reference strain *P. putida* KT2440 and the native strains, but did give high similarity levels among reference strain *P. fluorescens* WCS417r and the native strains. These rep-PCR techniques, applied together, thus provide a valuable tool for evaluating diversity of native fluorescent *Pseudomonas* strains isolated from the *M. piperita* rhizosphere, and most likely from rhizospheres of other aromatic species as well.

Comparison of Molecular Techniques Applied

In a comparison of the two rep-PCR techniques, ERIC-PCR showed better discrimination among native strains. The number of markers obtained was lower, resulting in relatively easy analysis and a high degree of clustering, i.e., lower similarity among native fluorescent *Pseudomonas* strains isolated from the *M. piperita* rhizosphere. At 80% similarity, 35 groups were obtained from ERIC-PCR analysis, vs. 30 groups obtained from BOX-PCR analysis. ERIC-PCR was useful for detailed study of differences in genomic DNA among closely related strains. BOX-PCR was useful for grouping strains while eliminating small differences among them, resulting in larger clusters. BOX-PCR produced a larger number (6) of duplicate profiles (Figure 4), i.e., a larger number of identical strains, than the number (3) obtained from ERIC-PCR (Figure 3). Previous studies gave similar degrees of heterogeneity by application of rep-PCR analysis in groups of isolated *Pseudomonas* strains (Selvakumar et al., 2009; Vyas et al., 2009; Wolska et al., 2011).

Simultaneous application of ARDRA and rep-PCR techniques could not distinguish among *P. fluorescens* and *P. putida* strains because of the 50% similarity with reference strains. However, both types of analysis revealed close relationships within the SJ09/SJ10/SJ13, SJ16/SJ20, and SJ31/SJ32 groups. The SJ17/SJ18, SJ21/SJ22, and SJ38/SJ39 groups showed high similarity in rep-PCR but lower similarity in ARDRA. Although BOX-PCR and ERIC-PCR sequences allow us to deeply probe genomic differences, they cannot be used as a tool to identify or determine phylogenetic positions of native fluorescent strains. Phylogenetic positioning is based on comparison of *16S rRNA* gene sequences. ARDRA identifies small differences among native strains by evaluating this particular gene of interest. The combination of ARDRA and *rDNA 16S* sequencing clearly identified *P. putida* as the major fluorescent species in *M. piperita* rhizospheric soil.

Plant Growth-Promoting Effects of mVOCs

Biomass

Plants were analyzed after 30 days incubation with mVOCs. A significant ($p \leq 0.05$) effect by mVOCs from native fluorescent Pseudomonas was observed for shoot fresh weight, which was 3.0 to 4.5-fold higher than control value. Values for shoot ramification number, node number, leaf number, root length, and dry weight did not differ significantly from control values ($p > 0.05$; Table 6). In one of our previous studies (Santoro et al., 2011), mVOCs from reference strain P. fluorescensWCS417r similarly increased shoot fresh weight of micropropagated *M. piperita*, but had no effect on leaf number or ramification number. The increase of shoot fresh weight can be attributed to greater leaf and stem size, leading to increased aerial biomass. We also observed a similar effect in *Ocimum basilicum* (basil) plants exposed to mVOCs from *B. subtilis* (Banchio et al., 2009). Zhang et al. (2007) explained the effect of mVOC exposure on plant biomass by demonstrating that stimulated genes were associated with cell wall structure, allowing expansion and reduced rigidity. They also observed reduced mVOC exposure in leaves, leading to enhanced shoot development through stimulation of elongation, and cellular differentiation (Zhang et al., 2008).

Table 6. Effects of mVOCs emitted by native fluorescent *Pseudomonas* strains on biomass parameters of micropropagated *M. piperita* plantlets.

Treatments	Root		Shoot				
	Length (cm)	Dry weight (mg)	Fresh weight (mg)	Length (cm)	Node number	Ramification number	Leaf number
Control	5.36 ± 1.20^{a}	2.21 ± 0.40^{a}	113.70 ± 12.34^{a}	4.02 ± 0.39^{a}	6.38 ± 0.19^{a}	1.07 ± 0.15^{a}	19.19 ± 1.62^{a}
SJ01	7.62 ± 1.25^{a}	4.89 ± 1.95^{a}	233.87 ± 29.47^{a}	4.45 ± 0.31^{a}	7.29 ± 0.49^{a}	1.25 ± 0.14^{a}	23.23 ± 2.40^{a}
SJ04	9.17 ± 1.25^{a}	3.34 ± 0.57^{a}	277.36 ± 26.52^{b}	4.78 ± 0.39^{a}	7.89 ± 0.45^{a}	1.80 ± 0.64^{a}	29.23 ± 3.71^{a}
SJ7b	8.66 ± 0.98^{a}	3.70 ± 0.60^{a}	386.02 ± 79.05^{b}	5.18 ± 0.17^{a}	8.48 ± 1.06^{a}	2.40 ± 1.20^{a}	49.87 ± 2.21^{a}
SJ08	7.37 ± 1.11^{a}	3.37 ± 0.60^{a}	227.23 ± 27.03^{a}	4.60 ± 0.04^{a}	7.00 ± 0.58^{a}	1.15 ± 0.08^{a}	25.44 ± 1.89^{a}
SJ10	7.05 ± 1.50^{a}	3.01 ± 0.65^{a}	301.36 ± 46.81^{b}	4.83 ± 0.31^{a}	7.44 ± 0.39^{a}	1.38 ± 0.39^{a}	29.52 ± 2.90^{a}
SJ13	7.77 ± 1.96^{a}	3.31 ± 0.19^{a}	298.82 ± 66.07^{b}	4.84 ± 0.18^{a}	7.81 ± 0.45^{a}	1.56 ± 0.51^{a}	27.88 ± 5.10^{a}
SJ22	8.19 ± 1.57^{a}	3.95 ± 0.93^{a}	258.88 ± 17.41^{a}	4.52 ± 0.20^{a}	7.77 ± 0.81^{a}	0.95 ± 0.05^{a}	25.40 ± 2.22^{a}
SJ24	9.92 ± 1.92^{b}	2.56 ± 0.49^{a}	251.72 ± 65.77^{a}	4.80 ± 0.70^{a}	7.53 ± 0.68^{a}	1.75 ± 0.60^{a}	30.56 ± 7.16^{a}
SJ25	9.16 ± 1.35^{a}	4.01 ± 0.77^{a}	326.18 ± 69.86^{b}	5.01 ± 0.30^{a}	7.60 ± 0.32^{a}	2.20 ± 0.41^{a}	32.41 ± 6.34^{a}
SJ27	9.15 ± 1.67^{a}	4.57 ± 1.02^{a}	279.09 ± 4.77^{b}	4.64 ± 0.41^{a}	7.68 ± 0.20^{a}	1.40 ± 0.28^{a}	27.17 ± 0.52^{a}
SJ28	10.05 ± 2.89^{a}	7.69 ± 3.54^{a}	321.17 ± 46.66^{b}	5.37 ± 0.43^{a}	8.45 ± 0.52^{a}	2.00 ± 0.71^{a}	32.20 ± 7.71^{a}
SJ31	7.66 ± 1.49^{a}	5.12 ± 1.33^{a}	346.38 ± 19.12^{a}	5.21 ± 0.51^{a}	8.15 ± 0.30^{a}	2.20 ± 0.57^{a}	31.10 ± 3.87^{a}
SJ33	6.61 ± 0.80^{a}	3.53 ± 0.48^{a}	253.79 ± 27.37^{a}	5.15 ± 0.21^{a}	7.40 ± 0.72^{a}	1.53 ± 0.35^{a}	28.35 ± 2.71^{a}
SJ36	8.75 ± 1.73^{a}	3.55 ± 0.74^{a}	221.97 ± 22.50^{a}	4.83 ± 0.41^{a}	7.00 ± 0.20^{a}	1.44 ± 0.16^{a}	22.94 ± 1.23^{a}
SJ38	9.50 ± 0.77^{a}	4.45 ± 1.06^{a}	268.52 ± 28.95^{b}	5.28 ± 0.19^{a}	8.11 ± 0.31^{a}	1.77 ± 0.36^{a}	27.69 ± 2.52^{a}
SJ40	9.13 ± 1.94^{a}	6.01 ± 2.66^{a}	277.08 ± 46.29^{b}	4.92 ± 0.31^{a}	7.60 ± 0.88^{a}	2.23 ± 0.93^{a}	28.75 ± 4.63^{a}
SJ41	6.55 ± 0.85^{a}	3.60 ± 0.83^{a}	228.41 ± 14.24^{a}	3.98 ± 0.19^{a}	7.68 ± 0.39^{a}	0.93 ± 0.39^{a}	22.85 ± 3.06^{a}
SJ44	7.76 ± 0.89^{a}	4.21 ± 0.79^{a}	238.85 ± 26.46^{a}	4.77 ± 0.42^{a}	7.30 ± 0.28^{a}	1.23 ± 0.39^{a}	23.86 ± 1.19^{a}
SJ45	9.37 ± 1.08^{a}	3.79 ± 0.33^{a}	261.12 ± 26.05^{a}	4.99 ± 0.13^{a}	7.86 ± 0.59^{a}	1.16 ± 0.10^{a}	26.19 ± 2.94^{a}
SJ46	8.06 ± 0.60^{a}	3.94 ± 0.70^{a}	315.16 ± 62.22^{b}	5.21 ± 0.34^{a}	8.22 ± 0.68^{a}	1.93 ± 0.40^{a}	32.69 ± 4.16^{a}
SJ48	8.23 ± 1.06^{a}	4.75 ± 1.10^{a}	265.52 ± 26.93^{b}	5.06 ± 0.50^{a}	8.03 ± 0.28^{a}	1.41 ± 0.21^{a}	26.44 ± 0.58^{a}
WCS417r	6.91 ± 1.00^{a}	3.62 ± 0.75^{a}	262.45 ± 28.97^{a}	4.66 ± 0.24^{a}	7.10 ± 0.63^{a}	1.61 ± 0.17^{a}	26.03 ± 2.00^{a}

[a,b] *Means followed by the same letter within a column are not significantly different according to Fisher's LSD test ($p < 0.05$).*

EO Production

We examined the effect of mVOCs on EO production in micropropagated *M. piperita* plantlets. The major components of extracted EOs were (−) limonene, terpineol, (−) menthone, (+) menthofuran, (−) menthol, (+) pulegone, and (−) menthyl acetate.

Exposure to mVOCs from strains SJ25 and SJ27 resulted in 86 and 57% increases, respectively, in total EO production relative to control (Figure 6). Limonene content was increased 2-fold by exposure to SJ04, SJ28, or SJ7b mVOCs, and 4-fold by SJ48mVOCs (Figure 7). Menthol content was increased three-fold by SJ04 mVOCs (Figure 8). Menthofuran Limonene content was reduced by exposure to mVOCs from all strains, most notably (20% reduction) by SJ48 mVOCs (Figure 7). Menthyl acetate content was increased four-fold by SJ7b mVOCs (Figure 9). The changes mentioned above were statistically significant at $p \leq 0.05$. Changes in menthone, terpineol, and pulegone content produced by exposure to mVOCs were not significant ($p > 0.05$; data not shown).

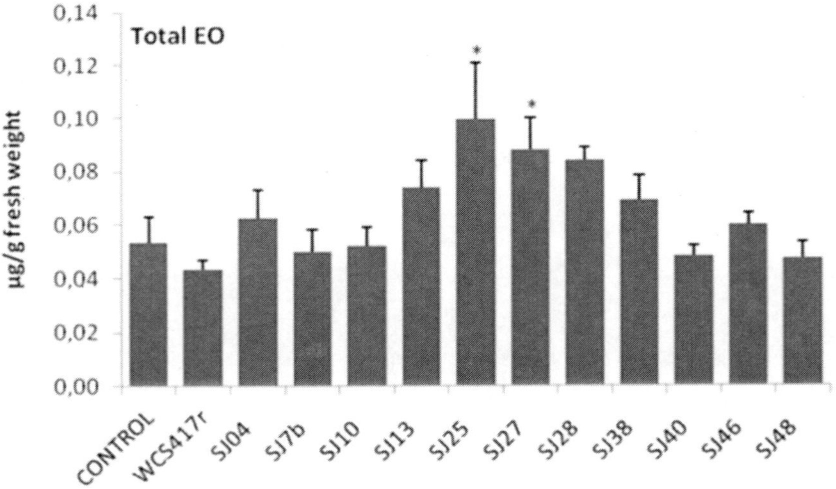

Figure 6. Effects of exposure to mVOCs of native and reference strains of fluorescent *Pseudomonas*on EO content of micropropagated *M. piperita* plantlets. *Significant difference ($p \leq 0.05$) from control value.

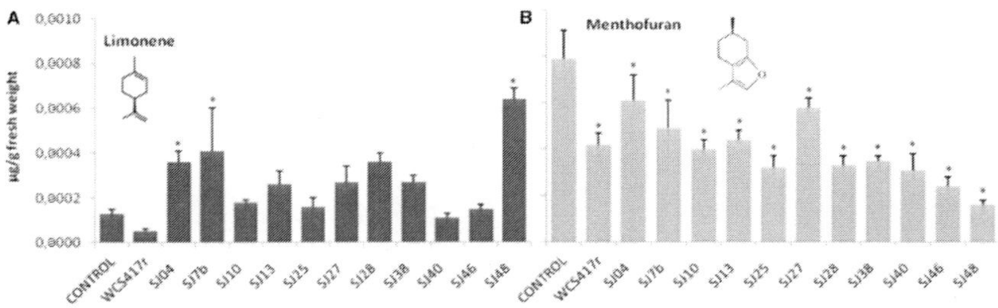

Figure 7. Effects of exposure to mVOCs of native and reference strains of fluorescent *Pseudomonas*on limonene and menthofuran content of micropropagated *M. piperita*. *Significant difference ($p \leq 0.05$) from control value.

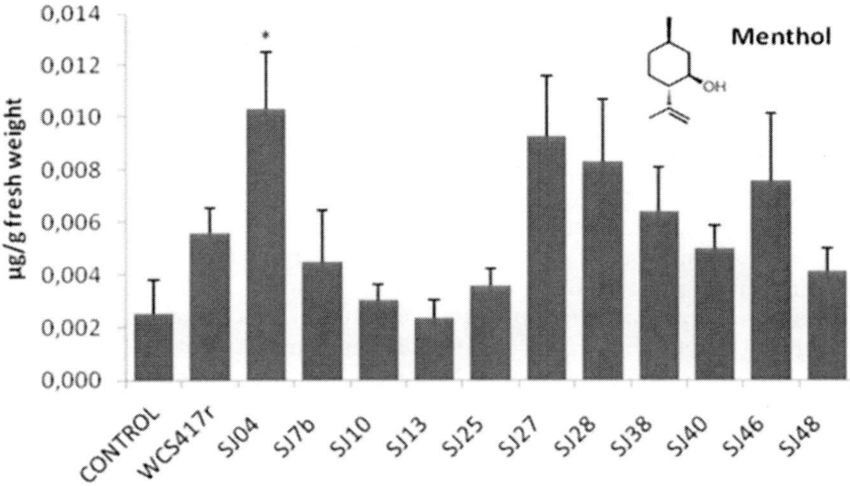

Figure 8. Effects of exposure to mVOCs of native and reference strains of fluorescent *Pseudomonas*on menthol content of micropropagated *M. piperita*. *Significant difference ($p \leq 0.05$) from control value.

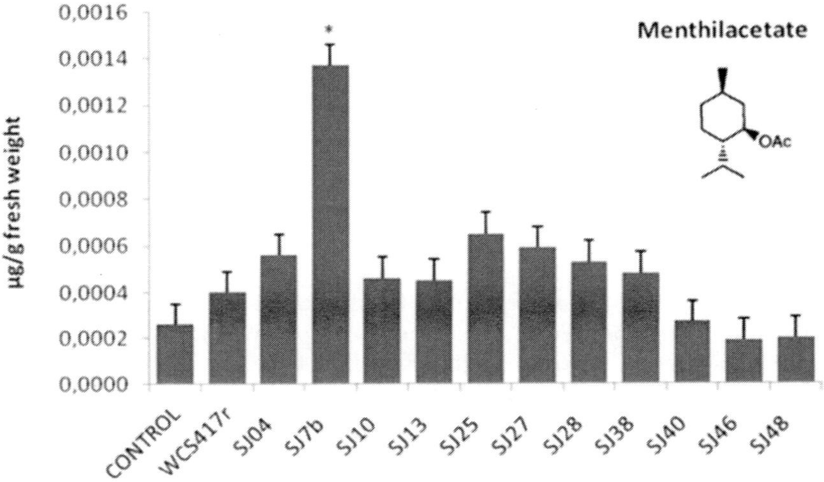

Figure 9. Effects of exposure to mVOCs of native and reference strains of fluorescent *Pseudomonas*on menthylacetate content of micropropagated *M. piperita*. *Significant difference ($p \leq 0.05$) from control value.

Content of menthol, the major monoterpene, was significantly elevated only by SJ04 mVOCs. Monoterpenes are the commercially valuable components among EOs (Gupta et al., 2002). The observed reduction of menthofuran content is related to the increase in menthol, because menthofuran represents a branch in the flux of terpene precursors to final production of menthol (Lange and Croteau, 1999; Turner and Croteau, 2004). The increased levels caused by SJ04, SJ7b, and SJ48 mVOCs may reflect a lack of maturation of EOs, considering that enzymes involved in later reactions are activated at ~30 days of foliar growth (McConkey et al., 2000). Thus, our findings suggest that mVOCs produced by native fluorescent strains stimulate monoterpene synthesis in micropropagated *M. piperita*.

Enhanced production of secondary metabolites may depend directly on improved nutritional status and on primary metabolism of plants following PGPR inoculation (Shulka et al., 1992; Giri et al., 2003; Farag et al., 2006). Studies by Zhang et al. (2007, 2008, 2009) suggest that plants exposed to mVOCs experience high availability of nutrients for growth and increased level of photosynthesis. These phenomena could lead to improved distribution of energy to support synthesis of secondary metabolites such as monoterpenes. Alternatively, they could reflect defensive responses of plants to bacterial colonization following PGPR inoculation, in view of the known antimicrobial properties of EOs (Hammerschmidt and Nicholson, 1999; Sangwan et al., 2001; Wittstock and Gershenzon, 2002). Studies by Radruppa et al. (2010) demonstrated induced systemic resistance (ISR) mediated by ethylene and salicylic acid in *A. thaliana* exposed to acetoin, a component of mVOCs produced by *B. subtilis* (Farag et al., 2006). Even in the absence of contact between a plant and bacterial mVOCs, enhanced EO production resulting from improved nutritional status could be mediated by improved defensive ability of the plant following mVOC exposure.

Chemical Analysis of mVOCs

mVOCs of native strains SJ04, SJ25, and SJ48, and reference strain *P. fluorescens* WCS417r, were analyzed using Super Q®filters. Only compounds not present in control samples (Table 7) were considered. Strain SJ48 showed the greatest mVOC production. Strains SJ25 and WCS417r had the same chromatographic profile. A total of 11 components were identified, belonging to 3 classes: hydrocarbons (cyclohexane, decane, dodecane, 2,6,10-trimethyl-tetradecane, dotriacontane, 11-decyl-docosane), aromatic compounds (methyl-benzene, 2-benzyloxy-ethanamine, 1-methy lnonadecyl- benzene, 1-(N-phenylcarbamyl)-2-morpholinocy clohexene), and halogenated compounds (1-chlorooctadecane). Hydrocar bons such as cyclohexane and dodecane were detected in strains of the genera *Bacillus* (Farag et al., 2006) and *Paenibacillus* (Lee et al., 2012), and were found to increase plant biomass in *A. thaliana*. They also induced ISR status, even though they were not the major mVOC components. The hydrocarbons decane and dodecane were detected in *Pseudomonas* strains, but did not display specific biocontrol activity against the plant pathogenic fungus *Sclerotinia sclerotiorum* (Fernando et al., 2005).

We conclude that the obtained chromatographic profiles are characteristic of the three native strains analyzed. This is interesting considering that the strains were isolated from the same location and presumably subjected to similar selective pressures.

It should be noted that low chemical diversity in a pool of mVOCs does not necessarily relfect low activity of the producer strain. Previous studies, even when focused on the major compound in the mVOC chemical profile, have consistently concluded that the degree of the observed effect reflects synergism among the compounds present. Even when investigators evaluated promoting activity using standards, they did not observe the same level of response in the parameters analyzed in comparison with the entire mVOC pool (Fernando et al., 2005; Farag et al., 2006; Radruppa et al., 2010; Groenhagen et al., 2013).

Table 7. mVOCs identified in headspace extracts of native and reference strains of fluorescent *Pseudomonas*.

rt (min)	Compound	Fluorescent *Pseudomonas* strains			
		SJ04	SJ25	SJ48	WCS417r
6.20	Cyclohexane		xxx		xxx
7.92	2-(benzyloxy)ethanamine			x	
8.02	Benzene, methyl		xxx		xxx
12.30	Decane			x	
13.28	1-(N-phenylcarbamyl)-2-morpholinocyclohexene	xx			
14.21	Dodecane			x	
15.24	Benzene, (1-methylnonadecyl)			x	
15.95	1-chlorooctadecane			x	
17.15	Tetradecane, 2,6,10-trimethyl			x	
20.59	Dotriacontane			x	
22.79	Docosane, 11-decyl			x	

Compounds were identified based on comparison of mass spectra with the NIST database. Chemical signals observed in both medium control and samples are not shown. Percentages of compounds are classified as 1–10% (x), 10–30 (xx), or 30–100% (xxx).

From a functional point of view, the various mVOC pools did not generate the same effects. Each of the native fluorescent bacterial strains analyzed had different PGPR activity. Although the strains all produced an increase in total fresh weight, their enhancing effect on EO content was variable, be causing of differing mixtures of emitted mVOCs. For example, mVOCs produced by strain SJ25 increased total EO content without improving the composition. In contrast, SJ04 and SJ48mVOCs improved the composition of EOs produced by the plants. Exposure of plants to SJ04 mVOCs resulted in decreased menthofuran and increased limonene and menthol contents. Exposure to SJ48 mVOCs similarly resulted in decreased menthofuran and increased limonene. These strains improved the quality of extracted EOs because of their ability to reduce menthofuran production, causing a diversion to the menthol production pathway, the primary indicator of EO quality (Lange and Croteau, 1999; Turner and Croteau, 2004). Clearly, differing mVOC pools emitted by bacteria induce differential responses in micropropagated *M. piperita*, and allow us to improve both quantity and quality of EO production.

CONCLUSION

We identified *Pseudomonas putida* as the predominant fluorescent PGPR species present in rhizospheric soil of *Mentha piperita* (peppermint), using a combination of phenotypic testing and molecular techniques (ARDRA, *rDNA 16S* sequence analysis). This species accounted for ~60% of native strains analyzed. Other *Pseudomonas* strains present in the rhizospheric soil were also identified. The whole genome of the fluorescent bacterial population isolated was evaluated using rep-PCR techniques (ERIC-PCR, BOX-PCR), which revealed the overall complexity of the population and the degree of dissimilarity among strains.

This is the first study focused on fluorescent bacterial populations in rhizospheric soil of aromatic plants in general and *M. piperita* in particular. The isolated native bacterial strains promoted *M. piperita* growth, through mVOC production. Exposure of plants to mVOCs results in increased shoot fresh weight and EO production, through improved quality of EO composition. The obtained chromatographic profiles were characteristic of the three native strains analyzed, and each mVOC pool generated a different level of PGPR activity in micropropagated *M. piperita* plantlets, which is interesting considering that the strains were isolated from the same location and presumably subjected to similar selective pressures. Our findings provide a basis for extensive further studies, and demonstrate the importance of PGPR properties in analysis of isolated native strains, in terms of phylogenetic position and adaptation to environmental and edaphic conditions. The combined analytical approach used represents a significant biotechnological advance in understanding of growth promotion of aromatic plants by rhizobacteria.

ACKNOWLEDGMENTS

This study was supported by grants from the Secretaría de Ciencia y Técnica de la Universidad Nacional de Río Cuarto, the Consejo Nacional de Investigaciones Científicas y Técnicas (CONICET), and the Agencia Nacional de Promoción Científica y Tecnológica (ANPCyT), Argentina. EB, WG, and PB are Career Members of CONICET. NN, LC, and MS have fellowships from CONICET. The authors are grateful to Dr. S. Anderson for English editing of the manuscript.

REFERENCES

1. Achouak, W., Thiery, J. M., Rouband, P., and Heulin, T. (2000). Impact of crop management on intraspecific diversity of *Pseudomonas corrugata* in bulk soil. *FEMS Microbiol. Ecol.* 31, 11–19. doi: 10.1111/j.1574-6941.2000.tb00666.x
2. Agaras, B., Wall, L. G., and Valverde, C. (2012). Specific enumeration and analysis of the community structure of culturable pseudomonads in agricultural soils under no-till management in Argentina. *Appl. Soil Ecol.* 61, 305–319. doi: 10.1016/j.apsoil.2011.11.016
3. Altschul, S., Madden, T., Schäffer, A., Zhang, J., Zhang, Z., Miller, W., et al. (1997). Gapped BLAST and PSI-BLAST: a new generation of protein database search. *Nucleic Acids Res.* 25, 3389–3402. doi: 10.1093/nar/25.17.3389
4. Balzarini, M. (2003). *Aplicaciones del Análisis Estadístico Multivariado.* Material teórico. Maestría en estadística aplicada; Universidad Nacional de Córdoba, Córdoba.
5. Banchio, E., Xie, X., Zhang, H., and Paré, P. W. (2009). Soil bacteria elevate essential oil accumulation and emissions in sweet basil. *J. Agric. Food Chem.* 57, 653–657. doi: 10.1021/jf8020305
6. Bogino, P., Abod, A., Nievas, F., and Giordano, W. (2013). Water-limiting conditions alter the structure and biofilm-forming ability of bacterial multispecies communities in the Alfalfarhizosphere. *PLoS ONE* 8:e79614. doi: 10.1371/journal.pone.0079614
7. Braun, B., Böckelmann, U., Grohmann, E., and Szewzyk, U. (2006). Polyphasic characterization of the bacterial community in an urban soil profile with *in situ* and culture-dependent methods. *Appl. Soil Ecol.* 31, 267–279. doi: 10.1016/j.apsoil.2005.05.003
8. Chen, F., Ro, D.-K., Petri, J., Gershenzon, J., Bohlmann, J., Pichersky, E., et al. (2004). Characterization of a root-specific *Arabidopsis* terpene synthase responsible for the formation of the volatile monoterpene 1,8-Cineole. *Plant Physiol.* 135, 1956–1966. doi: 10.1104/pp.104.044388
9. Di Renzo, J. A., Cas anoves, F., Gonzalez, L. A., Tablada, E. M., Díaz, M. P., Robledo, C. W., et al. (2001). *Estadística para las Ciencias Agropecuarias, Cuarta Edición.* Córdoba: Editorial Triunfar.
10. Farag, M. A., Ryu, C.-M., Sumner, L. W., and Paré, P. W. (2006). GC–MS SPME profiling of rhizobacterial volatiles reveals prospective inducers of growth promotion and induced systemic resistance in plants. *Phytochemistry* 67, 2262–2268. doi: 10.1016/j.phytochem.2006.07.021
11. Fernando, D. W., Ramarathnam, R., Krishnamoorthy, A., and Savchuk, S. (2005). Identification and use of potential bacterial organic antifungal volatiles in biocontrol. *Soil Biol. Biochem.* 37, 955–964. doi: 10.1016/j.soilbio.2004.10.021
12. Fuchs, R., Schäfer, M., Geoffroy, V., and Meyer, J.-M. (2001). Siderotyping – A powerful tool for the characterization of pyoverdines. *Curr. Top. Med. Chem.* 1, 31–57. doi: 10.2174/1568026013395542
13. Giri, B., Kapoor, R., and Mukerji, K. G. (2003). Influence of arbuscular mycorrhizal fungi and salinity on growth, biomass and mineral nutrition of *Acacia auriculiformis. Biol. Fertil. Soils* 38, 170–175. doi: 10.1007/s00374-003-0636-z
14. Gomila, M., Ramirez, A., and Lalucat, J. (2007). Diversity of environmental *Mycobacterium* isolates from hemodialysis water as shown by a multigene sequencing approach. *Appl. Environ. Microbiol.* 73, 3787–3797. doi: 10.1128/AEM.02934-06
15. Gordon, R. E., and Mihm, J. M. (1956). A comparative study of some strains received as *Nocardia. J. Bacteriol.* 73, 15–27.
16. Grimont, P. A. D. (2002). Risque bactériologique: repérage des souches. *C. R. Biol.* 325, 901–906. doi: 10.1016/S1631-0691(02)01499-3

17. Groenhagen, U., Baumgartner, R., Bailly, A., Gardiner, A., Eberl, L., Schulz, S., et al. (2013). Production of bioactive volatiles by different *Burkholderia ambifaria* strains. *J. Chem. Ecol.* 39, 892–906. doi: 10.1007/s10886-013-0315-y
18. Gupta, M. L., Prasad, A., Ram, M., and Kumar, S. (2002). Effect of vesicular-arbuscular mycorrhizal (VAM) fungus *Glomus fasciculatum* on the essential oil yield related characters and nutrient acquisition in the crops of different cultivars of menthol mint (*Mentha arvensis*) under field conditions. *Bioresour. Technol.* 81, 77–79. doi: 10.1016/S0960-8524(01)00109-2
19. Hammerschmidt, R., and Nicholson, R. (1999). "A survey of plant defense responses to pathogens," in *Induced Plant Defenses against Pathogens and Herbivores: Biochemistry, Ecology, and Agriculture*, eds A. Agrawal, S. Tuzun, and E. Bent (Minnesota, SD: APS Press, St. Paul), 55–57.
20. Harrewijn, P., van Oosten, A. M., and Piron, P. G. M. (2001). Natural terpenoids as messengers. A multidisciplinary study of their production, biological functions and practical applications. *Ann. Bot.* 90, 299–300. doi: 10.1093/aob/mcf187
21. Higgins, D., Thompson, J., Gibson, T., Thompson, J. D., Higgins, D. G., and Gibson, T. J. (1994). CLUSTAL W: improving the sensitivity of progressive multiple sequence alignment through sequence weighting, position-specific gap penalties and weight matrix choice. *Nucleic Acids Res.* 22, 4673–4680. doi: 10.1093/nar/22.22.4673
22. Hoagland, D. R., and Arnon, D. I. (1938). *The Water-Culture Method for Growing Plants without Soil.* Univ. Calif. Coll. Agric. Exp. Sta. Circ., Berkeley.
23. Houlden, A., Timms-Wilson, T. M., Day, M. J., and Bailey, M. J. (2008). Influence of plant developmental stage on microbial community structure and activity in the rhizosphere of three field crops. *FEMS Microbiol. Ecol.* 65, 193–201. doi: 10.1111/j.1574-6941.2008.00535.x
24. Hu, H. B., Xu, Y. Q., Chen, F., Zhang, X. H., and Hur, B. K. (2005). Isolation and characterization of a new fluorescent *Pseudomonas* strain that produces both phenazine-1-carboxylic acid and pyoluteorin. *J. Microbiol. Biotechnol.* 15, 86–90.
25. Kimura, M. (1980). A simple method for estimating evolutionary rate of base substitutions through comparative studies of nucleotide sequences. *J. Mol. Evol.* 16, 111–120. doi: 10.1007/BF01731581
26. King, E. O., Ward, M. K., and Randey, D. E. (1954). Two simple media for the demonstration of pyocyanin and fluorescein. *J. Lab. Clin. Med.* 44, 301–307.
27. Kloepper, J. W., and Schroth, M. N. (1978). "Plant growth-promoting rhizobacteria in radish," in *Proceeding of the 4th International Conference on Plant Pathogenic Bacteria*, Vol. 2. Angers: Station de Pathologic Vegetal et Phytobacteriologic, 879–882.
28. Lange, B. M., and Croteau, R. (1999). Genetic engineering of essential oil production in mint. *Curr. Opin. Plant Biol.* 2, 139–144. doi: 10.1016/S1369-5266(99)80028-4
29. Lee, B., Farag, M. A., Park, H. B., Kloepper, J. W., Lee, S. H., and Ryu, C.-M. (2012). Induced resistance by a long-chain bacterial volatile: elicitation of plant systemic defense by a C13 volatile produced by *Paenibacillus polymyxa*. *PLoS ONE* 7:e48744. doi: 10.1371/journal.pone.0048744
30. Loaces, I., Ferrando, L., and Scavino, A. F. (2011). Dynamics, diversity and function of endophytic siderophore-producing bacteria in rice. *Microb. Ecol.* 61, 606–618. doi: 10.1007/s00248-010-9780-9
31. Luria, S. E., and Burrous, J. W. (1955). Hybridization between *Escherichia coli* and *Shigella*. *J. Bacteriol.* 74, 461–476.
32. Lysenko, O. (1961). *Pseudomonas* - an attempt at a general classification. *J. Gen. Microbiol.* 25, 379–408. doi: 10.1099/00221287-25-3-379
33. Mavrodi, D. V., Bonsall, R. F., Delaney, S. M., Soule, M. J., Phillips, G., and Thomashow, L. S. (2001). Functional analysis of genes for biosynthesis of pyocyanin and phenazine-1-carboxamide from *Pseudomonas aeruginosa* PAO1. *J. Bacteriol.* 183, 6454–6465. doi: 10.1128/JB.183.21.6454-6465.2001

34. McConkey, M. E., Gershenzon, J., and Croteau, R. B. (2000). Developmental regulation of monoterpene biosynthesis in the glandular trichomes of peppermint. *Plant Physiol.* 122, 215–224. doi: 10.1104/pp.122.1.215

35. Medina, P., and Baresi, L. (2007). Rapid identification of gelatin and casein hydrolysis using TCA. *J. Microbiol. Meth.* 69, 391–393. doi: 10.1016/j.mimet.2007.01.005

36. Mehnaz, S., Weselowski, B., Aftab, F., Zahid, S., Lazarovits, G., and Iqbal, J. (2009). Isolation, characterization and effect of fluorescent pseudomonads on micropropagated sugarcane. *Can. J. Microbiol.* 55, 1007–1011. doi: 10.1139/W09-050

37. Mehri, I., Turki, Y., Chair, M., Chérif, H., Hassen, A., Meyer, J. M., et al. (2011). Genetic and functional heterogeneities among fluorescent *Pseudomonas* isolated from environmental samples. *J. Gen. Appl. Microbiol.* 57, 101–114. doi: 10.2323/jgam.57.101

38. Naik, P. R., Raman, G., Narayanan, K. B., and Sakthivel, N. (2008). Assessment of genetic and functional diversity of phosphate solubilizing fluorescent pseudomonads isolated from rhizospheric soil. *BMC Microbiol.* 8:230. doi: 10.1186/1471-2180-8-230

39. Nievas, F., Bogino, P., Nocelli, N., and Giordano, W. (2012). Genotypic analysis of isolated penaut-nodulating rhizobial strains reveals differences among populations obtained from soils with different cropping histories. *Appl. Soil Ecol.* 53, 74–82. doi: 10.1016/j.apsoil.2011.11.010

40. Palleroni, N. J. (2005). "Genus *Pseudomonas*," in *Bergey's Manual of Systematic Bacteriology*, 2nd Edn, Vol. 2., eds D. J Brenner, N. R. Krieg, and J. T. Staley, (New York, NY: Springer), 323–379.

41. Parejko, J. A., Mavrodi, D. V., Mavrodi, O. V., Weller, D. M., and Thomashow, L. S. (2012). Population structure and diversity of phenazine-1-carboxylic acid producing fluorescent *Pseudomonas* spp. from dryland cereal fields of central Washington State (USA). *Microb. Ecol.* 64, 226–241. doi: 10.1007/s00248-012-0015-0

42. Patten, C. L., and Glick, B. R. (2002). Role of *Pseudomonas putida* indole acetic acid in development of the host plant root system. *Appl. Environ. Microbiol.* 68, 3795–3801. doi: 10.1128/AEM.68.8.3795-3801.2002

43. Prosser, J. I. (2002). Molecular and functional diversity in soil microorganisms. *Plant Soil* 244, 9–17. doi: 10.1023/A:1020208100281

44. Radruppa, T., Biedrzycki, M. L., Kunjeti, S. G., Donofrio, N. M., Czymmek, K. J., Paré, P. W., et al. (2010). The rhizobacterial elicitor acetoin induces systemic resistance in *Arabidopsis thaliana*. *Commun. Integr. Biol.* 3, 130–138. doi: 10.4161/cib.3.2.10584

45. Ram, D., Ram, M., and Singh, R. (2006). Optimization of water and nitrogen application to menthol mint (*Mentha arvensis* L.) through sugarcane trash mulch in a sandy loam soil of semi-arid subtropical climate. *Bioresour. Technol.* 97, 886–893. doi: 10.1016/j.biortech.2005.04.047

46. Rameshkumar, N., Ayyadurai, N., Kayalvizhi, N., and Gunasekaran, P. (2012). Genotypic and phenotypic diversity of PGPR fluorescent pseudomonads isolated from the rhizosphere of sugarcane (*Saccharum officinarum* L.). *J. Microbiol. Biotechnol.* 22, 13–24. doi: 10.4014/jmb.1107.07025

47. Rangel-Castro, J. I., Levenfors, J. J., and Danell, E. (2002). Physiological and genetic characterization of fluorescent *Pseudomonas* associated with *Cantharellus cibarius*. *Can. J. Microbiol.* 48, 739–748. doi: 10.1139/w02-062

48. Saitou, N., and Nei, M. (1987). The neighbor-joining method: a new method for reconstructing phylogenetic trees. *Mol. Biol. Evol.* 4, 406–425.

49. Salisbury, W. A., and Lykos, J. J. (1972). Hydrolysis of casein: a differential aid for the indentification of *Serratia marcescens*. *J. Clin. Path.* 25, 1083–1085. doi: 10.1136/jcp.25.12.1083

50. Sangwan, V., Foulds, I., Singh, J., and Dhindsa, R. S. (2001). Cold–activation of *Brassica napus* BN115 promoter is mediated by structural changes in membranes

and cytoskeleton, and requires Ca^{2+} influx. *Plant J.* 27, 1–12. doi: 10.1046/j.1365-313x.2001.01052.x

51. Santoro, M. V., Zygadlo, J., Giordano, W., and Banchio, E. (2011). Volatile organic compounds from rhizobacteria increase biosynthesis of essential oils and growth parameters in peppermint (*Mentha piperita*). *Plant Physiol. Biochem.* 49, 177–1082. doi: 10.1016/j.plaphy.2011.07.016
52. Selvakumar, G., Piyush, J., Mishra, P. K., Bisht, J. K., and Gupta, H. S. (2009). Mountain aspect influences the genetic clustering of rhizospheric psychrotolerant phosphate solubilizing *Pseudomonads* in the Uttarakhand Himalayas. *Curr. Microbiol.* 50, 432–438. doi: 10.1007/s00284-009-9456-1
53. Shukui, Z., Lu, X., Dong, L., Xing, J., Su, X., Kong, H., et al. (2005). Quantitative determination of compounds in tobacco essential oils by comprehensive two-dimensional gas chromatography coupled to time-of-flight mass spectrometry. *J. Chromatogr.* 1086, 107–114. doi: 10.1016/j.chroma.2005.04.007
54. Shulka, A., Abad Farooqi, A. H., Shukla, Y. N., and Sharma, S. (1992). Effect of triacontanol and chlormequat on growth, plant hormones and artemisinin yield in *Artemisia annuaL. Plant Growth Regul.* 11, 165–171. doi: 10.1007/BF00024071
55. Sierra, G. (1957). A simple method for the detection of lipolytic activity of microorganisms and some observations on the influence of the contact between cells and fatty substrates. *Antonie Van Leeuwenhoek* 23, 15–22. doi: 10.1007/BF02545855
56. Sneath, P. H. A. (1956). Cultural and biochemical characteristics of the genus *Chromobacterium. J. Gen. Microb.* 15, 70–98. doi: 10.1099/00221287-15-1-70
57. Stanier, R. Y., Palleroni, N. J., and Doudoroff, M. (1966). The aerobic *pseudomonads*: a taxonomic study. *J. Gen. Microbiol.* 43, 159–271. doi: 10.1099/00221287-43-2-159
58. Tamura, K., Dudley, J., Nei, M., and Kumar, S. (2007). MEGA4: molecular evolutionary genetics analysis (MEGA) software version 4.0. *Mol. Biol. Evol.* 24, 1596–1599. doi: 10.1093/molbev/msm092
59. Torsvik, V., Sorheim, R., and Goksoyr, J. (1996). Total bacterial diversity in soil and sediment communities-a review. *J. Ind. Microbiol.* 17, 170–178. doi: 10.1007/BF01574690
60. Turner, G. W., and Croteau, R. (2004). Organization of monoterpene biosynthesis in *Mentha*. Immunocytochemical localizations of geranyl diphosphate synthase, limonene-6-hydroxylase, isopiperitenol dehydrogenase and pulegone reductase. *Plant Physiol.* 136, 4215–4227. doi: 10.1104/pp.104.050229
61. Urzí, C., Brusetti, L., Salamone, P., Sorlini, C., Stackebrandt, E., and Daffonchio, D. (2001). Biodiversity of *Geodermatophilaceae* isolated from altered stones and monuments in the Mediterranean basin. *Environ. Microb.* 3, 471–479. doi: 10.1046/j.1462-2920.2001.00217.x
62. Versalovic, J., Koeuth, T., and Lupski, J. R. (1991). Distribution of repetitive DNA sequences in eubacteria and application to fingerprinting of bacterial genomes. *Nucleic Acids Res.* 19, 6823–6831. doi: 10.1093/nar/19.24.6823
63. Versalovic, J. S. M., de Bruijn, F. J., and Lupski, J. R. (1994). Genomic fingerprinting of bacteria using repetitive sequence-based polymerase chain reaction. *Methods Mol. Cell Biol.* 5, 25–40.
64. Viti, C., and Giovannetti, L. (2005). Characterization of cultivable heterotrophic bacterial communities in Cr-polluted and unpolluted soils using Biolog and ARDRA approaches. *Appl. Soil Ecol.* 28, 101–112. doi: 10.1016/j.apsoil.2004.07.008
65. Vyas, P., Rahi, P., and Gulati, A. (2009). Stress tolerance and genetic variability of phosphate-solubilizing fluorescent *Pseudomonas* from the cold deserts of the trans-Himalayas. *Microb. Ecol.* 58, 425–434. doi: 10.1007/s00248-009-9511-2
66. Wang, M. C., Liu, Y. H., Wang, Q., Gong, M., Hua, X. M., Pang, Y. J., et al. (2008). Impacts of methamidophos on the biochemical, catabolic, and genetic characteristics of soil microbial communities. *Soil Biol. Biochem.* 40, 778–788. doi: 10.1016/j.soilbio.2007.10.012

67. Weisburg, W. G., Barns, S. M., Pelletier, D. A., and Lane, D. J. (1991). 16S Ribosomal DNA amplification for phylogenetic study. *J. Bacteriol.* 173, 697–703.
68. Willems, A., Coopman, R., and Gillis, M. (2001). Phylogenetic and DNA–DNA hybridization analyses of *Bradyrhizobium* species. *Int. J. Syst. Evol. Microbiol.* 51, 111–117. doi: 10.1099/00207713-51-1-111
69. Wittstock, U., and Gershenzon, J. (2002). Constitutive plant toxins and their role in plant defense. *Curr. Opin. Plant Biol.* 5, 300–307. doi: 10.1016/S1369-5266(02)00264-9
70. Wolska, K., Kot, B., Jakubczak, A., and Rymuza, K. (2011). BOX-PCR is an adequate tool for typing of clinical *Pseudomonas aeruginosa* isolates. *Folia Histochem. Cytobiol.* 49, 734–738. doi: 10.5603/FHC.2011.0098
71. Wolska, K., and Szweda, P. (2008). A Comparative evaluation of PCR ribotyping and ERIC PCR for determining the diversity of clinical *Pseudomonas aeruginosa* isolates. *Pol. J. Microbiol.* 57, 157–163.
72. Wu, M., Zhang, X., Zhang, H., Zhang, Y., Li, X., Zhou, Q., et al. (2009). Soil *Pseudomonas* community structure and its antagonism towards *Rhizoctonia solani* under the stress of acetochlor. *Bull. Environ. Contam. Toxicol.* 83, 313–317. doi: 10.1007/s00128-009-9731-7
73. Zhang, H., Kim, M. S., Krishnamachari, V., Payton, P., Sun, Y., and Crimson, M. (2007). Rhizobacterial volatile emissions regulate auxin homeostasis and cell expansion in *Arabidopsis*. *Planta* 226, 839–851. doi: 10.1007/s00425-007-0530-2
74. Zhang, H., Sun, Y., Xie, X., Kim, M. S., Dowd, S. E., and Paré, P. W. (2009). A soil bacterium regulates plant acquisition of iron via deficiency-inducible mechanisms. *Plant J.* 58, 568–577. doi: 10.1111/j.1365-313X.2009.03803.x
75. Zhang, H., Xie, X., Kim, M. S., Kornyeyev, D. A., Holaday, S., and Paré, P. W. (2008). Soil bacteria augment *Arabidopsis* photosynthesis by decreasing glucose sensing and abscisic acid levels in planta. *Plant J.* 56, 264–273. doi: 10.1111/j.1365-313X.2008.03593.x

CHAPTER 21

Different Responses of Rhizosphere and Non-Rhizosphere Soil Microbial Communities to Consecutive *Piper Nigrum* L. Monoculture

Zhigang Li, Chao Zu, Can Wang, Jianfeng Yang, Huan Yu & Huasong Wu

Spice and Beverage Research Institute, Chinese Academy of Tropical Agricultural Sciences & Key Laboratory of Genetic Resources Utilization of Spice and Beverage Crops, Ministry of Agriculture & Hainan Provincial Key Laboratory of Genetic Improvement and Quality Regulation for Tropical Spice and Beverage Crops, Wanning 571533, Hainan, P. R. China.

ABSTRACT

Soil microorganisms have important influences on plant growth and health. In this study, four black pepper fields consecutively monocultured for 12, 18, 28 and 38 years were selected for investigating the effect of planting age on rhizosphere and non-rhizosphere soil microbial communities and soil physicochemical properties. The results revealed that the relative abundance of the dominant bacterial phyla in rhizosphere soil increased considerably with long-term consecutive monoculture but decreased in non-rhizosphere soil with a significant decline in *Firmicutes*. For fungi, an increasing trend over time was observed in both rhizosphere and non-rhizosphere soils, with the abundance of the pathogenic fungi *Fusarium* increasing significantly accompanied by a decrease in the bacteria *Pseudomonas* and *Bacillus* that is beneficial for black pepper. Consecutive monoculture, especially for 38 years, considerably decreased soil microbial diversity. Additionally, the rhizosphere soil pH and organic matter and available K contents decreased with increasing planting duration, though available N and P increased. All soil nutrient contents and microbial diversity indices were higher in rhizosphere soil compared to non-rhizosphere soil. The results suggest that long-term consecutive monoculture leads to variations in soil microbial community composition and physicochemical properties in both rhizosphere and non-rhizosphere soils, thus inhibiting the black pepper growth.

INTRODUCTION

Black pepper (*Piper nigrum* L.), of the Piperaceae family, is a perennial woody climber plant that yields the commercial product known as "black pepper"[1,2]. Considered to be the "King of Spices", black pepper has a wide range of uses, from a simple dietary constituent to compelling pharmacological benefits[3,4]. The plant has great economic importance in many tropical countries, and Vietnam, India, Indonesia, Brazil and China are the world's major producers and exporters of the black pepper commodity[5]. Hainan Province, the major producer and exporter in China, cultivates 22,000 ha of *P. nigrum* and produces 3.6×10^7 kg of dried peppercorns annually, comprising 90% of the pepper production in China. This constitutes an important tropical crop industry in China, with an output value of more than 15 million USD6.

However, consecutive monoculture problems are among the key actors that affect the productivity and quality of black pepper in China[7,8]. Although consecutive cropping is common in agricultural production, it can cause issues such as soil quality degradation[9], crop yield reduction[10], and aggravated crop pest-induced plant diseases[11]. For black pepper, long-term monoculture results in poor growth, low yield and serious disease, severely hindering the black pepper industry. There are a certain number of researches on black pepper consecutive monoculture problems, mainly focused on fruit quality[12], soil nutrition and physicochemical properties[13,14].

Slow decline/wilt, Phytophthora foot rot disease and root-knot nematodes are the major black pepper diseases. These three kinds of diseases occur and spread more frequently in long-term consecutive monocultured black pepper fields[5,7]. *Fusarium* is reported to be the pathogenic fungus of black pepper slow decline/wilt[5], and *Pseudomonas* and *Bacillus* has antagonist effects against *Phy. Capsici*[15] and root-knot nematode[16]. However, the changes of the abundance of the pathogenic fungus *Fusarium* and the antagonistic bacteria *Pseudomonas* and *Bacillus* in the black pepper soil with the planting age has not been reported.

The rhizosphere is a hot spot of microbial interactions, as exudates released by plant roots are the main food source for microorganisms and a driving force of their population density and activities[17,18]. Plants produce root exudates that stimulate heterotrophic growth, which leads to local competition between roots and microorganisms for inorganic nutrients[19,20], and the root-associated microbial community plays an important role in the soil ecosystem, influencing many soil biochemical processes and impacting the growth and health of plants[21]. Although a large number of studies have shown that continuous cropping results in changes in the rhizosphere soil microbial structure, very few studies have involved black pepper. Conversely, non-rhizosphere soil is not (or only lightly) affected by plants roots and root exudates, with a consequent lower level of microbial activity and soil fertility. Nonetheless, this part of the soil is necessary for the stability of soil aggregates and the resistance of soil to negative phenomena (e.g., soil erosion, leaching of nutrients)[22]. There are few studies to date on the effect of continuous cropping on non-rhizosphere or bulk soil microbial community composition. Furthermore, comprehensive comparison of rhizosphere and non-rhizosphere soil microbial communities and the physicochemical property response to continuous cropping are rarely reported.

Soil physicochemical characteristics, especially soil nitrogen (N) availability and soil pH, have a large effect on the composition of soil microbial communities[23,24]. Our previous study reported that long-term black pepper consecutive monoculture led to a significant decline in soil pH, organic matter (OM), and enzymatic activities, resulting in a decrease in soil bacterial abundance[25]. In the present study, we focused in comparing the effect of long-term consecutive monoculture on the composition of bacterial and fungal communities as well as the physicochemical characteristics between rhizosphere and non-rhizosphere soils.

RESULTS

Microbial community composition

After filtering reads according to basal quality control and singleton operational taxonomic unit (OTU) removal, 1,477,241 sequences comprising 9,665 of bacterial OTUs were obtained from 24 samples. The number of high-quality sequences per sample varied from 36,084 to 89,822. Classified sequences across all samples were affiliated with 25 bacterial phyla, and the remaining sequences were unclassified. The dominant phyla (average relative abundance [ARA] >1%) across all samples
were *Acidobacteria* (23.41%), *Proteobacteria* (18.83%), *Firmicutes* (6.75%), *Chloroflexi* (6.37%), *Verrucomicrobia* (5.57%), *Actinobacteria* (3.91%), *P lanctomycetes* (3.87%), *Bacteroidetes* (3.09%) and *Gemmatimonadetes* (1.09%) (Fig. S1a). In addition, *Nitrospira, Armatimonadetes*, *Cyanobacteria* and *WS3* were present in most soils but at relatively low abundance (ARA: 0.1–1%). Twelve other rarer phyla belonged to Others (ARA <0.1%).

Regarding fungi, after filtering reads as indicated above, pyrosequencing-based analysis of ITS sequences recovered 2,039,925 high-quality sequences across all samples, with 47,817–139,906 high-quality sequences per sample clustering into 3,249 OTUs. Based on 97% species similarity, 2,921 fungal OTUs from *Ascomycota* were predominantly observed, accounting for 89.92% of the total fungal sequences obtained (Fig. S1b). The other four phyla observed were *Basidiomycota* (2.52%), *Zygomycota* (0.91%), *Chytridiomycota* (0.44%) and *Glomeromycota* (0.01%); 6.21% of fungi were unclassified.

The relative abundance (RA) of different phyla and genera from rhizosphere and non-rhizosphere soils were compared (Figs 1 and 2). As shown in Fig. 1a, long-term monoculture resulted in considerable increases in RA of the dominant bacteria phyla in the rhizosphere soil but decreases in the non-rhizosphere soil. In rhizosphere soil, RA of *Acidobacteria* and *Firmicutes* decreased, whereas that of *Proteobacteria, Chloroflexi, Verrucomicrobia, Actinobacteria, Planctomycetes, Bacteroidetes* and *Gemmatimonadetes* increased. In non-rhizosphere soil, RA of *Firmicutes* decreased from 20.31% (12 years) to 5.37% (38 years). In contrast, a rising trend of fungal RA was observed in both the rhizosphere and non-rhizosphere soils as planting duration increased (Fig. 1b), with the RA of *Ascomycota, Basidiomycota, Zygomycota* and *Glomeromycota* increasing in both soil types. However, greater *Zygomycota* abundance was detected in rhizosphere soil compared to non-rhizosphere soil.

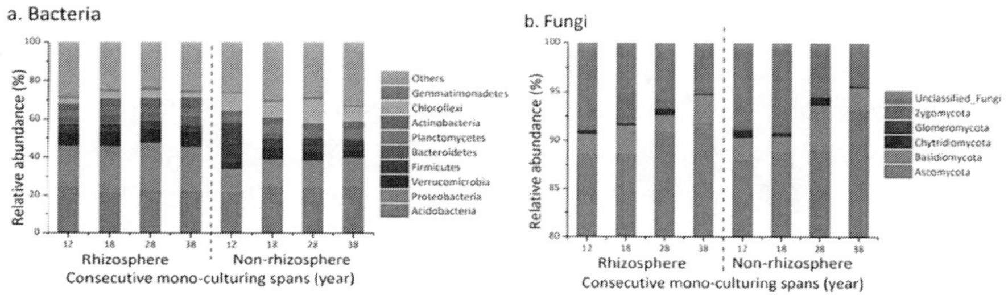

Figure 1 Relative abundance of main bacterial (**a**) and fungal (**b**) phylum in rhizosphere and non-rhizosphere soil at four consecutive monoculture spans.

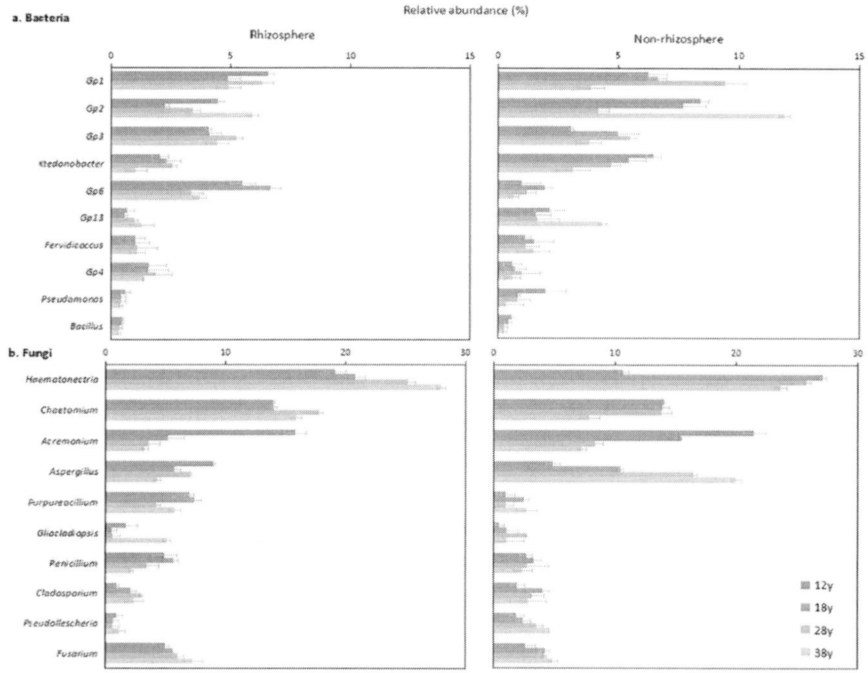

Figure 2 Relative abundances of the top 10 bacterial (**a**) and fungal (**b**) genera in rhizosphere and non-rhizosphere soil at four consecutive monoculture spans. The "12y", "18y", "28y" and "38y" refers to the four consecutive monoculture spans of 12, 18, 28 and 38 years, respectively.

At the genus level (Fig. 2), *GP4* and *GP6* genera had higher RA in rhizosphere soil than in non-rhizosphere soil, whereas the opposite trend was observed for the other eight genera. In non-rhizosphere soil, the RA of *Ktedonobacter* decreased as planting duration increased; however, the RA of other genera exhibited no obvious variation trend. For fungi, the tendencies of almost all genera in the rhizosphere soil were similar to those in the non-rhizosphere soil, except for *Aspergillus,* which decreased in the rhizosphere soil and increased in the non-rhizosphere soil.

Strikingly, the pathogenic fungus *Fusarium* increased significantly as planting duration increased in both the rhizosphere and non-rhizosphere soils. In contrast, *Pseudomonas* and *Bacillus*, beneficial for the growth of black pepper, showed a considerable decline.

BACTERIAL AND FUNGAL A-DIVERSITY

The microbial community richness and diversity indices are listed in Table 1. The coverage of all samples was above 96%, indicating that the depth of the sequencing met the needs of our experiments. With regard to fungi, the Shannon index revealed a considerable decline in both rhizosphere and non-rhizosphere soils as the planting duration increased. In particular, the soil fungi diversity significantly decreased after 38 years of black pepper cultivation. For bacteria, the soil bacterial richness and diversity decreased with planting duration in both the rhizosphere and non-rhizosphere soils, even though the Chao1 and Shannon indices did not reveal a significant difference.

Soil chemical properties and effects on bacterial or fungal taxa

As shown in Table 2, the soil OM and available K (AK) contents decreased significantly with time of monoculture in both the rhizosphere and non-rhizosphere soils. Changes in the pH value for both soils showed fluctuation with a downward trend. Furthermore, available N (AN) and available P (AP) contents increased with time in the rhizosphere soil, and the AP value of the rhizosphere soil was approximately 10 times higher than that of the non-rhizosphere soil.

Table 1: Bacterial and fungal α-diversity indices in rhizosphere and non-rhizosphere soil.

Microbial community	Soil sample	Consecutive monoculturing spans	Coverage	Chao 1	Shannon
Bacteria	Rhizosphere Soil	12 y	96.49 ± 0.24 a	4930.32 ± 421.96 ab	6.72 ± 0.40 a
		18 y	96.36 ± 0.22 a	5248.75 ± 349.74 a	7.09 ± 0.16 a
		28 y	97.10 ± 1.30 a	4756.31 ± 269.36 abc	6.82 ± 0.22 a
		38 y	96.65 ± 0.21 a	4676.70 ± 376.47 abc	6.59 ± 0.26 a
	Non-rhizosphere Soil	12 y	97.33 ± 0.93 a	3854.48 ± 255.05 cd	5.92 ± 0.95 a
		18 y	96.55 ± 0.50 a	4143.82 ± 184.53 bcd	6.63 ± 0.39 a
		28 y	98.01 ± 0.93 a	3502.97 ± 482.39 d	6.22 ± 0.08 a
		38 y	97.49 ± 0.45 a	3620.26 ± 360.17 d	6.06 ± 0.34 a
Fungi	Rhizosphere Soil	12 y	97.43 ± 0.04 a	904.09 ± 23.05 a	4.46 ± 0.05 a
		18 y	97.46 ± 0.41 a	876.56 ± 31.13 a	4.20 ± 0.25 abc
		28 y	97.59 ± 0.67 a	854.45 ± 14.16 ab	4.15 ± 0.09 abc
		38 y	97.65 ± 0.04 a	822.09 ± 22.37 abc	3.88 ± 0.07 c
	Non-rhizosphere Soil	12 y	97.95 ± 0.53 a	764.67 ± 17.44 bcd	4.15 ± 0.12 abc
		18 y	97.54 ± 0.08 a	887.06 ± 3.00 a	4.29 ± 0.17 ab
		28 y	98.11 ± 0.25 a	689.43 ± 80.69 d	3.96 ± 0.12 bc
		38 y	97.98 ± 0.68 a	732.71 ± 26.32 cd	4.07 ± 0.06 bc

Table 2: Soil characteristics of black pepper fields.

Soil sample	Consecutive monoculturing spans	OM (g/kg)	AN (mg/kg)	AP (mg/kg)	AK (mg/kg)	pH
Rhizosphere soil	12 y	21.64 ± 0.69 a	117.48 ± 3.46 bc	204.60 ± 14.63 c	206.85 ± 13.54 b	5.90 ± 0.62 ab
	18 y	17.36 ± 0.57 b	126.93 ± 4.26 ab	342.87 ± 42.62 b	131.03 ± 9.11 cd	5.92 ± 0.20 a
	28 y	15.59 ± 0.26 c	129.85 ± 2.30 a	462.18 ± 20.24 a	168.81 ± 9.07 bc	5.83 ± 0.24 ab
	38 y	15.27 ± 0.40 c	131.03 ± 4.45 a	467.15 ± 17.87 a	141.37 ± 2.75 cd	5.35 ± 0.02 abc
Non-rhizosphere soil	12 y	14.61 ± 0.27 cd	107.96 ± 0.94 c	48.42 ± 4.01 d	253.06 ± 32.34 a	5.17 ± 0.18 bc
	18 y	12.03 ± 0.24 ef	120.44 ± 7.62 ab	49.32 ± 0.70 d	118.32 ± 3.37 d	4.71 ± 0.12 c
	28 y	13.48 ± 1.06 de	129.49 ± 3.44 a	48.20 ± 0.12 d	58.32 ± 1.41 e	4.75 ± 0.10 c
	38 y	11.91 ± 0.29 f	128.25 ± 0.98 ab	48.36 ± 0.54 d	127.53 ± 9.83 d	4.64 ± 0.01 c

Monte Carlo tests based on soil chemical properties and the numbers of the most abundant bacterial and fungal phyla revealed that the selected soil chemical properties were significantly correlated with variations in the selected phyla ($P = 0.01$). Redundancy analysis (RDA) showed that the first and second RDA components explained 52.76% and 30.27% of the total bacterial and fungal variations, respectively (Fig. 3). For bacteria, the first component (RDA1), which explained 37.57% of the total variation, separated the rhizosphere soil samples obtained from the four field plots of different planting duration from the non-rhizosphere soil samples. As shown by their close groupings and vectors, the dominant bacterial phylum in the rhizosphere soil samples was *Proteobacteria* and was related to AP, pH, and OM and AN contents, whereas the dominant bacterial phylum in the non-rhizosphere soil was *Acidobacteria*. Regarding fungi, the first component (RDA1), which explained 15.65% of the total variation of fungal phyla, separated the 38-year rhizosphere and non-rhizosphere soil samples from the other samples. As shown by their close grouping and vectors, the top fungal phylum in the 38-year treatments was *Ascomycota* and was related to the AN content.

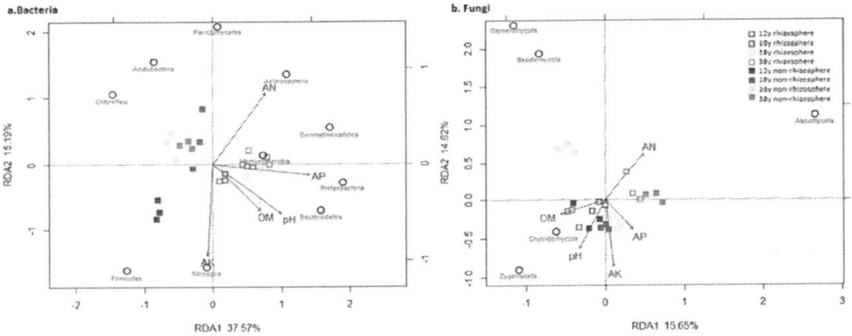

Figure 3 Redundancy analysis (RDA) of abundant soil bacterial (**a**) and fungal (**b**) phyla and soil physicochemical properties for individual samples from four consecutive monoculture in black pepper fields. "12y", "18y", "28y" and "38y" refers to the four consecutive monoculture spans of 12, 18, 28 and 38 years, respectively.

Spearman's rank-order correlation was used to evaluate relationships between abundant phyla and soil physicochemical properties (Table S1). The results indicated different correlations of soil microbial phyla and soil physicochemical properties in the rhizosphere and non-rhizosphere soils. For example, the RA of the non-rhizosphere soil microbial phyla was negatively correlated (phylum *Firmicutes* was positively correlated) to soil pH, but there was no significant correlation with AP. In contrast, there was no significant correlation with pH and a positive correlation with AP in the rhizosphere soil.

DISCUSSION

In this study, Illumina MiSeq sequencing analysis of 16S rRNA and ITS sequences revealed that long-term consecutive black pepper monocropping altered the structure of the soil microbial community. The variations observed for bacteria were consistent with our previous results showing that *Acidobacteria* and *Proteobacteria* are the main phyla, with decreases in *Firmicutes*, under long-term continuous cropping[25]. In non-rhizosphere soil, the RA of *Acidobacteria* considerably increased with continuous planting, whereas *Bacteroidetes* and *Firmicutes* decreased. The variations in these three phyla were consistent with previous research on consecutive cropping[26]. With regard to fungi, *Ascomycota* was the most abundant phylum, accounting for 89.92%, and its RA increased significantly with planting duration. In rhizosphere soil, the RA of the top five phyla all increased with time. However, the *Chytridiomycota* RA decreased slightly in the rhizosphere soil, whereas the RA of the other four phyla increased. In addition, a significant decrease in the Shannon index of the rhizosphere soil was observed in the 38-year old field, suggesting that long-term consecutive cropping may cause black pepper rhizosphere soil fungal diversity to decrease significantly.

Our experimental results demonstrate that at the genus level, the abundance of the pathogenic fungus *Fusarium* increased significantly with time in both rhizosphere and non-rhizosphere soils. In contrast, *Pseudomonas* and *Bacillus*, beneficial bacteria for black pepper, showed a considerable decline. In this study, *Fusarium*, the pathogenic fungus of black pepper slow decline/wilt, significantly increased in both rhizosphere and non-rhizosphere soils. Thus, intensifying pathogen might cause black pepper to be more vulnerable to slow decline/wilt disease in long-term monoculture. Phytophthora foot rot disease and root-knot nematodes are the other two major black pepper diseases in China. *Phytophthora capsici* Leonian, which causes foot and root rot of black pepper, also referred to as "sudden wilt", is a destructive and economically important pathogen of black pepper, occurring wherever the crop is grown[27]. *Phy. capsici* can infect all plant parts during all growth stages, and black pepper yield losses caused by *Phy. capsici* can reach 40–50%[28]. Root-knot nematodes, *Meloidogyne* spp., not only cause root damage that results in rot but also increase black pepper susceptibility to Phytophthora foot rot disease[5]. In our study, the RA of *Pseudomonas* and *Bacillus*, which have been reported to be antagonistic to *Phy. capsici*[15] and root-knot nematode[16], revealed a considerable decline with planting duration in both rhizosphere and non-rhizosphere soils. Increasing pathogen and decreasing antagonistic bacteria might cause black pepper to be more vulnerable to diseases in long-term monoculture fields.

In accordance with our previous study results[23], the soil OM and AK contents and the non-rhizosphere soil pH decreased significantly but AN and AP increased with the duration of consecutive monoculture. This may have been caused by the fertilizer regime applied to the black pepper fields. In China, to obtain higher yields, growers use large amounts of N and P fertilizers, whereas organic and K fertilizers have been ignored for a long time. Applying excessive amounts of N and P fertilizers not only causes the fertilizer to be ineffective but also causes the soil pH to decrease significantly[29,30]. In addition, the rhizosphere soil AP content increased and was approximately 10 times higher than that in the non-rhizosphere soil; this may have been caused by root exudates, which play a major role in P bioavailability *via* several mechanisms[31,32]. Such conditions may enhance the activity of phosphate-solubilizing bacteria, thus increasing the plant's P supply[33]. However, the decline in OM content in both the rhizosphere and non-rhizosphere soils suggests that more organic fertilizers should be applied in long-term consecutive monoculture of black pepper. We also found that in each black pepper field plot, the rhizosphere soil pH as well as the AK (except in the 12-year field), AN, AP and soil OM contents were all higher than in the non-rhizosphere soil. This suggests that long-term consecutive monoculture results in considerable soil nutrient unbalance and soil acidification. Although this deterioration might be a certain ease for rhizosphere soil, this would still inhibit the growth of black pepper.

The biological properties of rhizosphere soil can differ greatly from those of bulk soils[34], and a similar phenomenon was observed in this study. First, the average RA of *Proteobacteria* in the rhizosphere soil (RA of 23.55%) was much higher than that in the non-rhizosphere soil (RA of 14.11%). Second, the RA of *Firmicutes* decreased significantly from 20.31% (12 years) to 5.37% (38 years) in the non-rhizosphere soil, but no significant difference was observed for the rhizosphere soil. Finally, at the genus level, the RA of most genera were different between the rhizosphere and non-rhizosphere soils. These results may be due to the complex interactions among root exudates, nutrient elements and soil microorganisms[35]. Moreover, soil nutrient bioavailability and microbial community richness and diversity were higher in the rhizosphere soil than in the non-rhizosphere soil. Overall, to understand the problems of consecutive monoculture of black pepper, more attention should be given to interactions among soil nutrient elements, root exudates and the microbial community, especially the function of root exudates in rhizosphere soil.

METHODS

Sampling sites

The study was conducted at the Spice and Beverage Research Institute (18°72′N–18°76′N, 110°19′E–110°22′E), Wanning City, Hainan Province, P. R. China. This area is characterized by a tropical monsoon climate with an average temperature of 24.6 °C, average annual precipitation of 2,150 mm, and annual mean relative humidity of 85%.

The experimental soil samples were collected on July 15, 2015, from four different black pepper (*Piper nigrum* L. *cv.* Reyin No. 1) consecutive monoculture fields (12, 18, 28 and 38 years) with the same agronomic management and fertilization regime. For each black pepper field, nine black pepper plants were randomly selected, and soil sampled at random from each of three plants was mixed together as a single soil sample. For each plant, carefully dug up clods with black pepper roots. Soil farther than 1 cm from the roots was collected and considered as non-rhizosphere soil. Then, carefully remove the soil 5 mm away from the roots with brush and tweezers. Soil adjacent to the root segments, at 1–5 mm from the root surface, was shaken off and defined as rhizosphere soil. All soil samples were placed into sterile plastic bags, placed in an ice box, and transported to the laboratory. After passing through a 2-mm sieve, each sample was divided into two subsamples: one portion was air-dried for soil characteristic analysis, and the remainder was stored at −80 °C for DNA extraction.

DNA EXTRACTION AND ILLUMINA MISEQ SEQUENCING

Total DNA was extracted from 24 soil samples using a MoBio PowerSoil™ DNA Isolation Kit (Mo Bio Laboratories Inc., Carlsbad, CA, USA) according to the manufacturer's instructions. The Genomic DNA concentration and purity were measured using a NanoDrop ND-2000 spectrophotometer (NanoDrop Technologies, Wilmington, DE).

Bacterium-specific primers 515F (GTGCCAGCMGCCGCGGTAA) and 806R (GGACTACVSGGGTATCTAAT) were used to amplify the V4 hyper-variable regions of the bacterial 16S rRNA gene. Fungus-specific primers ITS1F (CTTGGTCATTTAGAGGAAGTAA)[36] and ITS2 (GCTGCGTTCTTCATCGATGC)[37] were selected to target the ITS1 region. The polymerase chain reaction (PCR) was performed according to our previous publication[23,38]. Finally, paired-end sequencing was carried out using an Illumina MiSeq sequencer at Personal Biotechnology Co., Ltd (Shanghai, China).

OTU-based sequence analysis

OTU-based sequence analysis was carried out using our previous methods[38]. Briefly, after removing the adaptors and primer sequences, raw sequences were assembled for each sample according to the unique barcode using QIIME[39]. Split sequences for each sample were merged using FLASH V1.2.7[40], and low-quality sequences were discarded by QIIME. The sequences retained for each sample were processed following the established UPARSE pipeline[41]. Briefly, sequences with a quality score lower than 0.5 or a length shorter than 200 bp were removed. After discarding singletons, the remaining reads were assigned to OTUs with a threshold of 97% identity. The chimera removal processes were then performed. Finally, bacterial representative sequences were searched against the RDP database[42,43], and fungal representative OTUs were classified using the UNITE database[44].

Determination of soil physicochemical propertiesh

Soil physicochemical properties were determined in the same manner as reported in our previous paper[23]. Briefly, soil pH was measured using a glass electrode meter in a soil-water suspension (1:5 w/v). OM and AN (hydrolyzable N) were determined using the potassium dichromate external heating method. AP was determined using the molybdenum blue method. AK was determined by flame photometry.

Statistical analyses

Coverage, richness (Chao1 index) and diversity (Shannon index) were used to estimate the α-diversity of each sample using Mothur (version 1.36.0). Correlations between abundant soil microbial phyla and soil characteristics were determined by RDA using the vegan package of R (version 3.2.3). One-way analyses of variance (ANOVA) with Tukey's HSD multiple range tests were performed for multiple comparisons using SPSS V20.0 (SPSS Inc., USA), and Spearman's rank correlations were calculated using SPSS V20.0 (SPSS Inc., USA).

ACKNOWLEDGEMENTS

We would like to thank the staff of the National Engineering Research Center for Organic-based Fertilizers and Nanjing Agricultural University (Nanjing, China) for their tremendous help in analyzing the data and improving the manuscript. This study was financially supported by National Natural Science Foundation of China (31301857) and Natural Science Foundation of Hainan Province (20153102).

REFERENCES

1. Sharangi, A. B., Kumar, R. & Sahu, P. K. *Survivability of black pepper (Piper nigrum L.) cuttings from different portions of vine and growing media. J. Crop Weed* **6**(1), 52–54 (2010).
2. Tilahun, D., Tefera, W. & Weldetsadik, K. *Macro Propagation of Black Pepper (Piper Nigrum L.). As Influenced by Growth Media and Physiological Conditions of the Stock Plant* (ed. Tilahun, D.) Ch. 1, 1 (LAP Lambert Academic Publishing, 2011).
3. Nair K. P. P. *The agronomy and economy of black pepper (Piper nigrum L.) —The "king of spices". Adv. Agron.* **82**, 271–389 (2004).
4. Nair K. P. P. *Agronomy and Economy of Black Pepper and Cardamom* (ed. Nair K. P. P.) Ch. 1, **1** (Elsevier, 2011).
5. Lai, K. F. & Paulus, A. D. *Pepper Production Technology in*Malaysia (eds Lai, K. F. & Sim, S. L.) Ch. 1, 1–6 (Malaysian Pepper Board, 2011).
6. Zu, C. *et al. Acid soil is associated with reduced yield, root growth and nutrient uptake in black pepper (Piper nigrum L.). Agr. Sci.* **5**, 466–473 (2014).
7. Zheng, W. Q. *et al. Chief factors influencing pepper succession cropping and prevention measures. Chinese J. Trop. Agr.* **30**(3), 52–55 (2010).
8. Yang, J. F. *et al. Status and development countermeasures of pepper industry in China. Chinese J. Trop. Agr.* **30**(10), 13–18 (2010).
9. Vargas, G. S. *et al. Field assessment of soil biological and chemical quality in response to crop management practices. World J. Microb. Biot.* **25**(3), 439–448 (2009).
10. Wu, L. K. *et al. Plant-microbe rhizosphere interactions mediated by Rehmannia glutionosa root exudates under consecutive monoculture. Sci. Rep.-UK* **5**, 15871, doi: 10.1038/srep15871 (2015).
11. He, J. Z., Zheng, Y. & Chen, C. R. *Microbial composition and diversity of an upland red soil under long-term fertilization treatments as revealed by culture-dependent and culture-independent approaches. J. Soil Sediment.* **8**, 349–358 (2010).
12. Yu, H., ZU, C., WU, H. S., Yang, J. F. & Zheng, W. Q. *The difference analysis of quality in pepper from different planting years. Chinese J. Trop. Crop* **34**(5), 850–854 (2013).
13. LI, Z. G. *et al. Soil physicochemical properties and microbial dynamics in the fields with the different aging black pepper. Chinese J. Trop. Crop* **33**(7), 1245–1249 (2012).
14. Yang, J. F. *et al. Effects of the planting period on the soil chemical fertility parameters in pepper garden. Chinese J. Trop. Crop* **32**(4), 592–597 (2011).
15. Tran, H. & Kruijt, M. J. *Diversity and activity of biosurfactant-producing Pseudomonas in the rhizosphere of black pepper in Vietnam. J. Appl. Microbiol.* **104**(3), 839–851 (2008).

16. Devapriyanga, R., Jonathan, E. I., Meena, K. S. & Kavitha, P. G.*Bioefficacy of Pseudomonas and Bacillus isolates against root-knot nematode, Meloidogyne incognita in black pepper cv. Panniyur 1. Indian J. Nematol.* **42**(1), 57–62 (2012).
17. Raaijmakers, J. M., Paulitz, T. C., Steinberg, C., Alabouvette, C. & Moënne-Loccoz, Y. *The rhizosphere: a playground and battlefield for soilborne pathogens and beneficial microorganisms. Plant Soil***321**, 341–361 (2009).
18. Berendsen, R., Pieterse, C. & Bakker, P. *The rhizosphere microbiome and plant health. Trends Plant Sci.* **17**, 478–486 (2012).
19. Klemedtsson, L., Svensson, B. H. & Rosswall, T. *Relationships between soil moisture content and nitrous oxide production during nitrification and denitrification. Biol. Fert. Soils* **6**(2), 106–111 (1988).
20. Mukerji, K. G., Manoharachary, C. & Singh, J. *Microbial Activity in the Rhizosphere* (eds Mukerji, K. G. *et al.*). (Springer publishing, 2006).
21. Zhao, Y. Z. *et al. Interaction of Pseudostellaria heterophylla with Fusarium oxysporum f. sp. heterophylla mediated by its root exudates in a consecutive monoculture system. Sci. Rep.-UK* **5**, 8197, doi: 10.1038/srep08197 (2015).
22. Jakub, E. & Jaroslav, Z. *The comparison of microbial activity in rhizosphere and non-rhizosphere soil stressed by drought. Mental. Net.* 234–240 (2014).
23. Sari, S., Anu, E. & Minna, K. M. *Regulation of microbial community composition and activity by soil nutrient availability, soil pH, and herbivory in the Tundra. Ecosystems* **15**, 18–33 (2012).
24. Jennifer, K. C., Deirdre, R., Deirdre, B. G. & Nicholas, C. *Altering the mineral composition of soil causes a shift in microbial community structure. Fems, Microbiol. Ecol.* **61**(3), 414–423 (2007).
25. Xiong, W. *et al. The effect of long-term continuous cropping of black pepper on soil bacterial communities as determined by 454 pyrosequencing. PLoS ONE* **10**(8), e0136946, doi: 10.1371/journal. Pone. 0136946 (2015).
26. Lauber, C. L., Hamady, M., Knight, R. & Fierer, N. *Pyrosequencing-based assessment of soil pH as a predictor of soil bacterial community structure at the continental scale. Appl. Environ. Microbiol.* **75**, 5111–5120 (2009).
27. George, C. K., Abdullah, A. & Chapman, K. *Pepper production guide for Asia and the Pacific. Joint effort of theInternational Pepper Community (IPC) and Food and Agricultural Organization of United Nations (FAO).* Bangkok, Thailand:IPC (2005).
28. Drenth, A. & Sendall, B. *Economic impact of Phytophthora disease in Southeast Asia. Diversity and Management of Phytophthora in Southeast Asia* (eds. Drenth, A. & Guest, D. I.) Ch. 2, 10–28 (Australian Centre for International Agricultural Research, 2004).
29. Malhi, S. J., Nyborg, M. & Harapiak, J. T. *Effects of long term N fertilizer induced acidification and liming on micronutrients in soil land in brome grass hay. Soil Till. Res.* **48**, 91–101 (1998).
30. Jackie, L. S. *et al. Soil acidification from long-term use of nitrogen fertilizers on winter wheat. Soil Sci. Soc. Am. J.* **75**, 957–964 (2011).
31. Hinsinger, P. *Bioavailability of soil inorganic P in the rhizosphereas affected by root-induced chemical changes: a review. Plant Soil* **237**, 173–195 (2001).
32. Vance, C. P., Uhde-Stone, C. & Allan, D. L. *Phosphorus acquisition and use: critical adaptations by plants for securing a nonrenewable resource. New Phytol.* **157**, 423–447 (2003).
33. Richardson, A. E. *Prospects for using soil microorganisms to improve the acquisition of phosphorus by plants. Aust. J. Plant Physiol.* **28**, 897–906 (2001).
34. Hinsinger, P., Bengough, A. G., Vetterlein, D. & Young, I. M.*Rhizosphere: biophysics, biogeochemistry and ecological relevance. Plant Soil* **321**, 117–152 (2009).
35. Jason, A. P. *et al. Diversity and heritability of the maize rhizosphere microbiome under field conditions. PNAS* **110**(16), 6548–6553 (2013).
36. Gardes, M. & Bruns, T. D. *ITS primers with enhanced specificity for basidiomycetes-application to the identification of mycorrhizae and rusts. Mol. Ecol.* **2**, 113–118, doi: 10.1111/j.1365-294X. 1993. tb00005. x (1993).

37. White, T. J., Bruns, T., Lee, S. & Taylor, J. W. *Amplification and direct sequencing of fungal ribosomal RNA genes for phylogenetics. PCR Protoc. Guide Methods Appl.* **18**, 315–322, doi: 10.1016/b978-0-12-372180-8.50042-1 (1990).

38. Xiong, W. *et al. Comparison of fungal community in black pepper-vanilla and vanilla monoculture systems associated with vanilla Fusarium Wilt Disease. Front. Microbiol.* **7**, 117, doi: 10.3389/fmicb. 2016. 00117 (2016).

39. Caporaso, J. G. *et al. QIIME allows analysis of high-through put community sequencing data. Nat. Methods* **7**, 335–336 (2010).

40. Magoč, T. & Salzberg, S. L. *FLASH: fast length adjustment of short reads to improve genome assemblies. Bioinformatics* **27**, 2957–2963, doi: 10.1093/ bioinformatics /btr507 (2011).

41. Edgar, R. C. *UPARSE: highly accurate OTU sequences from microbial amplicon reads. Nat. Methods* **10**, 996–998, doi: 10.1038/nmeth.2604 (2013).

42. Kõljalg, U. *et al. Towards a unified paradigm for sequence-based identification of fungi. Mol. Ecol.* **22**, 5271–5277, doi: 10.1111/mec.12481 (2013).

43. Cole, J. R. *et al. The ribosomal database project: improved alignments and new tools for rRNA analysis. Nucl. Acids Res.* **37**, 141–145 (2009).

44. Wang, Q., Garrity, G. M., Tiedje, J. M. & Cole, J. R. *Naive bayesian classifier for rapid assignment of rRNA sequences into the new bacterial taxonomy. Appl. Environ. Microbiol.* **73**, 5261–5267 (2007).

CHAPTER 22

Highly Efficient Removal of Lead and Cadmium During Wastewater Irrigation using a Polyethylenimine-Grafted Gelatin Sponge

Bingbing Li[1], Feng Zhou[1], Kai Huang[1], Yipei Wang[2], Surong Mei[1], Yikai Zhou[1] & Tao Jing[1]

[1]State Key Laboratory of Environment Health (Incubation), Key Laboratory of Environment and Health, Ministry of Education, Key Laboratory of Environment and Health (Wuhan), Ministry of Environmental Protection, School of Public Health, Tongji Medical College, Huazhong University of Science and Technology, Wuhan, Hubei, 430030, China. [2]Institute of Environmental Pollution and Health, School of Environmental and Chemical Engineering, Shanghai University, Shanghai 200444, China.

ABSTRACT

Wastewater irrigation is a very important resource for heavy metal pollution in soil and then accumulation in vegetable crops. In this study, a polyethylenimine (PEI)-grafted gelatin sponge was prepared to effectively adsorb heavy metals during wastewater irrigation. Based on the strong water adsorption ability, wastewater remained in the PEI-grafted gelatin sponge for a sufficient time for the heavy metals to interact with the sorbents. The binding capacities of Pb(II) ions and Cd(II) ions on the PEI-grafted gelatin sponge were 66 mg g^{-1} and 65 mg g^{-1}, which were much more than those on the gelatin sponge (9.75 mg g^{-1} and 9.35 mg g^{-1}). Subsequently, the PEI-grafted gelatin sponge was spread on the surface of soil planted with garlic and then sprayed with synthetic wastewater. The concentrations of cadmium and lead in the garlic leaves were 1.59 mg kg^{-1} and 5.69 mg kg^{-1}, respectively, which were much lower than those (15.78 mg kg^{-1} and 27.98 mg kg^{-1}) without the gelatin sponge, and the removal efficiencies were 89.9% and 79.7%. The PEI-grafting gelatin sponge could effectively remove heavy metals during wastewater irrigation, which improved the soil environment and reduced human exposure to heavy metals.

INTRODUCTION

Water shortages in the world have led to the reuse of municipal wastewater for irrigation. In many developing countries, wastewater irrigation is a very important way in agricultural production, however, it would increase human exposure to the pollutants. Amin *et al.* found that vegetables grown in the Mardan District of Pakistan were highly polluted by heavy metals due to the irrigation of untreated wastewater, which would lead to serious health risks for local people[1]. Jiang *et al.* reported that agricultural soil was potentially at risk of heavy metal accumulation from irrigation water, which was higher than those from atmospheric deposition and fertilizer[2]. Thus, decrease the contaminants in wastewater system was an important way to avoid their entrance into agricultural soil and then food chain.

Various technologies were proposed to decrease the heavy metal concentration in wastewater, such as ion exchange, adsorption, chemical precipitation and membrane separation[3-7]. Among them, adsorption was considered as a flexible, convenient and low cost technology. After simple desorption, the adsorbents are regenerated for multiple uses and it still has a high efficiency[4,8]. Therefore, it was a preferred process for wastewater pollution control. However, developing a specific sorbent with high adsorption capacity is an interesting yet still challenging task. Currently, carbon materials (such as carbon foam, carbon nanotubes, graphene and activated carbon)[9-11], nanosized metal oxides (such as ferric oxides, manganese oxides and zinc oxides)[4], chitosan composites[12-13] and low-cost adsorbents (such as zeolites and weathered coal)[14-16] are all considered in the literatures to be promising choices for the heavy metal pollution control. Unfortunately, it is impossible to recover these materials and they are often used in a sewage station based on the special operating devices. Wastewater irrigation is different from wastewater emission, which is an unpalatable choice in water scarce countries. They weren't even sure about the sources of irrigated wastewater and dry season. Due to the economic costs, it is difficult for farmers to develop a purification system by themselves to further decrease the amounts of heavy metals using these materials as the sorbents.

The easiest and most direct way is to spread a cheap material on the soil surface to remove heavy metals during wastewater irrigation. In other words, this technique implies that the following two conditions are necessary: this material has high adsorption capacities for the contaminants and allows wastewater to flow through at a slow flowing rate to provide sufficient time for the removal of heavy metals. A gelatin sponge is an easily available biopolymer that not only has good chemical, physical, and biological stability but also contains different types of functional groups, which can interact with heavy metals by different forces[7,17,18]. Most importantly, although the water absorption of a gelatin sponge is very high, the volume does not noticeably increase and a highly porous structure is still maintained for the fast adsorption kinetics[19].

However, due to its crosslink with the amino groups, adsorption capacity of a gelatin sponge is not very high. Surface modification can be used to further improve its adsorption capacity and mechanical stability. Branched poly(ethyleneimine) (PEI) is a water soluble polymer and the amine groups exist in primary, secondary and tertiary forms in any given chain segment[20-22]. Thus, it can be used for heavy metals adsorption in real water samples due to the formation of chelation complexes[23]. In this study, a PEI-grafted gelatin sponge was prepared and spread on the surface of soil planted with garlic to decrease the heavy metal concentrations during wastewater irrigation, because of its following properties: (1) the PEI-grafted gelatin sponge with a high adsorption capacity could be used for the heavy metal pollution control; (2) the high water adsorption ability can provide sufficient time to decease the amounts of heavy metals in wastewater.

RESULTS AND DISCUSSION

Synthetization and characterizations of PEI-grafted gelatin sponge In this study, preparation of PEI-grafted gelatin sponge and its application are shown in Fig. 1. Due to the introduction of 3,3′-dithiodipropionic acid, the amino groups of gelatin sponge were converted to the carboxylic groups by the formation of amide bonding between the carboxylic acid groups of the 3,3′-dithiodipropionic acid and the amino groups of gelatin. Then, the target product PEI-grafted gelatin sponge was further prepared based on the ionic interactions between the amine groups of PEI and the carboxylic groups (3,3′-dithiodipropionic acid and gelatin) based on the NHS/EDC coupling. The number of amino groups increased significantly and the mechanical stability was improved based on the modification of PEI.

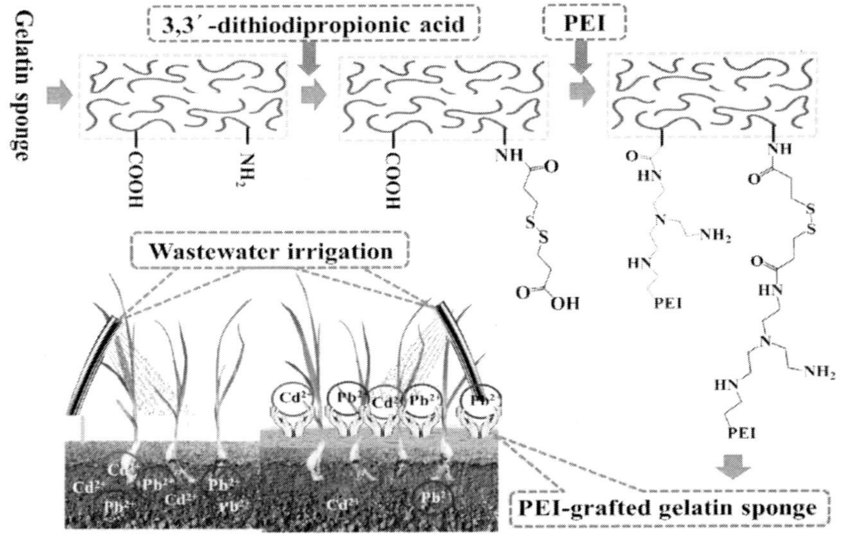

Figure 1: Preparation of the PEI-grafted gelatin sponge and its application in the removal of heavy metals during wastewater irrigation.

Supplementary Fig. S1 indicates the SEM micrographs of the gelatin sponge and PEI modified-gelatin sponge. The surface morphologies were considerably different. The porous three-dimensional (3D) structure of the gelatin sponge was a fibrous network, whereas a lamellar membrane was observed on the PEI modified-gelatin sponge. Zeta potential measurement was used to study the modified mechanism. It was shown that the potential of the gelatin sponge was +12.14 mV in aqueous solution (pH = 5.0), since the gelatin had a net positive charge on the surface at low pH. But, after modified by the 3,3′-dithiodipropionic acid, it changed to −13.78 mV, because of the presence of carboxyl groups. Finally, this value of PEI modified-gelatin sponge was +27.43 mV, implying a PEI layer was created around gelatin sponge.

FT-IR spectra were obtained to verify the change of function groups on the gelatin sponge (see Supplementary Fig. S2). The infrared spectrum of the gelatin sponge exhibited broad bands assigned to O-H and N-H stretching centered at 3429 cm^{-1}. The characteristic peak of C = O and C-N at 1629 cm^{-1} and N-H stretching at 1537 cm^{-1} were attributed to amide I and amide II. The peak emerging at 1336 cm^{-1} implied the presence of amide III[24]. After modification with PEI, the band intensity peak at

3429 cm^{-1} was greatly increased and the peaks of the peptidic bond of the protein were not changed, which indicated that the enhancement of amino groups resulted from PEI.

The surface chemical compositions of the gelatin sponge and PEI-grafted gelatin sponge were studied based on the XPS spectra. It is shown that the nitrogen atom concentration was increased from 11.91% to 17.35%, implying the successful modification of PEI on the gelatin sponge. For the gelatin sponge, the N1s peak at 399.75 eV appeared due to the NH (NH_2) groups, and the O1s peak at 531.7 eV attributed to the C-OH/C-O-C groups (see Supplementary Fig. S3)[24,25]. After PEI modification, a new S1s peak at 167.4 eV was assigned to the -S-S- groups due to the introduction of 3,3′-dithiodipropionic acid.

Effect of pH The pH value was an important parameter for the removal efficiency of the sorbents, because it not only changed the form of heavy metals, but also had a significant influence on the species of functional groups. As shown in Supplementary Fig. S4, the adsorption capacities increased with the increasing of pH value and reached a plateau when this value changed from 4.0 to 7.0. Subsequently, the adsorption capacities were significantly decreased in basic solution because the concentration of OH^- increased and the lead ions or cadmium ions might exist in the forms of $X(OH)^+$ and $X(OH)_2$ (X = Pb or Cd)[26,27]. Thus, the percentage removal of Cd^{2+} (or Pb^{2+}) by precipitation was much greater than that by adsorption at a pH of above 7.0. From the perspective of electrostatic interaction, at low pH values, hydrogen ions could compete with target ions for the active sites, resulting in the decrease of binding capacities. When the pH value was increased, competition from hydrogen ions and the electrical repulsion force were all decreased, and the removal efficiency was gradually increased[28]. Thus, the proposed sorbents can be used for the adsorption of heavy metals in surface water or wastewater due to the formation of chelation complexes[28]. Furthermore, PEI could be used as a proton acceptor for the carboxyl group of gelatin, which would facilitate the binding of the carboxylate to metal ions. Thus, a pH of 6.0 was fixed to remove lead ions and cadmium ions from the real water samples.

Adsorption for lead and cadmium The PEI-grafted gelatin sponge was applied as a sorbent to adsorb heavy metals in wastewater and a common gelatin sponge was also studied for comparison. The evaluated methods are shown in the Supporting Information. The adsorption capacities of lead ions and cadmium ions on the PEI-grafted gelatin sponge were 66.1 mg g^{-1} and 65.2 mg g^{-1}, respectively, which were more than those of the gelatin sponge (9.35 mg g^{-1} and 9.75 mg g^{-1}) (Fig. 2). Gelatin as a biological macromolecule had different types of functional groups, such as -COOH, $-NH_2$, -OH and -SH, which could be applied to remove target ions (see Supplementary Fig. S5). The amino groups could be used not only to remove the heavy metals in wastewater due to the formation of chelation complexes, but also as a proton acceptor for the carboxyl group of gelatin, which would facilitate the binding of the carboxylate to metal ions. Furthermore, the gelatin could also form complexes with metal ions through various side groups such as imidazole and thiol. Further modification of PEI could significantly increase the number of amino groups to enhance the removal efficiency of sorbents.

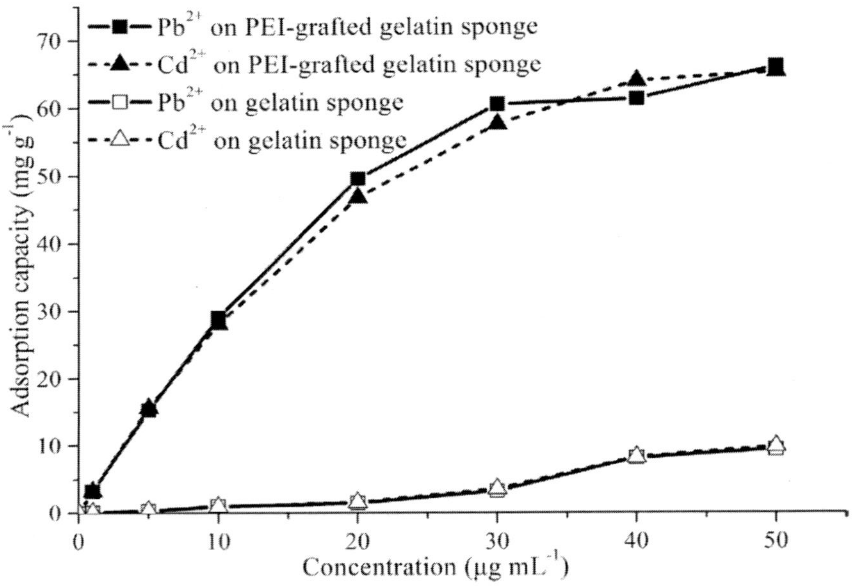

Figure 2: Adsorption isotherms of heavy metals on the gelatin sponge and the PEI-grafted gelatin sponge.

The Freundlich isotherm and Langmuir isotherm are often used to study the adsorption mechanism of the PEI-grafted gelatin sponge. In this study, the adsorption parameters are shown in Table 1. The adsorption capacities of Pb(II) ions and Cd(II) ions onto the PEI-grafted gelatin sponge were increased with the increasing of original concentration, and the experimental data was good fit with the Langmuir model. It was shown that the removing mechanism was the controlled monolayer adsorption. Based on the Langmuir model, the Q_{max} values of the PEI-grafted gelatin sponge were 80.6 mg g^{-1} for lead ions and 79.9 mg g^{-1} for cadmium ions, respectively. Dimensionless constant named as equilibrium parameter was applied to indicate the essential characteristics of adsorption model (Supplementary R_L). These values for both of the test ions on the PEI-grafted gelatin sponge were 0.24, which demonstrated that the experimental parameters were favorable for the removal of heavy metals.

Table 1: The isotherm constants of for PEI-grafted gelatin sponge for the adsorption of heavy metals.

	Langmuir model				Freundlich model		
Test metal ions	Q_{max} (mg g^{-1})	K_L (L mg^{-1})	R_L	R^2	K_F (mg g^{-1})	n	R^2
Pb(II)	80.6	0.0404	0.024	0.9932	6.11	1.27	0.9716
Cd(II)	79.9	0.0404	0.024	0.9896	6.05	1.28	0.9763

Adsorption kinetic studies for Pb(II) ions and Cd(II) ions The adsorption kinetics of the PEI-grafted gelatin sponge for the target ions was studied in this study, especially in wastewater irrigation. Due to the water transmission characteristics of soil, the adsorption of heavy metal ions must be completed when the wastewater flows from the surface of the sponge to the bottom layer. Otherwise, it would lead to soils polluted by heavy metals and toxicological effects for vegetable crops. The high water adsorption ability of PEI-grafted gelatin sponge could slow the transit of wastewater to the soil and provide sufficient time for the adsorption of target pollutants on the sponge. Unlike general removing procedure at a fixed flow rate, remain of wastewater on the sponge was a complex kinetics procedure and there were a complex interaction among water molecules, PEI-grafted gelatin sponge, heavy metal ions and soil. Therefore, shaking adsorption procedure was performed to investigate the interaction between the sponge and heavy metals and water losing experiment was used to evaluate the kinetics procedure of water.

The results of the shaking adsorption experiments indicated that more than 54.7% of the total metals were remained on the PEI-grafted gelatin sponge in the first 15 minutes (Fig. 3), which demonstrated that the removal rate was quite rapid, which was beneficial to decrease heavy metal pollution during wastewater irrigation. The results of the water losing experiments indicated that the outflow of water on the wet PEI-grafted gelatin sponge was changed considerably within the first 30 minutes and then gradually slowed over time. The results indicated that 45.5% water still remained on the sponge after 15 min (Fig. 3). Furthermore, the water molecules organized around the PEI-grafted gelatin sponge could be competitively adsorbed by sodium chloride (NaCl) and then induced a salting-out effect. The curve of the remaining water was not changed even if the concentration of NaCl was 0.1 g mL^{-1}, which implied that the PEI-grafted gelatin sponge had high water adsorption ability.

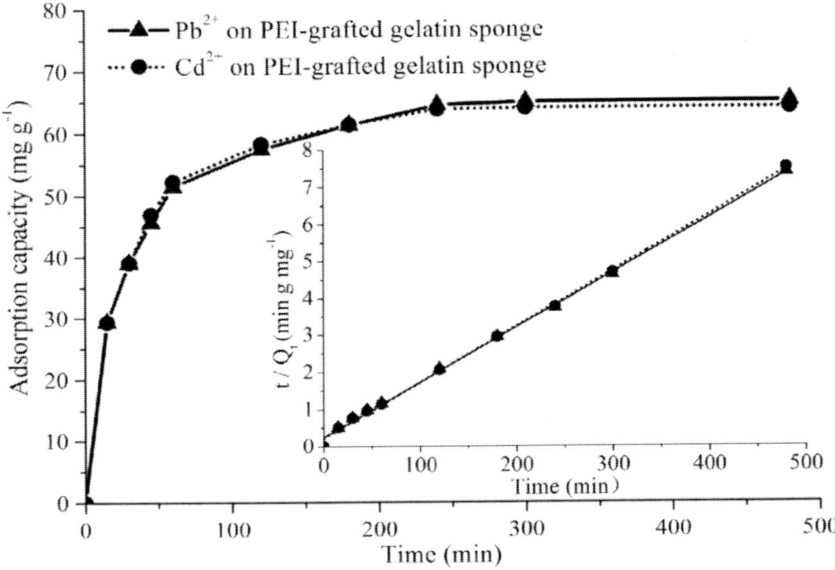

Figure 3: Rebinding kinetic behavior of of heavy metals on the PEI-grafted gelatin sponge. Inset: the pseudo-second-order mode for the adsorption of heavy metals.

To further study the mechanism of adsorption kinetics, intra-particle diffusion, pseudo-first-order and pseudo-second-order were performed to investigate the adsorption dynamics of heavy metals on the PEI-grafted gelatin sponge. The kinetic equations are given in the Supporting Information. It was shown that high correlation coefficients (R^2) were obtained, and the calculated values of the binding capacity ($Q_{e,calc}$) were the same as the experimental values ($Q_{e,exp}$) (Fig. 4, Table 2). Therefore, the adsorption kinetic procedure of heavy metals on the PEI-grafted gelatin sponge was the pseudo-second-order kinetic procedure. Furthermore, the intra-particle diffusion model was further used to describe the adsorption process of metal ions. The plots exhibited double linearity, implying there were two steps in the adsorption process (see Supplementary Fig. S6).

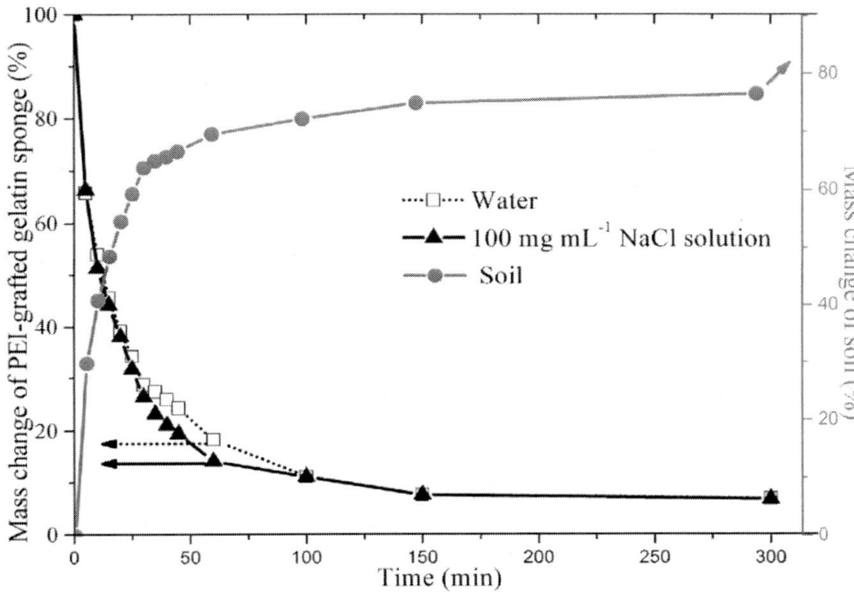

Figure 4: Retention behavior of water on the PEI-grafted sponge and the effect of sodium chloride on the retention behavior.

Table 2: The kinetic constants of PEI-grafted gelatin sponge for the adsorption of heavy metals. Notes: *Q_e denotes the adsorbed amount of heavy metals at the equilibrium concentration and $Q_{e,\ calc}$ is the calculated value of Q_e according to the kinetic model. The experimental values of Q_e of Pb(II) and Cd(II) are 61.31 mg g^{-1} and 60.46 mg g^{-1}, respectively.

	Pseudo-Frist order			Pseudo-Second order			Intra-particle diffusion model		
Test metal ions	k_1 (min^{-1})	$Q_{e,calc}$* (mg g^{-1})	R^2	k_2 (g mg^{-1} min^{-1})	$Q_{e,calc}$* (mg g^{-1})	R^2	k_i (mg g^{-1} min$^{-1/2}$)	C	R^2
Pb(II)	0.023	57.90	0.9695	0.00082	64.52	0.9968	2.0578	25.739	0.7674
Cd(II)	0.018	50.32	0.9854	0.00073	63.70	0.9964	2.0848	24.059	0.7674

Stability and regeneration The stability of PEI-grafted gelatin sponge was studied by investigate the adsorption capacities of heavy metals after 90 d of storage in aqueous solution. It is shown that removal efficiency was almost unchanged, implying the excellent time stability. Since the adsorption of heavy metals onto the sorbent was a reversible process, a sponge was repeatedly putted into the wastewater samples and taken out for regeneration using an acetic acid solution (0.1 mmol L^{-1}). The results showed that the removal ability was decreased with the increasing of usage, over 80% efficiency was received after the fifth adsorption-desorption cycles, implying that the PEI-grafted gelatin sponge was cost-effective. However, performances of the sponge were affected by several factors in the natural environment, such as sunshine, wind and so on. Thus, PEI-grafted gelatin sponge was spread on the soil surface in the Taiyuan city of china for two weeks (Weekly mean of temperature: 17 ± 1.1–31 ± 2.4 °C, Weekly mean of relative humidity: 57%; Rainy Day: 2 days; Precipitation: 11 mm; breeze). Then, the weight and the adsorption capacities were investigated to indicate the degradation of PEI-grafted gelatin sponge in natural condition. It was shown that the loss weight was $6.3 \pm 1.2\%$ and the adsorption capacities for lead ions and cadmium ions were 57.8 mg g^{-1} and 58.2 mg g^{-1}, which were 87.4% and 89.2% of the original adsorption capacities, respectively. Thus, we recommended that the PEI-grafted gelatin sponge was changed once a week, due to ease of operation and low cost.

Removal of metal ions during wastewater irrigation Dietary intake is the main route of heavy metal exposure, especially in vegetable crops irrigated with metal-contaminated wastewater. Excessive uptake of heavy metals could result in serious health problems in humans. Garlic (*Allium sativum* L.) was widely used as food, spice and medicine, which has some beneficial effect in preventing heavy metal induced alterations, but people seldom focus on the accumulation of pollutants[29].

Therefore, the PEI-grafted gelatin sponge was spread onto the surface of soil planted with garlic to effectively remove target ions during wastewater irrigation. The average length of garlic leaves decreased significantly in the positive control (23.4 and 19.5 cm) and the mixed contamination group (17.2 cm) as compared to the blank control (37.2 cm) (Fig. 5). Compared with the treatment groups and the blank control, leaf drooping of garlic was found in the positive control and mixed contamination group, which demonstrated a significantly toxic effect on the plant. In addition, the results demonstrated that the average lengths of garlic leaves in the positive control were much lower than those in the treatment groups ($P < 0.05$). There was no statistically difference of garlic growth among blank control (I), treatment groups (III and V) and mixed treatment groups (VII) ($F = 2.27$, $P > 0.05$), implying a high removal efficiency of heavy metals by the PEI-grafted gelatin sponge during wastewater irrigation.

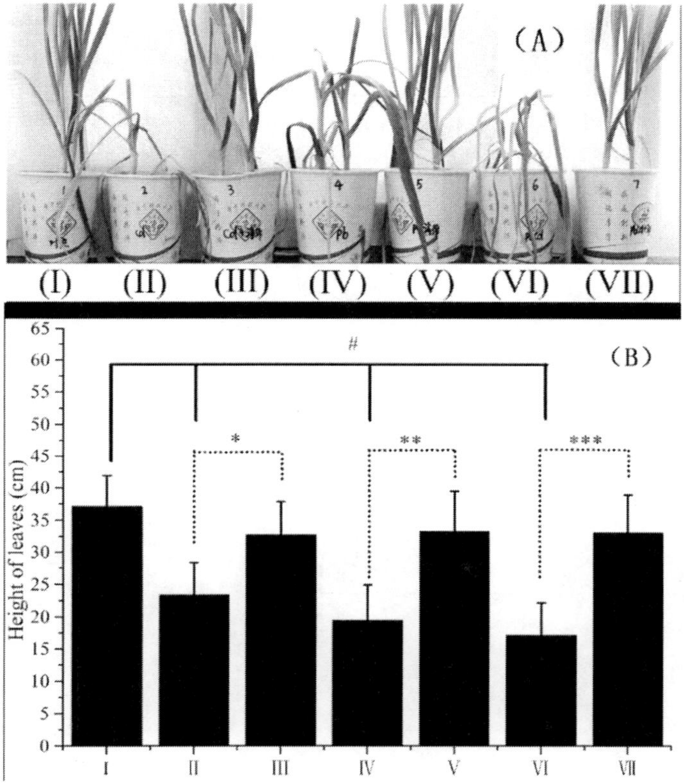

Figure 5 Photograph of garlic one month after treatment (**A**); leaf growth of garlic after one month (**B**). (I) Blank control (spraying with ultrapure water); (II) positive control of Cd(II); (III) treatment group of Cd(II); (IV) positive control of Pb(II); (V) treatment group of Pb(II); (VI) positive control of mixed contamination; and (VII) treatment group of mixed contamination. Notes: *Denotes a significant difference using a T-test. #Denotes a significant difference based on analysis of variance.

Figure 6 shows that the concentrations of Cd(II) ions and Pb(II) ions in the soil were 2.91 mg kg^{-1} and 5.92 mg kg^{-1}, respectively, which is much lower than those (8.82 mg kg^{-1} and 21.37 mg kg^{-1}) without the protection of PEI-grafted gelatin sponge, and the removal efficiencies were 67.0% and 72.3% (Fig. 6). However, there was no difference of Cd(II) content in the roots with and without the protection of the sponge, which should attract sufficient attention for cadmium pollution. For the positive control of Cd(II), 57.7% and 41.1% of Cd(II) was accumulated in the roots and leaves of garlic, respectively. By using the PEI-grafted gelatin sponge, the amount of Cd(II) in the leaves changed from 15.78 to 1.59 mg kg^{-1}, and the removal efficiency was 89.9%. When the soil was simultaneously contaminated by Pb(II) and Cd(II), the PEI-grafted gelatin sponge with high adsorption capacity could still effectively remove Cd(II) from the wastewater and then decrease the risk of heavy metals. The removal efficiencies were 71.4% and 92.4% for the soil and leaves, respectively.

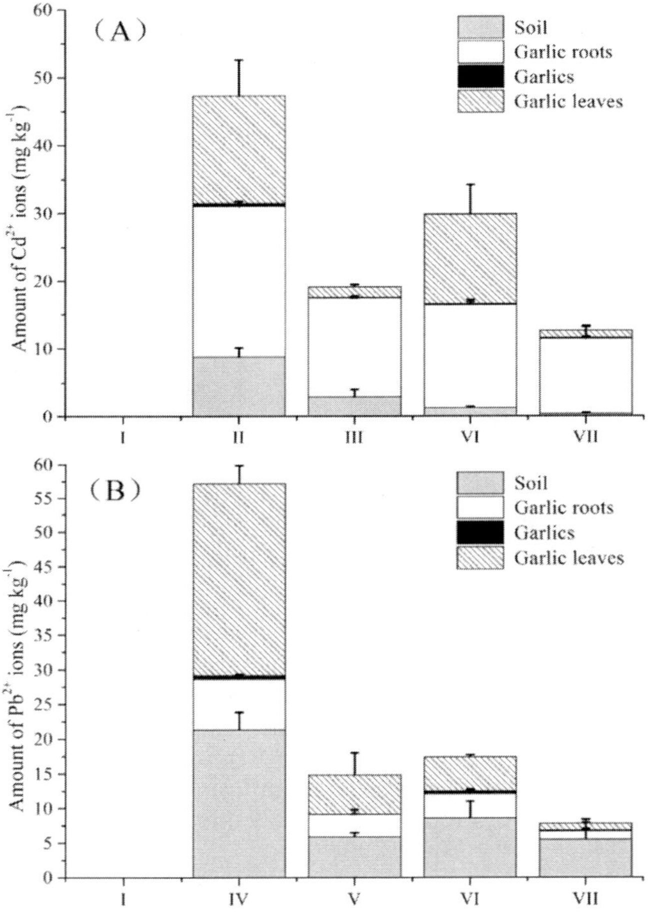

Figure 6: Bioaccumulation of Cd(II) (A) and Pb(II) (B) in the soil and edible parts of vegetable crops with or without the spreading of the PEI-grafted sponge. Notes: (I-VI) denote the same contents as in Fig. 5. Each group has three cups and two garlic plants were planted in each cup. The concentration in the sprayed solution were 500 mg L^{-1}(signal contamination) and 250 mg L^{-1} (mixed contamination), respectively. The sprayed volume was 10 mL day^{-1} and the treatment time was one month.

For the Pd(II) in the positive control (IV), 19.2% and 73.7% of Pd(II) was accumulated in the root and leaves of garlic, respectively. Most of the Pd(II) ions remained in the soil and also accumulated in the leaves. After using the PEI-grafted gelatin sponge, the amount of Pd(II) in the leaves changed from 27.98 to 5.69 mg kg^{-1} and the removal efficiency was 79.7% (Fig. 6). The amount of Pd(II) in the roots also decreased from 7.3 mg kg^{-1} to 3.23 mg kg^{-1} and the removal efficiency was 55.75%. Similar results were obtained when the soil was simultaneously contaminated by Pb(II) ions and Cd(II) ions.

It was shown that PEI-grafted gelatin sponge could be applied for the pollution control during wastewater irrigation to improve the soil quality and then reduce the exposure of crops to heavy metals. This study provided a simple, effective and economic technology to prevent further contamination of heavy metals in the soil. For the contaminated soil, the removal ability of PEI-grafted gelatin sponge for heavy metals was very weak. The removal efficiency was only 5.3%, when the dry PEI-grafted gelatin sponge was spread onto the surface of wet soil.

CONCLUSION

In this study, a PEI-grafted gelatin sponge was prepared and spread onto the surface of soil planted with garlic to decrease the risk of heavy metals during wastewater irrigation. According to the strong water adsorption ability and the high removal efficiency of the sponge for heavy metals, wastewater remained in the PEI-grafted gelatin sponge for a sufficient period of time for the heavy metals to interact with the sorbents. Garlic was used as the test plant to study the removal efficiency of the PEI-grafted gelatin sponge during wastewater irrigation. The removal efficiencies for lead ions and cadmium ions were 89.9% and 79.7%, respectively. Furthermore, the PEI-grafted gelatin sponge had the advantages of excellent mechanical stability, environmental friendliness, ease of operation, low operating costs and high adsorption capacity. We believe that this study provide an effective, practical and direct technology to improve the soil quality and decrease the accumulation of heavy metals.

METHODS

Materials Absorbable gelatin sponges (60 mm × 20 mm × 5 mm) were purchased from Nanchang Xiangentang medical equipment Co. Ltd.

(Xixiang, China). N-hydroxysuccinimide (NHS), branched polyethylenimine (PEI) with a molecular weight of 800 Da, 3,3′-dithiodipropionic acid, ethanol, dimethylformamide and 1-ethyl-3-(3-dimethylaminopropyl) carbodiimide hydrochloride (EDC) were purchased from Sigma-Aldrich (St. Louis, MO, USA). Stock aqueous solutions of lead and cadmium (1 mg mL^{-1}) were purchased from the National Institute of Metrology of China and then dissolved with ultrapure water.

Preparation of PEI-grafted gelatin sponge A PEI-grafted gelatin sponge was synthesized according to the literature with a modification30. 3,3′-dithiodipropionic acid solution (0.2 mol L^{-1}) was prepared by using aqueous dimethylformamide and then mixed with a 0.1 mol L^{-1} of EDC solution. The mixture was stirred at 40 °C for 1 h and then slowly added dropwise into 10 mL of acetic acid solution (pH = 5.0). A piece of absorbable gelatin sponge was introduced into the resulting solution. The reaction was performed at 40 °C for 24 h with continuous shaking (200 rpm) to convert the amino groups to the carboxyl groups. The sponge was removed from solution using a tweezers and placed on a paper to remove the remaining water. Subsequently, 0.5 mmol EDC, 0.5 mmol NHS and 0.5 mmol PEI were dissolved by 20 mL of acetic acid solution (pH = 6.5) and the resulting sponge was introduced to the mixture, which was incubated at 40 °C for 24 h with continuous shaking (200 rpm). The final product was received using a tweezers and washed three times with water.

Adsorption experiments The adsorption abilities of the gelatin sponge and PEI-grafted gelatin sponge were investigated with batch shaking adsorption experiments. Approximately 0.2 g of each sponge was placed into 20 mL of aqueous solution (pH = 6.0) spiked with different concentrations of heavy metals (lead and cadmium as the test metal ions). After shaking for 12 h (at 200 rpm), the amounts of lead ions and cadmium ions in the supernatant were detected by an inductively-coupled plasma atomic emission spectroscope (ICP-AES, Perkin Elmer Plasma 3200RL). Kinetic studies of PEI-grafted gelatin sponge were the same as the adsorption procedure, however the testing samples were taken at scheduled time intervals.

Removal of lead ions and cadmium ions during wastewater irrigation Garlic growth assays were carried out using an indoor cultivation method (temperature was 25 °C and sunshine hours was 8 h) to verify the application perspective of the PEI-grafted gelatin sponge. Garlic is a bulbous plant and garlic growth is easy. Thus, it was used to investigate the bioaccumulation of pollutants in the fruit, leaves and roots of garlic after spreading with a gelatin sponge. Soils for the incubation studies were sampled from the city park soils in September 2015 (sand 120 g

kg^{-1}, silt 410 g kg^{-1}, and clay 470 g kg^{-1}). The soils were then packed to a paper cup (I.D. = 15 cm and the packed height = 6 cm) for the garlic growth.

Fresh garlic bulbs were chosen from the local farmers market and then each clove was pushed into the soil to a depth of approximately 1.5 cm. Subsequently, the PEI-grafted gelatin sponge was spread on the soil surface and surrounded the garlic clove. Lead and cadmium are commonly present in different types of industrial effluents and mainly responsible for environmental pollution and public health. Thus, it was sprayed with lead (or cadmium) solutions (0.5 mg mL^{-1}, 50 mL) at a flow rate of 1 mL min^{-1}. The positive controls without the sponge spreading and a blank control (spraying with water) were also included in the experiments. After 30 days of growth, all of the plants were harvested (leaf, garlic, and root), washed with tap water to remove the soil, and then dried in an oven (105 °C, 30 min). The concentrations of target ions in the dried plants and soil were determined by ICP-AES after digestion with the mixture of HNO_3 and $HClO_4$ (1/4, v/v)[31].

ACKNOWLEDGEMENTS

This work was supported by the National Nature Science Foundation of China (No. 21207044), the Fundamental Research Funds for the Central Universities (No. 2015TS097) and Young talent project of health and Family Planning Commission of Hubei Province (No. WJ2015Q011). We also thank the Analytical and Testing Center of Huazhong University of Science and Technology for XPS, FT-IR and SEM analysis.

REFERENCES

1. Noor-ul-Amin Hussain, A., Alamzeb, S. & Begum, S. *Accumulation of heavy metals in edible parts of vegetables irrigated with waste water and their daily intake to adults and children, District Mardan, Pakistan*. *Food Chem* **136**, 1515–1523 (2013).
2. Jiang, W. *et al. Annual input fluxes of heavy metals in agricultural soil of Hainan Island, China. Environ Sci Pollut R* **21**, 7876–7885 (2014).
3. Fu, F. & Wang, Q. *Removal of heavy metal ions from wastewaters: A review. J Environ Manage* **92**, 407–418 (2011).
4. Hua, M. *et al. Heavy metal removal from water/wastewater by nanosized metal oxides: A review. J Hazard Mater* **211**, 317–331 (2012).
5. Moosavirad, S. M., Sarikhani, R., Shahsavani, E. & Mohammadi, S. Z. *Removal of some heavy metals from inorganic industrial wastewaters by ion exchange method. J Water Chem Techno* **37**, 191–199 (2015).

6. Qiu, J., Dong, S., Wang, H., Cheng, X. & Du, Z. *Adsorption performance of low-cost gelatin-montmorillonite nanocomposite for Cr(III) ions*. *Rsc Adv* **5**, 58284–58291 (2015).
7. Zhang, Y. *et al. Fabrication of chitosan/gelatin foams with ordered porous structures for use in drug release and metal ion adsorption*. *Rsc Adv* **4**, 33840–33847 (2014).
8. Pan, B. *et al. Development of polymeric and polymer-based hybrid adsorbents for pollutants removal from waters*. *Chem Eng J* **151**, 19–29 (2009).
9. Tofighy, M. A. & Mohammadi, T. *Adsorption of divalent heavy metal ions from water using carbon nanotube sheets*. *J Hazard Mater* **185**, 140–147 (2011).
10. Lee, C. *et al. Application of carbon foam for heavy metal removal from industrial plating wastewater and toxicity evaluation of the adsorbent*, *Chemosphere* **153**, 1–9 (2016).
11. Inagaki, M., Qiu, J. & Guo, Q. *Carbon foam: Preparation and application, Carbon* **87**, 128–152 (2015).
12. Liu, B., Wang, D., Yu, G. & Meng, X. *Adsorption of heavy metal ions, dyes and proteins by chitosan composites and derivatives - A review*. *J Ocean U China* **12**, 500–508 (2013).
13. Ngah, W. S. W., Teong, L. C. & Hanafiah, M. A. K. M. *Adsorption of dyes and heavy metal ions by chitosan composites: A review*. *Carbohyd Polym* **83**, 1446–1456 (2011).
14. Kumar, K. Y., Muralidhara, H. B., Nayaka, Y. A., Balasubramanyam, J. & Hanumanthappa, H. *Low-cost synthesis of metal oxide nanoparticles and their application in adsorption of commercial dye and heavy metal ion in aqueous solution*. *Powder Technol* **246**, 125–136 (2013).
15. Hegazy, E. Z., Abdelmaksod, I. H. & Kosa, S. A. *Removal of Heavy Metal Quaternary Cations Systems on Zeolite A and X Mixtures Prepared from Local Kaolin*. *Clean-Soil Air Water* **42**, 775–778 (2014).
16. Nguyen, T. A. H. *et al. Applicability of agricultural waste and by-products for adsorptive removal of heavy metals from wastewater*. *Bioresource Technol* **148**, 574–585 (2013).
17. Jeevika, A. & Shankaran, D. R. *Visual colorimetric sensing of copper ions based on reproducible gelatin functionalized silver nanoparticles and gelatin hydrogels*. *Colloid Surface a* **461**, 240–247 (2014).
18. Wang, W., Huang, D., Kang, Y. & Wang, A. *One-step in situ fabrication of a granular semi-IPN hydrogel based on chitosan and gelatin for fast and efficient adsorption of Cu^{2+} ion*. *Colloid Surface B* **106**, 51–59 (2013).
19. Huang, F. *et al. Preparation of three-dimensional macroporous chitosan–gelatin B microspheres and $HepG_2$-cell culture*. *J Tissue Eng Regen M* (2014).
20. Jia, J., Wu, A. & Luan, S. *Spectrometry recognition of polyethyleneimine towards heavy metal ions*. *Colloid Surface a***449**, 1–7 (2014).
21. Sun, X. *et al. Enhanced Cu(II) and Cr(VI) biosorption capacity on poly(ethylenimine) grafted aerobic granular sludge*. *Colloid Surface B* **82**, 456–462 (2011).
22. Xu, H., Liu, D., He, L., Liu, N. & Ning, G. *Adsorption of Copper(II) from an Wastewater Effluent of Electroplating Industry by Poly(ethyleneimine)-Functionalized Silica*. *Iran J Chem Chem Eng***34**, 73–81 (2015).
23. He, L., Liu, D., Wang, B. & Liu, N. *Poly(ethyleneimine)-functionalized Silica-supported Organic-Inorganic Hybrid Sorbent Prepared by Combining Sol-Gel Method and Hydrothermal-assisted Process*. *Chem Lett* **43**, 579–581 (2014).
24. Sun, X. *et al. Enhanced Cu(II) and Cr(VI) biosorption capacity on poly(ethylenimine) grafted aerobic granular sludge*. *Colloid Surface B* **82**, 456–462 (2011).
25. Zhou, Y. *et al. Sorption of heavy metals on chitosan-modified biochars and its biological effects*. *Chem Eng J* **231**, 512–518 (2013).
26. Weng, C. H. *Modeling Pb(II) adsorption onto sandy loam soil*. *J Colloid Interf Sci* **272**, 262–270 (2004).

27. Luo, C., Wei, R., Guo, D., Zhang, S. & Yan, S. *Adsorption behavior of MnO2 functionalized multi-walled carbon nanotubes for the removal of cadmium from aqueous solutions. Chem Eng J* **225**, 406–415 (2013).
28. Pang, Y. *et al. PEI-grafted magnetic porous powder for highly effective adsorption of heavy metal ions. Desalination* **281**, 278–284 (2011).
29. Nwokocha, C. R., Owu, D. U., Nwokocha, M. I., Ufearo, C. S. & Iwuala, M. O. E. *Comparative study on the efficacy of Allium sativum (garlic) in reducing some heavy metal accumulation in liver of wistar rats. Food Chem Toxicol* **50**, 222–226 (2012).
30. Li, Z. *et al. Chitosan-graft-polyethylenimine with improved properties as a potential gene vector. Carbohyd Polym* **80**, 254–259 (2010).
31. Luo, C. *et al. Heavy metal contamination in soils and vegetables near an e-waste processing site, South China. J Hazard Mater* **186**, 481–490 (2011).

Index